Human Factors in the Chemical and Process Industries

Human Factors in the Chemical and Process Industries

Making it work in practice

Edited by

Janette Edmonds

AMSTERDAM • BOSTON • HEIDELBERG • LONDON • NEW YORK • OXFORD
PARIS • SAN DIEGO • SAN FRANCISCO • SINGAPORE • SYDNEY • TOKYO

Elsevier
Radarweg 29, PO Box 211, 1000 AE Amsterdam, Netherlands
The Boulevard, Langford Lane, Kidlington, Oxford OX5 1GB, United Kingdom
50 Hampshire Street, 5th Floor, Cambridge, MA 02139, United States

Notices
Knowledge and best practice in this field are constantly changing. As new research and experience broaden our understanding, changes in research methods, professional practices, or medical treatment may become necessary.

Practitioners and researchers must always rely on their own experience and knowledge in evaluating and using any information, methods, compounds, or experiments described herein. In using such information or methods they should be mindful of their own safety and the safety of others, including parties for whom they have a professional responsibility.

To the fullest extent of the law, neither the Publisher nor the authors, contributors, or editors, assume any liability for any injury and/or damage to persons or property as a matter of products liability, negligence or otherwise, or from any use or operation of any methods, products, instructions, or ideas contained in the material herein.

British Library Cataloguing-in-Publication Data
A catalogue record for this book is available from the British Library

Library of Congress Cataloging-in-Publication Data
A catalog record for this book is available from the Library of Congress

ISBN: 978-0-12-803806-2

For Information on all Elsevier publications
visit our website at https://www.elsevier.com/

www.elsevier.com • www.bookaid.org

Publisher: Joe Hayton
Acquisition Editor: Fiona Geraghty
Editorial Project Manager: Maria Covey
Production Project Manager: Debbie Clark
Designer: Mark Rogers

Typeset by MPS Limited, Chennai, India

This book is dedicated to the staff at The Keil Centre who were all involved in the writing and reviewing of the chapters.

Contents

SECTION I INTRODUCTION TO HUMAN FACTORS WITHIN THE CHEMICAL AND PROCESS INDUSTRIES

CHAPTER 5 Management frameworks for human factors 59
J. Edmonds

SECTION II MANAGING HUMAN FAILURE

CHAPTER 6 Human factors in risk management 73
R. Scaife

CHAPTER 8 Human factors in incident investigation 131
E. Novatsis and J. Wilkinson

SECTION III HUMAN FACTORS WITHIN DESIGN AND ENGINEERING

CHAPTER 9 Overview of human factors engineering 153
J. Edmonds

CHAPTER 17   Human factors in the design of procedures.............291
E. Novatsis and E.J. Skilling

SECTION IV UNDERSTANDING AND IMPROVING ORGANIZATIONAL PERFORMANCE

CHAPTER 19 Managing organizational change.................................335
K. Gray and J. Wilkinson

CHAPTER 20 Staffing the operation 357
J. Edmonds

Author Profiles

Chiara Amati—Occupational Psychologist MA, MSc, DBA, C.Psychol

Chiara is a Chartered Occupational Psychologist whose main area of expertise is the relationship between the individual and their work. She has experience of working with individuals, teams, and organizations to promote resilience, well-being, and engagement at work. Chiara also has a special interest in management and leadership development, having recently completed a doctorate study in this area. In her roles as a coach and an experienced facilitator of leadership development, Chiara also promotes an understanding of how personality shapes individuals' experience in work contexts.

Janette Edmonds—Director/Principal Consultant Ergonomist

Janette has a bachelor's degree in psychology and a master's degree in ergonomics. Janette is a Chartered Ergonomics and Human Factors Specialist (CErgHF), a Fellow of the Chartered Institute of Ergonomics and Human Factors (FIEHF), and a Chartered Member of the Institution of Occupational Safety and Health (CMIOSH). Janette has been a practitioner since 1994 working in a broad range of industries including oil and gas, petrochemical, chemicals and plastics, nuclear, manufacturing, food production, utilities, energy, the emergency services, defense, rail, telecoms, medical products, and other applications. Janette has been a human factors manager for several engineering projects and has considerable experience of human factors applied to operational assets, incident investigation, and human reliability. Janette is the course director of the IChemE *Human Factors in Health and Safety* training program in the United Kingdom and Europe.

Jenny Foley—Clinical Service Manager/Counseling Psychologist

Jenny has a bachelor's degree in psychology, a master's degree in counseling psychology, and a postmaster diploma in counseling psychology. Jenny is a chartered counseling psychologist and member of The British Psychological Society and registered with the Health and Care Professions Council. As well as assessing and treating a range of conditions, Jenny is also a qualified Eye Movement Desensitisation and Reprocessing (EMDR) practitioner. EMDR is recognized as particularly effective in the treatment of trauma.

Jenny has a special interest in absence management and the return to work, as well as working with promoting effective communication and healthy relationships.

Ken Gray—Director/Occupational Psychologist

Ken has an Honors Degree in Psychology and has been Chartered by the British Psychological Society as an Occupational Psychologist since 1995 (CPsychol). He is registered as a professional practitioner with the Health and Care Professions Council (HCPC), holds the Qualifications in Test Use Register (RQTU), and is on the Occupational Safety and Health Consultants Register (OSCHR). Ken is an Associate Fellow of the British Psychological Society (AFBPS) and a member of the BPS Division of Occupational Psychology. Ken was shortlisted in 2010 for the BPS Practitioner of the Year. With his expertise in behavioral assessment, skills and competency advancement, team dynamics, and high impact leadership development, Ken works with both individuals and management teams to maximise individual potential and embed safe behaviors within organizations. He is particularly interested in psychological well-being and building resilience; both of which provide a solid basis for those leading organizational change or building future leadership capability.

Kirsty McCulloch—Principal Human Factors Advisor
Kirsty has a bachelor's degree in psychology and a PhD
in Applied Behavioral Science, specializing in fatigue
risk management. Kirsty is an adjunct research fellow
with Central Queensland University, and a member of
the Human Factors and Engineering Society of Australia
(HFESA). Kirsty is considered a world expert in fatigue
risk management, and her methods have been adopted as
regulatory models in several industries, across multiple
countries. Kirsty also has extensive experience in understanding human failure, human factors in incident investi-

gation, task analysis, critical procedure reviews, and safety culture. She has worked
in many safety-critical industries, including petroleum, mining, power generation,
medicine, emergency services, transport, and manufacturing.

Johnny Mitchell—Principal Occupational Psychologist
Johnny is a Chartered Occupational Psychologist with considerable experience in human factors, safety culture, and
psychological health. Over the last 10 years he has worked
in every continent on global safety culture development
projects and related topics. Johnny regularly presents at
industry conferences and delivers the safety culture module on the IChemE Human Factors Course. He recently
developed a model of Human Factors Maturity which featured in SHP magazine and a set of tools (TEAVAM) that
support the proactive detection and management of human

error. Johnny provides training in The Keil Centre's Human Factors Analysis Tools
and he is also frequently involved in supporting organizations to investigate major
incidents in the oil and gas, pharmaceutical, chemical, and construction industries.

Colin Munro—Consultant Ergonomist

Colin is qualified to master's degree level with merit in ergonomics (human factors) and with a Bachelor of Science degree with honors in physiotherapy. Colin commenced his professional career in 2002 and his experience includes 6 years within occupational health and safety for a safety critical organization. Colin has particular interests in ergonomics and human factors applied to design, control room analysis and design, human error analysis, safety culture, and safer systems of work. He has worked in the oil and gas industry, nuclear, rail, biohazard, and medical sectors.

Emily Novatsis—Organizational Psychologist/Principal Human Factors Advisor

Emily holds a Bachelor of Science (Honors) degree and a Doctor of Psychology (Organizational) degree, where her research focus was organizational culture. Emily is a Registered Organizational Psychologist and has been a practitioner since 2000, working in the fields of human factors and organizational development. She has worked in both internal and external consulting roles and in a range of hazardous industries. Her experience includes 6 years leading Woodside's human factors team, where she was instrumental in establishing the organization's human factors programs and internal capability. Emily's specific human factors areas of expertise are organizational and safety culture, competency frameworks, incident investigation, learning and development, procedure design, and stress management and resilience.

Richard Scaife—Director/Occupational Psychologist

Richard has a BSc(Tech) in Applied Psychology and a MSc in Occupational Psychology. He is a Chartered Ergonomics and Human Factors Specialist (CErgHF), Fellow of the Chartered Institute of Ergonomics and Human Factors (FIEHF), European Ergonomist (Eur.Erg), Chartered Psychologist (CPsychol), Registered Psychologist with the Health and Care Professions Council, Chartered Scientist (CSci), and Associate Fellow of the British Psychological Society (AFBPsS). Richard has been a practitioner since 1989, beginning his career as a psychologist with the Royal Air Force. His early career included Human Factors specialist in the defense industry and air traffic control, latterly as the head of human safety of the NATS Human Factors Unit. Richard joined The Keil Centre 14 years ago, and has worked in a broad range of industry sectors including aviation, oil and gas, construction, petrochemicals, logistics and distribution, rail, nuclear, manufacturing, food production, defense, and energy. Richard specializes in all aspects of human factors, particularly human factors engineering, organizational safety, human safety analysis (including human error), and incident investigation. Richard has extensive experience of major project management in industry, and of managing teams of human factors and associated professionals.

Elaine J. Skilling—Principal Consultant Ergonomist

Elaine has a bachelor's (Honors) degree in physiotherapy and a master's degree in ergonomics. Elaine is a Chartered Ergonomics and Human Factors Specialist (CErgHF), a Chartered Physiotherapist, a Registered Member of the Health and Care Professions Council (HCPC), and Registered Member of the Association of Chartered Physiotherapists in Occupational Health and Ergonomics (ACPOHE). Elaine has provided ergonomic interventions within a variety of sectors since 2000, including oil and gas, chemical processing, manufacturing pharmaceuticals, and public sectors. Elaine is skilled in safety critical task analysis, human reliability analysis, procedure development, and control room design. She has also developed and implemented behavioral safety culture programs for several international clients. She is a qualified trainer and has extensive experience in delivering training on multiple human factors topics in many locations throughout the world.

Karen Smith—Interior Designer (guest author for Chapter 11)

Karen has a Bachelor's Degree in Interior Design and has focused her career on creating ergonomically compliant and aesthetically pleasing spaces for the 24-hour environment. Karen cofounded BAW Architecture in 1992, with the vision of pioneering best practices in control building design. Recognized globally as a notable speaker and roundtable chair, she is also an accomplished publisher of relevant articles addressing the challenges associated with designing state-of-the-art control buildings. BAW Architecture and The Keil Centre first collaborated in 2012 to develop the ISO 11064 compliant design of one of the largest iconic centralized control buildings yet to be constructed in the world (2018). The two organizations have since collaborated on several control room and building projects. Karen served as the ASM® Consortium's designated representative as an associate member, bringing the unique interior perspective to the development of Best Practices in Control Building design. Here she received the NIST Advanced Technology Program's Certificate of Appreciation for her 10-year contribution to R&D development and implementation. Recently a graduate of the Institute for Integrative Nutrition, Karen will continue to champion the voice of operations through '*to Designwell*': translating seasoned expertise into skillfully designed control rooms that incorporate proper ergonomics with precision focus on mind, body, and space.

John Wilkinson—Principal Human Factors Consultant

John has a Psychology and Philosophy dual degree and a postgraduate diploma in health and safety. John is a Chartered Ergonomics and Human Factors Specialist (CErgHF), a Fellow of the Chartered Institute of Ergonomics and Human Factors (FIEHF), a Chartered Member of the Institution of Occupational Safety and Health (CMIOSH), and a Member of the British Psychological Society. John worked in the UK's Health and Safety Executive (HSE) as a regulator for 22 years and as Human and Organizational Factors (HOF) Team Leader and Principal Specialist Inspector (Human Factors) from 2003. He helped develop the very successful "HSE Key Topics" approach to HOF including the HSE web page guidance, regulator training, safety report assessment, inspection, and investigation. He led the HOF strand of the HSE investigation at Buncefield, and was an expert reviewer for the US CSB reports on Texas City and Macondo. He has worked with the EU Major Accident Hazards Bureau as a regulator and consultant. Since 2011 John has worked as a practitioner with the Keil Centre on a wide range of HOF projects in the United Kingdom, EU, and further afield. He has presented widely at conferences and written many papers.

Foreword

This book focuses on the principle that there is a fundamental need to understand humans and their behaviors, and apply this to chemical engineering and the process industries.

In writing the foreword it is unusual to begin by recommending another book to the reader, but that is where I want to start. In 2015 I read Professor Steve Hilton's book entitled "More Human—Designing a World Where People Come First." In this "big systems" focused book there were some fascinating case studies of where well-meaning government and other attempts at solving social problems had failed because the systems that were put in place were "inhuman"—governments, bureaucrats, and others had failed to take account of the needs of people and their behaviors. All too often the "problem solvers" had not bothered to ask people what they needed or how they would like things to be—they had assumed people would behave in a certain way only to find later, after the money had been spent and the "solution" had not worked, that the people did not behave in the assumed way at all. Hilton's book is a good read for anyone of any discipline to remind us all that human beings lie at the heart of everything and we ignore that at our peril.

During my time as a member of IChemE Council and my Presidency of IChemE (2013–14), I spoke at many events including Hazards Conferences around the globe about the need for us to revisit process safety and put it at the heart of what all of us as chemical engineers do—whether we are design engineers, process engineers, or plant managers. I am delighted that the profession is reawakening its interest in this important subject, *and* realizing that Human Factors are integral and fundamental to success.

If the title "More Human—Designing a World Where People Come First" had not already been used, it would have been a very apt subtitle for this book. Engineers design and operate processes which bring benefits to people but which also have people at the heart of the processes themselves. We must not allow ourselves to be seduced into thinking we can engineer everything to avoid human error or behaviors which we have not prescribed or considered.

I commend the authors for bringing together all of the vital material which engineers need to absorb and integrate into their thinking and into their work to put Human Factors where they need to be—at the core.

We often say that "systems thinking" is what defines chemical engineering but let us all be clear that systems are human not just chemical, mechanical, or digital.

January 2016

Dame Judith Hackitt DBE

Preface

REASON FOR THE BOOK

The Institution of Chemical Engineers (IChemE) joined forces with The Keil Centre in 2009 to develop a human factors training program tailored specifically to meet the needs of the chemical and process industries. Managing human factors was identified as the next major step change in safety and business performance. The training program was intended to develop and up skill industry professionals, many of whom had no formal qualifications or training in behavioral or human sciences. By early 2016, the human factors training program had been delivered in four continents, with delegates from over 150 leading international companies. The IChemE also responded to the growing focus on human factors within its "Hazards" conference program, which now has a well-established human factors topic stream. The high level of interest in both the training program and the human factors topic stream at the IChemE Hazards conferences prompted the production of this book.

The increased emphasis on managing human factors has come from regulators, industry bodies, and individual businesses in recognition of the significant potential to improve health, safety, and performance. Major incidents in particular, including those at Texas City, Buncefield, and elsewhere in the world have highlighted the importance of addressing this crucial aspect of performance.

The book is written for those who want a comprehensive overview of the subject, focusing on the practical application of human factors. It complements the training program as it provides a narrative to multiple sources of human factors material. This is expanded upon, illustrated, and used to provide explanation for the case studies and practical exercises. However, the book is certainly not limited to the training program audience alone as it actually provides a comprehensive narrative to the topic of human factors within the chemical and process industries for anyone with an interest in this topic.

The authors are human factors specialists whose daily work involves resolving real world human factors issues and achieving higher standards in the chemical and process industries. The focus is on helping the reader develop a greater understanding of key human factors issues to enable them to identify and implement more effective solutions.

TEAM OF AUTHORS

The Keil Centre is a leading practice of chartered ergonomics and human factors specialists, chartered safety specialists, registered occupational psychologists and registered clinic psychologists. It provides consultancy and training services in three business areas: human factors in health and safety; organizational assessment and development; and clinical psychology. Eleven specialists from the three business areas have contributed to the book.

Established in 1983 and based in Scotland and Australia, the Keil Centre team has a unique combination of expertise in the human and behavioral sciences. The team works across all continents and has extensive experience of applying human factors within the oil and gas, and petrochemicals industries, as well as chemicals and plastics, pharmaceuticals, nuclear, mining, and other process-related industries.

The Keil Centre has worked extensively for the UK Health and Safety Executive and other regulatory bodies to provide advice on the application of human factors and psychology to safety. In the offshore oil and gas sector, the company was involved in the Step Change in Safety initiative following Piper Alpha.

So, the authors of this book have a wide range of practical, 'real world' experience of supporting organizations to manage and resolve human factors issues. The emphasis of the book reflects their collective experience in the practical application of Human Factors.

THE EMERGENCE AND DEVELOPMENT OF HUMAN FACTORS IN THE CHEMICAL PROCESS AND OTHER INDUSTRIES

Human factors is primarily concerned with health, safety, and the performance of humans within the work context. As a discipline it has strong scientific and methodological underpinnings and these are applied in industrial settings to improve the overall design of the work system.

The topics which collectively form the subject of human factors have developed at different rates over the last century. It has grown from the first time-and-motion and behavioral studies within manufacturing settings to a much broader application in the modern day. For example, safety culture in particular became a key focus within the offshore oil and gas industry after the Piper Alpha disaster in 1989. Human safety analysis has been a key area of focus for the nuclear industry since the 1980s. Human factors engineering has established much of its body of knowledge from military applications starting during the World War II, and usability engineering gained momentum in the 1970s with consumer products such as those produced by Apple™.

The rate of development of human factors has accelerated over recent years, mostly as a result of increasing complexity of industrial systems, major accidents, regulatory influence, and societal expectations. It has also been recognized that "engineering the human out" approach through "blanket" automation does not work: what is increasingly required is a proportionate consideration of human capabilities and limitations within any work system.

As would be expected given the current level of development in human factors across industry, the chemical and process industries have made great strides in some areas, whilst others require significantly more development. There has also been a variable rate of progress in the different industries which are part of the chemical and process industries, and indeed even between different organizations within each of these industries. A general statistic that is often cited is that humans contribute to 80% of all accidents, which is true for the chemical and process industries (HFI DTC, 2006). In the 80% human-related accidents, 64% occurred during operations. This presents an opportunity and a challenge for these industries. There is clearly a need to do more to improve the reliability of the human element, and the more that the industry understands about human factors, the more effective it will be at making these improvements.

This book brings together relevant human factors topic areas within a framework of interconnected influences and learning from across industry, to support its development within the chemical and process industries.

OVERVIEW OF THE CONTENT

The topics within this book have been selected on the basis of their relevance to the chemical and process industries and their level of priority in achieving safe and reliable human performance. The book has been divided into chapters, but the topics are related to each other. Organizational and operational contexts and the design of the human interface all have significant impacts on each other and ultimately on human performance, health, and safety. A comprehensive approach to human factors requires consideration of the depth of the different topics discussed, whilst recognizing the breadth of coverage and the need to approach the subject from a holistic perspective.

The book is presented in four sections: (I) introducing and positioning human factors within the chemical and process industries; (II) managing human failure; (III) managing human factors in design and engineering; and (IV) managing organizational performance.

The introduction section (I) provides a working definition of what human factors is about in the context of work systems, what it aims to achieve, and the general approach for its application. Relevant human factors issues for the chemical and process industries are discussed, taking account of how these industries are generally organized and operate. Significant major incidents are reviewed to demonstrate and provide learning opportunities about where human factors vulnerabilities typically reside and where they have contributed to major accidents and injuries. The regulatory and governmental focus on human factors is summarized, providing examples of different approaches around the world and the relative merits of different types of approach. The section ends with a chapter on a framework for measuring and managing human factors within organizations.

The managing human failure section (II) presents a review of human factors within risk management and the necessity for, and benefits of, an integrated approach with other risk management tools and techniques. The proactive approach to predicting and managing risks related to human failure is discussed, and includes specific subsections on managing human error (unintentional behaviors) and managing violations (intentional behaviors). An effective approach to human factors in incident investigation is presented, with a discussion on methods to enable a comprehensive understanding of how human behavior has contributed to an incident. The human factors related to the incident investigation process are also considered. This section provides the reader with the opportunity to understand how to improve human reliability within their own organization.

The managing human factors in design and engineering section (III) specifically focuses on relevant human factors issues within design and how it can be effectively integrated and managed within engineering programs. Specific applied topics that are relevant to the chemical and process industries are discussed including: building and control room design; control system interface design; work area and workspace design; plant side aspects; materials handling; environmental ergonomics; and procedures. The focus for these chapters is to explain how human performance is directly influenced by design. The key human factors issues are discussed, and the means to optimize the design for the end user.

The understanding and improving organizational performance section (IV) provides a comprehensive review of the key organizational factors that affect performance within the chemical and process industries. Safety culture is discussed in relation to its effect on safety performance and the process by which it can be assessed, developed, and sustained. Organizational change has a significant impact on individuals, teams, and the organization and is often insufficiently managed, to the detriment of the whole organization. An appreciation of the key issues and the process which needs to be applied is presented to explain an effective means of managing organizational change. A chapter is presented on staffing the operation, which discusses how to determine the optimal number and type of people required to operate the work system, how to develop and maintain competence, and how to achieve effective supervision. Fatigue management is discussed in relation to health and safety performance, taking account of the sometimes challenging shifts and routines worked within the chemical and process industries. The final chapter provides guidance on managing performance under pressure and discusses psychological well-being and performance in relation to acute and chronic states of strain and distress.

WRITING STYLE

The book has been written in an "industry" style, rather than an academic style. This could be described as:

- More direct—considering various implications, but directing the reader to the heart of the issue relatively quickly;
- Simple language—avoiding overly complex language to ensure that the book is accessible and easy to digest for a wide audience;
- Applied—making use of case studies and practical examples to explain points being made;
- Practical references—directing the reader to industry texts, websites, and guidance which are easy to access and understand.

Janette Edmonds

Acknowledgment

The Keil Centre would like to acknowledge and thank Cheryl MacKenzie, Investigations Team Lead, US Chemical Safety Board, for her advice and guidance on defining the regulatory framework in the United States of America which is discussed in the chapter on the current regulatory and government focus on human factors. Any errors or omissions are the author's. Cheryl MacKenzie is a world renowned expert in the field of Human Factors and her expertise has been instrumental in understanding the human contribution to significant and well-known major accidents.

Acknowledgment and gratitude are extended to BAW Architecture, Denver, Colorado, USA, for the provision of figures and support in writing the chapters related to control room and workstation design. BAW Architecture is a world leading specialist in control building design with over 100 built projects on five continents including within the chemical and processing control industries. The company is exemplary in how it embraces the human-centered design approach and uses a multidisciplinary team to achieve the highest standards and compliance in design. The Keil Centre has worked with BAW Architecture on several control (and other) building projects where the benefits of human factors in design are clearly understood. Further information can be obtained at www.bawarchitecture.com.

Introduction to human factors within the chemical and process industries

What is human factors?

J. Edmonds

LIST OF ABBREVIATIONS

PPE   Personal Protective Equipment

Despite significant technological developments and investment within the chemical and process industries, systems do not always work or function as effectively as planned. More often than not this can be attributed to the failure to consider human interactions or because inaccurate assumptions are made about how people will behave. At best this is inconvenient, at worst it is catastrophic. There is therefore a fundamental need to recognize that people are at the heart of any system, even the most highly automated of systems.

Human beings differ quite considerably from the hardware, software, or chemical elements. Nevertheless, the human is an essential, beneficial, and integral aspect of the work system and needs to be effectively considered and defined. It is therefore important to understand the context and the interfaces within the work system where human interactions occur. Within this chapter, and indeed the book, the term "system" includes all the technical engineering (chemical, software, and hardware) and human aspects.

The primary focus of human factors is to optimize health, safety, and performance and is undertaken by focusing specifically on the human interactions that occur within a work system. This chapter intends to provide a definition of its scope and application.

Humans vary considerably in their capabilities and limitations. It is not always easy to predict or preempt how they may behave. However, a large body of human science-based knowledge has developed over several decades, which helps understand and preempt how human interactions may occur. Using this knowledge to understand the specific work system context enables optimization of the design of systems and processes that depend on human interaction.

The application of human factors knowledge may be predictive, such as when designing a new system. It may also be used to understand and optimize existing systems of work. In either case, human factors provides the key to unlock many of these "unknowns" to provide a suitable solution.

THE WORK SYSTEM

The work system, as referred to in this book, relates to the complete work context. This could be, a hospital, an oil refinery, an airport, or any other workplace with its various functions, infrastructure, work areas, equipment, and importantly, the people who work in it, maintain it, support it, and so on.

The work system has a particular purpose, and various functions are performed within the work system to meet the objective of the system goal. An offshore oil installation, for example, has the purpose of extracting crude oil. Its functions include drilling and well construction, production, with its incumbent processes, such as separation, compression, heating/cooling, and export to a terminal where other work systems take over. There will be other functions that enable the primary purpose to be accomplished.

At the start of developing a new work system, it does not actually exist. It is merely a concept. It will then go through three distinct phases:

- *Design*: where it is conceptualized, designed, developed, and built;
- *Operation*: where it performs the function it was designed to perform;
- *Decommissioning*: where it is taken back to its starting point where it no longer exists as a system (Fig. 1.1).

The definition is simplistic and the boundaries between the phases do in reality overlap with each other as the system continues to dynamically evolve through

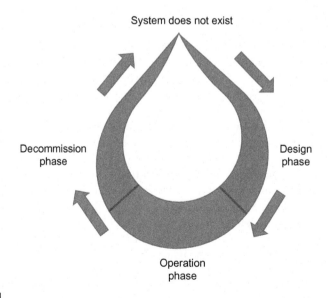

FIGURE 1.1

Work system phases.

the operational phase. This means that it is important for human factors to remain a core consideration to ensure continued health, safety, and performance during the evolution.

Within the work system, most functions require the people within it to use the inanimate "machine" aspects, such as hardware and software to achieve the functional goal. Different roles perform different tasks and interact with different interfaces within the system (including other roles) but ultimately the tasks are undertaken to achieve the system goal.

Human factors applies to any and all aspects of the work system where there are human interactions. It also applies to all work system phases. Ideally, much of the effort in preempting the human interactions for operations and decommissioning will occur within the design phase to avoid issues at a later date. There is a more in-depth discussion about human factors in design and engineering in Section III, Human Factors Within Design and Engineering.

HUMANS WITHIN THE WORK SYSTEM

Within the work system, there are human interactions that are essential for the system goal to be achieved. These are illustrated in Fig. 1.2 and this is intended to act as a framework to illustrate how people interact with various elements of a work system, including other people within that system.

The different types of human interactions within the work system are described in more detail as follows:

> *Work tasks*: the work system requires certain functions to be performed. They are typically subdivided and allocated either to the "machine," the "human," or both (dependent on the level of automation). The human tasks

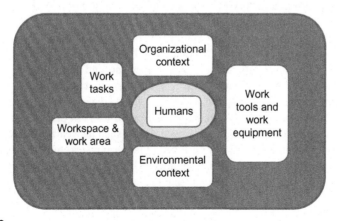

FIGURE 1.2

Human interactions within the work system.

related to the system functions are typically grouped into different job roles, such as an operator, maintainer, and so on. The performance of these tasks will include physical (or manual) elements as well as cognitive elements (the psychological aspects of the task). Human factors is applied to each of these aspects of task and work design. There is a need to achieve an optimal allocation of functions between the human and the machine, and appropriate design and grouping of tasks for different job roles. The purpose is to ensure the tasks are achievable, cohesive, and within the limitations and capabilities of the people undertaking them.

Work tools and equipment: work tools and equipment are used by people to achieve their tasks. Even in more primitive times, people developed and used tools to achieve their goals and it is one of the aspects that separate human beings from the rest of the animal kingdom. Tools and equipment can take many forms, from a simple spanner, through to a complex control system, even including the written procedures and signs that people use or refer to. Work tools and equipment have become increasingly complex and technologically advanced. However, it is important to recognize that humans have not evolved at the same pace. This means that tools and equipment must be designed to meet the needs of the end user, whilst taking account of their capabilities and limitations. Otherwise, there is potential to induce human failure. If the human fails, there is potential for the system performance to be adversely affected.

Workspaces and work areas: physical space is allocated and designed to provide the area for humans to perform their tasks. There is the immediate workspace or workstation where a person or people perform their tasks, and there is a wider work area that accommodates a multitude of work and machine/equipment spaces. The immediate workspace could, for example, be the cockpit of a crane, a controller's console, or a work bench. The wider work area may be a plant area, a workshop, a factory floor, or a control building. For a workspace or workstation to be effective, it needs to be designed to support the tasks that are performed and provide the right amount of space for the human to fit within and access and use the equipment. There are two basic physical space needs; the person needs to have sufficient clearance (space around them), and be able to reach and see the equipment interfaces in a posture that is least likely to injure them. This is relatively simple if it is a tailor-made design for an individual, but workspaces need to accommodate a variety of users of different sizes and shapes, sometimes with additional space needs (such as wearing Personal Protective Equipment (PPE), or performing larger physical movements). In addition to physical accommodation, the workspace needs to support the functionality required which means that the layout and arrangement of the space must be optimized. Likewise, the wider work area also needs to optimize task performance and provide the optimal amount of space.

Environmental context: the most common environmental considerations within the scope of human factors are noise; vibration; thermal comfort; and lighting, although inherent hazards, such as chemicals, hot or cold surfaces, and other hazards may also be included within this category. There are two aspects that are considered; protecting people from the adverse effects of the environmental conditions, and designing the environmental conditions to enhance performance. The environmental context can have negative effects on the physical and psychological capability, and well-being.

Noise has potential to cause permanent hearing loss, but it is important to ensure that the acoustic environment is conducive to performing the relevant auditory/verbal tasks. This might include responding to auditory signals and equipment communications, as well as human to human communications.

Vibration has different effects on the body at different frequencies and amplitudes and there is a limited physical tolerance before injury occurs. The task performance can also be affected. Although the main emphasis is to protect people from vibration, it can be used within design to provide information, for example, in the form of vibrotactile feedback.

Temperature—thermoregulation is the biological mechanism which maintains the internal body temperature at 36.8°C. There is a small tolerance range before impairment and death occur. Thermal comfort is achieved where there is a balance between the external thermal environment and the internal thermal processes. It is affected by ambient temperature (air temperature and radiant temperature), air movement, humidity, human physical activity, and insulation of the body through clothing. In the work system, there is a need to protect the human from thermal conditions that can place an excessive demand on thermoregulation, but also to design the thermal environment to achieve thermal comfort and thereby enhance physical and mental performance.

Light is required to enable humans to perform visual tasks in the work environment. The lighting design requirements are dependent on the tasks that need to be achieved whilst ensuring that adverse lighting conditions, such as glare and reflection are avoided. The lighting environment is affected by the uniformity and type of light (artificial and natural), the level of illumination, the eyesight of the human, the reflectance of light from work surfaces and a number of other factors.

Organizational context: within a work system there are a number of people performing a multitude of different job roles who work to certain work patterns. They form an organizational structure, which develops its own culture, with implicit and explicit ways of working. These factors have a strong influence on human behavior, motivation, and performance. The organizational factors discussed in this book are not an exhaustive presentation of organizational factors, but represent key aspects affecting organizational performance and safety. The organizational factors include:

Organizational culture: groups of people develop their own culture with shared attitudes, beliefs, and ways of behaving. A safe culture is one where safety is a core value. There are several elements that positively or negatively affect culture and these elements can be identified, measured, and developed to improve and mature the safety culture.

Management of organizational change: change is an inevitable aspect of any organization. However, mismanaged change can adversely impact behavior, the ability to perform jobs, and the well-being of the individuals and teams involved. Well-managed organizational change can have a positive impact on safety and avoid poorly managed risks.

Staffing and workload: staffing is concerned with having the optimal number and type of people to perform tasks in all operational scenarios, at all levels of the organization. People perform at their best when workload is optimal, not too much or too little, and when it is evenly shared between roles within a team. The people selected to perform job roles need to have the right knowledge, skills and attitudes, and maintain the required level of competence through training and experience. Supervisors have a key role in maintaining high standards of performance and safe behavior and require specific leadership capabilities to enable this to happen.

Safety critical communication: there are many and various communications that occur within an organization; some more critical than others. The methods of communication between the transmitter(s) and receiver(s) of information can affect the reliability of the communication. Particular areas of communication vulnerability, such as shift handover and permit to work systems, are heavily reliant on good communication and are key factors that can affect safe performance.

Fatigue: sleep is a basic biological need. The way shifts and work hours are organized can impact sleep opportunity, and adversely affect human performance. Hours of work, sleeping conditions, and certain tasks and environments can exacerbate the effects of fatigue, and ultimately performance.

Psychological well-being: there are common organizational factors which can lead to chronic or acute stress. There is a need to manage these factors to avoid mental ill health and to maintain safe behaviors and performance.

The "human factors within the work system" are not independent of each other. For example, a glare issue caused by poor lighting design may be resolved by a rearrangement of the workspace layout. Equipment interface design can be compromised if the work layout is poor. The organizational context sets up the behavior and motivations of people in their interactions with equipment.

The "humans within the work system" (also known as the sociotechnical system) represent a complex web of interactions with all aspects of the system, whether it is an interaction with the machine elements or other people. Human factors is about understanding how all of these factors individually and collectively interact to influence human performance, and provide the opportunity to optimize system performance.

SCIENTIFIC BACKGROUND

The human being is itself a system in its own right and it is the job of the human scientist to provide the applied knowledge about how humans work to enable the work system to be designed accordingly. Human factors draws on both the psychological and physical human sciences, and is used to understand human capabilities and limitations.

Psychology: two main areas of psychology feature heavily in human factors: cognitive psychology and organizational psychology. Cognitive psychology is focused on how people process information received through the senses, how it is perceived, how the memory works, how decisions are made, and then factors affecting performance of actions. Organizational psychology is the study of people in groups and organizations, and includes topics such as motivation, team work, group dynamics, psychological well-being, and job design. Clinical psychology features in relation to the treatment of people with psychological ill health, and developing resilience.

Physiology: physiology is concerned with the biological processes of the body, particularly in relation to physical work, including physical endurance and work capacity, thermoregulation, strength, and movement control. It also includes biological processes, such as circadian rhythms which have an influence on restorative functions, that is, sleep, and levels of alertness.

Anatomy and biomechanics: anatomy and biomechanics are used to understand the musculoskeletal system, and functions including static and dynamic posture, balance, and movement dynamics. It includes knowledge about how to optimize musculoskeletal performance and avoid injury.

Anthropometry: anthropometry is the study of body size. Every person has a unique dimensional profile but each dimension, such as stature, arm length, and torso girth, aligns with a normal distribution for a population. This means that there will be few people at the extreme smaller and larger ends of the size range with a "bell-shaped" distribution curve for dimensions in between, as shown in Fig. 1.3.

Different dimensions are used for different purposes. For example, if there is a need to determine seat height, the popliteal range (back of the knee to the floor) is used. The vertical position of a visual display needs to be determined using either the sitting or standing eye height as the basic reference point. There are static and dynamic body size dimensions, where static is the dimension measurement taken in a static posture. Dynamic measurements take account of the movement of body members. Anthropometry is used to accommodate as wide a range of people in the end user population as possible. It is typically focused on the limiting user, so for clearance, there is benefit in designing for the largest person by using more than the 95th or preferably 99th percentile dimension so that smaller people also have sufficient space. For reach, the limiting user is the smallest, using say the 5th percentile functional grip reach dimension. By designing to this person, everyone with a longer grip reach can also reach the item. There is often benefit in incorporating an adjustable range, in which case, the 5–95th percentile range (or more) would be used. This is relatively simple for a single dimension, but more complex when there are several dimensions to accommodate.

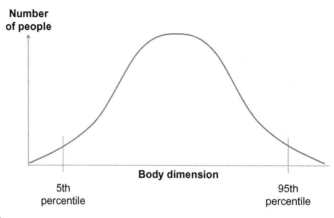

FIGURE 1.3

Normal distribution for body dimensions.

SCIENTIFIC APPROACH

It is necessary to understand the human interactions within the specific work context to direct the application of the most relevant scientific information for the system. There are many tools and techniques to support this process throughout the system lifecycle, from analyzing the tasks required, through to testing the human interaction. Tools and techniques are described throughout the book as relevant to the subject area being discussed but the following small, nonexhaustive list is presented as an illustration:

- Task analysis is used to understand and aid the design of tasks;
- Link analysis is used to understand human interactions (e.g., role to role and role to equipment);
- Human reliability analysis is used to understand how humans may fail and reduce the risk of human failure;
- Workload analysis is used to assess a person's ability to manage the amount of work being demanded of them;
- Staffing assessment is used to determine the type and numbers of people required within the system;
- Training needs analysis is used to determine skill gaps and support the identification of the optimal means to reduce the training gap.

The scientific approach may also include calculation, such as anthropometric or visual angle/distance calculations to determine workstation dimensions and equipment positioning. It may also include a combination of tools and techniques for performance measurement, such as would be used during user trials and testing.

The suite of human factors tools and techniques is constantly evolving and so it is necessary to keep abreast of the latest developments.

SUMMARY: WHAT IS HUMAN FACTORS?

Without adequate attention to the human interactions that occur within a work system, there is potential that performance and safety may be compromised. This chapter introduced a simple model to explain those interactions and outline aspects where human factors can be used to optimize design, whether it is a new system or an existing one.

Human factors is based on well-established underpinning scientific knowledge and tools and techniques that enable that knowledge to be applied to system design, development, and optimization.

In the next chapter, the explanation provided within this definition of human factors is made more specific to the chemical and process industries.

KEY POINTS

- Human factors focuses specifically on the human interactions within a work system, whatever that system is, such as an oil platform, chemical process plant, or a hospital.
- It is applied during all aspects of a system lifecycle, including design, operations, and decommissioning.
- The key human interactions relate to the tasks people perform; the tools and equipment they use to perform the task; the workspace/work area where the work is undertaken; and the influence of the environmental and organizational context.
- Humans vary in their capabilities and limitations, and this is informed by the applied human sciences.
- The scientific information about humans is facilitated using tools and techniques to understand and measure the human interactions within the work system.

FURTHER READING

Chartered Institute of Ergonomics and Human Factors: www.ergonomics.org.uk.
Energy Institute: www.energyinst.org/technical/human-and-organisational-factor.
Health and Safety Executive: www.hse.gov.uk/humanfactors/humanfactors/index.htm.
Human Factors and Ergonomics Society of Australia: www.ergonomics.org.au/.
Human Factors and Ergonomics Society (US): www.hfes.org/web/Default.aspx.
Step Change in Safety: www.stepchangeinsafety.net/.

What part does human factors play within chemical and process industries?

2

J. Edmonds and J. Wilkinson

LIST OF ABBREVIATIONS

HFE	Human Factors Engineering
HRO	High Reliability Organization
HSE	Health and Safety Executive
ISO	International Organization for Standardization
PPE	Personal Protective Equipment
PSV	Pressure Safety Valve

Chapter 1, What is Human Factors? provided a definition of human factors, describing the concept of a work system and the human interfaces within it. This chapter uses the same definition to reflect on how work is organized in the chemical and process industries; and the relevance of human factors for these work systems. The purpose of the chapter is to outline typical human factors deficiencies and how human factors can be used to improve human, and therefore overall system performance. The current level of maturity of human factors within the industry is discussed to focus attention on areas of human factors where substantial gains may be made.

WORK SYSTEMS IN THE CHEMICAL AND PROCESS INDUSTRIES

The chemical and process industries include (but are not limited to): oil and gas production; mid-stream oil and gas processing; refining; petrochemicals; chemical and plastic production; food and drink processing; pharmaceuticals; water and waste water; paper production; nuclear power and other forms of power generation; and mining.

The common denominator for these work systems is that they involve large industrial scale processes. They typically manage huge volumes of chemicals or materials, in batch or continuous processes, which may be highly toxic, unstable, or flammable.

There are different examples of work systems but in general they include a variety of processes in which materials undergo chemical reactions and/or physical changes to produce products. The range of products include: chemicals for agriculture, manufacturing, and construction; plastics (and eventually consumer goods); medicines; food and drink; and energy and water.

Understanding and valuing the human contribution to these processes is vital for enabling efficient and safe operation. As they become increasingly complex, managing the human contribution becomes more important and more challenging.

As well as having predictable properties, systems also have emergent properties (often not identified until an accident occurs). The emergent properties are unpredictable because they "emerge" from the complex interactions within the system. In other words, such systems are more than the sum of their parts. It is therefore important to be vigilant for emerging warning signs that may herald unexpected events. This is consistent with the "chronic unease" said to be exhibited by High Reliability Organizations (HROs), organizations that manage and sustain almost error-free performance despite operating in hazardous conditions where the consequences of errors could be catastrophic (Health and Safety Executive (HSE), 2011). It is also about preparing to deal with the unexpected and building resilience through preparing staff to make optimal decisions under uncertain conditions. Improvement opportunities may also be generated by unexpected but useful emergent properties of the system. Most operational workarounds are undertaken to take advantage of these useful emergent properties, and make the system work in practice. Only occasionally do these lead to unexpected failures.

TYPICAL HUMAN FACTORS ISSUES

Human factors is relevant to the whole of the work system, because the human is intrinsic to the system. People design it, install it, commission it, operate it, maintain it, and support it.

It is useful to make a distinction between human factors in *design* and human factors within *operations*, as the emphasis can be quite different. Human factors in *design* provides the opportunity to improve human performance through effective design of the tasks, tools, equipment, workspaces, work environment, and organizational arrangements. Human factors can also be used to improve existing *operations*, not just through design modification but through other interventions. These might include: the improvement of safety culture; shift systems; control of work systems; improvements in procedures; safety critical communications; and other arrangements.

This section provides a simple explanation of the work organization within these industries and relevant human factors considerations. The description is generic, although it is recognized that there are variations from one type of work system to another and from one organization to another. An illustration is provided in Fig. 2.1.

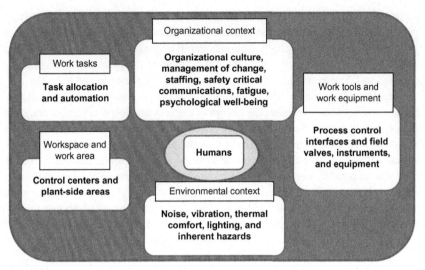

FIGURE 2.1

Human interactions within industrial work systems.

WORK TASKS

The processes within these industries typically require a high degree of control and management to ensure process variables (such as flow, temperature, level, pressure, and demand or supply) and supporting utilities, remain within design and safety limits. Within the processes and associated support functions there are "machine" tasks and there are "human" tasks. Human tasks are combined and allocated to roles within operations, maintenance, and support functions.

The control systems tend to have a high level of automation managed by a process control system that monitors data and controls the plant equipment. There are specific safety functions built in, such as interlocks that are linked with the process variables. For example, if there is a danger of overfilling a vessel, the interlock uses logic controllers to prevent further material additions. There are also relief systems to enable material to, for example, flow to a "catch tank," or route to the flare, where it can be safely managed in an over-fill or over-pressure situation. The safety systems are typically automated and triggered by process variables outside safety limits. More detail on safety systems is discussed in Chapter 3, Review of Common Human Vulnerabilities and Contributions to Past Accidents. It is entirely appropriate to incorporate automation in this way, and much of the improvements in safety over recent decades are owed to the engineering development of process control and safety systems.

However, engineering out the human through automation is not always the right decision as it can reduce performance effectiveness in different parts of the system or leave humans with a loss of situational awareness or motivation. The allocation of tasks to the "human" and/or the "machine" needs to ensure a good balance and optimize the strengths of both aspects to achieve optimal overall system performance. It is important to remember that skilled people get things right most of the time, often "completing" and sometimes "compensating" for the design through experience and practice.

Regardless of the extent of automation used or the allocation of tasks to job roles, there is still a need to ensure that the design of the task and job is commensurate with human capabilities and limitations. Task and job design has evolved, driven by factors such as technology changes, drivers for efficiency, and the availability of skilled resources. However, some tasks in these industries are designed remotely from the people who ultimately conduct them. For example, elements of the process control system or plant equipment are often purchased either mostly or partly "off the shelf" from a particular supplier, and there is a significant cost involved to deviate from the "standard" product. So in reality, task design for a whole system is unlikely to be a "bottom up" design where usability options and concepts are analyzed in detail, but people needs, capabilities, and limitations still need to be taken into account. People should not be expected to adapt and "make it work" as this can set them up to fail. Many assumptions are made about error-free human performance. However, the implications of human error must be considered to avoid undermining the technical design of the system. The performance of any safety system, for example, can be undermined if it is incorrectly maintained. This topic is discussed in more detail in Section II, Managing Human Failure.

WORK TOOLS AND WORK EQUIPMENT

The work tools and equipment are the direct interfaces that system users have with the system. They provide the window into the system and the means by which to control it. The forms of these interfaces are multiple and variable from digitized complex control systems to mechanical tools. These are discussed in relation to process control interfaces, and field valves, instruments, and equipment.

Process control interfaces: The process control system is typically operated from a central control room, although there may be local control stations, local equipment rooms, and manual equipment in the field. The control system includes software and hardwired control elements and presents the human interface via graphical screen displays and control panels.

There have been many advances in the design of the control system interfaces, particularly software interfaces. The first generation transitions from hardware to software were not overly successful from a human factors perspective. For example, the "system overview," which was readily transparent with a wall of dials, suddenly became nested within deep layers of a complex set of screens. This often led to a loss of situational awareness. In addition, the user–system dialogue was often "unforgiving" of errors, leading to the control of the system being onerous, frustrating, and unnecessarily difficult. The incorporation of features, such as system overviews, fewer screen levels, trend displays, and good alarm presentation (amongst other aspects), has provided the opportunity for the control operator to preempt and predict what the system is doing and proactively control and monitor, rather than being reactive to events.

Although there are good examples of effective control system interface design within the industry, there are plenty of examples where the design falls short of acceptable. There are also many examples of modifications and partial improvements to process control and interfaces, leaving operators managing two "not-very-well-integrated" systems, such as an older analogue system side by side with a digital system. One notable omission is the lack of user trials to objectively measure performance (and errors) in the use of the interface. It is important to recognize the connection between poor control system interface and human error.

Field valves, instruments, and equipment: With regard to the main process plant, field operators interact with various interfaces, such as manual valves and field instruments. They are involved in the plant-side aspects of the process, including activities such as material loading/receiving and pig launching/receiving. The control room and field operators typically work as a team with both roles undertaking aspects of the same function or operation. Maintenance teams interact with the control systems and plant equipment to ensure their correct operation. Although the same equipment may be interacted with by operations and maintenance, the actual interfaces are often different from the maintenance interactions which are typically more invasive. An example is provided in Fig. 2.2.

The maintenance teams use a large variety of tools and test equipment for use on the plant. They will also use lifting and handling equipment ranging from trolleys and lifting beams to requiring the support of a crane operations team. Different work teams often need to work together, such as in this example and so coordination, management, and cross-team communication are required to ensure that critical work activities are not missed.

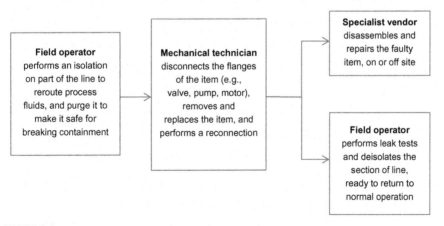

FIGURE 2.2

Example of how different teams interact with the same equipment.

The "ergonomic" design of plant equipment is often overlooked both for operability and maintainability. Sometimes this is because equipment is standard "off the shelf" and changes to the standard design are prohibitively expensive. Sometimes it is just accepted that equipment has been designed this way for years and the need for improvements is not recognized (unless there has been an incident).

There are plenty of examples where the basic principles of interface design have not been implemented. This includes aspects such as equipment being oriented so the labels are upside down, gauges not being readable because they are too small given the visual distance away from them, control types being used which are prone to accidental activation, and valves being mounted so the valve handles cannot be operated. It is necessary to remember that these interfaces are the human "window" to the system and that they can be a critical source of error. They therefore present considerable system vulnerability.

Maintenance interfaces are, in some regards, even more critical than process operation interfaces given that they may involve risks associated with breaking containment, overriding safety systems and are typically more invasive. There are huge opportunities for improvement in this area.

The chemical and process industries do not currently have a single comprehensive standard for the human factors design of plant and equipment, although one or two individual organizations have filled this gap by generating their own. There are some good resources available which are discussed in Chapter 14, Plant and Equipment Design.

WORKSPACE AND WORK AREA

In this chapter, work areas have been distinguished as control centers (which includes the control room and other functional areas within a control building), and plant-side areas (which includes non-control center areas).

Control centers: Most process and production plants are coordinated via a central control room, often within a central control building that accommodates associated support functions. A control room typically includes one or more control consoles with process control screens; communications equipment; and hardwired safety system control panels. Some control rooms use off-console displays such as large process or activity overview screens that are wall mounted. The control room is a critical area of any site to design well, particularly as the functions performed are typically safety critical.

There have been some great advances in control building and control room design over recent years, and many organizations have made the connection between human factors and control center design. It is a common aspect of design that human factors specialists are asked to support. This has been driven, to a large extent by the international standard for the ergonomic design of control centers (International Organization for Standardization (ISO 11064, 1999). Indeed, some organizations have developed their own standards for control "room" design, which reflect the guidance provided in ISO 11064. The critical human factors aspects include the design of control consoles, control room layout, and building arrangement (as well as the process control interfaces), and the primary focus is to enhance human performance. Refer to Chapter 11, Building and Control Room Design for a comprehensive overview of the topic area.

Plant-side areas: Many activities are performed in external plant areas, including operation and maintenance of the plant-side process-related equipment, materials handling, facilities management, possibly helideck operations, fire and emergency response, and security. Plant-side areas often suffer from a lack of human factors attention. The layout of the plant is not always logical, accessible, or well routed and can fail to guard against common errors such as interacting with the wrong (similar) equipment. Space can be a constraint, so access to equipment and the layout of equipment interfaces within a workspace can adversely affect how well the tasks are performed and increase the potential for human failure. The spaces required for maintenance tasks in particular can often be overlooked. Manual handling injuries are a frequent occurrence and mechanical handling activities can present major safety issues if the tasks and plant design have not been well designed. These issues are discussed in Chapter 12, Building and Control Room Design, Chapter 14, Plant and Equipment Design, and Chapter 15, Human Factors in Materials Handling.

ENVIRONMENTAL CONTEXT

The key human factors issues are noise, vibration, thermal comfort, lighting, and inherent hazards in the workplace. These are discussed in more detail in Chapter 16, Environmental Ergonomics.

Noise and vibration: Chemical and process plants often have noisy vibrating equipment such as generators and compressors, which present a risk of noise-induced hearing loss, musculoskeletal injury or other injury risk from vibration. While control rooms are generally quieter than plant, even relatively low noise and vibration levels can still present a stressor hazard, leading to an inability to concentrate during highly cognitive and safety critical task conduct. Communication can also be adversely affected.

Thermal comfort: In site buildings the thermal environment is generally controlled within a small range (whether or not this is comfortable is a topic for debate). The external environment varies more widely, so that field operators and maintainers may work in more extreme hot and cold conditions, which may be exacerbated by other factors such as wind velocity, radiant heat from equipment, wet conditions, physical work rate, and the use of Personal Protective Equipment (PPE). An inadequate thermal environment can present performance shaping factors that make human error more likely. It can also lead to personal injury and death in extreme situations.

Lighting: Internal and external lighting can enhance someone's work ability or hinder it. Poor lighting is a performance shaping factor that can make human error more likely. This not only includes inadequate illumination for the type of task, but also poor lighting arrangements that cause issues such as glare and reflection.

Inherent hazards: Field operators, maintainers, and other roles which are external may be exposed to inherent hazards such as chemicals, mechanical impact, radiation, and other hazards specific to the site.

Some of the environmental topics are well controlled as they fall within the remit of general health and safety, and safety engineering. However, not all of these aspects are sufficiently well considered, such as defining the limitations for physically demanding work under different environmental conditions, or identify and reducing psychosocial stressors. These factors can present a significant influence on safe task performance, not just the health of the person.

ORGANIZATIONAL CONTEXT

There has been a focus on organizational safety issues to varying degrees across these industries, but gaps and challenges still exist. These are discussed in detail in Chapter 4, Current Regulatory and Government Focus on Human Factors and summarized below:

Organizational culture has been a focus for some organizations and sectors since Chernobyl in 1986 and Piper Alpha in 1988. Various models, tools, and techniques for improving culture have emerged from the body of knowledge that has grown over this time. There are day-to-day techniques for use by operational personnel to enhance safety culture, such as positive safety conversations. Despite the increased focus, organizational culture remains an area for improvement, as investigations continue to highlight examples where poor safety culture has contributed to incidents.

Management of organizational change: Most sites have management of change policies and procedures. They are often focused on the "technical" change (including plant and engineering), missing the "organizational" change component, and the significant impact this often has on safe operation. The human factors focus is on assessing and managing the human risks of the change itself and transition to the change. This may include, for example, ensuring balanced workload and adequate staffing, sufficient competence, and reducing the risks associated with potential human failure. It also includes managing the psychological adjustment for change.

Staffing: The staffing profile over the operational life of a facility may remain fairly stable but in some facilities such as an offshore platform, the profile may vary depending on the nature of the platform and/or the phase of operation. For example, planned maintenance shutdowns can cause the staffing to swell enormously. Maintenance requirements also generally increase over time as plant and equipment age.

Ensuring the right numbers with the right competencies to manage workload is a major challenge. Some sites use a structured system to identify the right staffing levels, but many sites analyze this in far less detail. A common oversight (also linked with the management of change) is an untested belief that a site can run just as effectively with fewer people. Staff cuts can have enormous implications for the ability of the remaining teams to cope, especially in high workload scenarios, such as plant upsets. This ultimately has the potential to undermine safety performance.

A number of sites are developing competency management systems. This is a big undertaking and one that needs to incorporate a systematic and thorough process for continuous improvement.

Safety critical communication: The chemical and process industries rely on safe communication, particularly in relation to shift handovers, for communicating accurate operational status. There are many critical communications between: members of the same team, such as control room and field operators; between teams such as between operations and maintenance; between shift workers and day staff; and in multi-party work situations, such as lifting operations, over-side working, managing the receipt and transfer of materials; and so on. Another type of safety critical communication is control of work (including permit to work) systems to assess and communicate the hazards relating to parts of the plant under maintenance or other non-normal operation.

Fatigue: Chemical and process plants often have 24-hour operations and it is common practice to use 12-hour shift rotations. Other operations use callouts to respond to trips and upsets. There can sometimes be a "culture" to work longer hours especially by management teams who are typically on "day" turns, or for covering staff shortages. Any work system that requires long work hours, long commutes, time zone changes, or working at times of the day (or night) when it is normal to be sleeping has the potential to impact on sleep opportunity, and cause fatigue-related risk. There is no such thing as a perfect shift pattern or rota, but factors such as shift duration, number of shifts worked in a row, speed, and direction of shift rotation, length and timing of breaks/days off can be planned to optimize sleep opportunity, and thereby performance. This concept is rarely well understood within organizations, and is an opportunity for improvement.

Psychological well-being: As with any industry, there are opportunities for chronic stress caused by work-related factors, such as having too much work to cope with, poor relationships and support, lack of control over work, poor management of change or ill-defined roles. There are also opportunities for acute stress, especially in relation to being involved in or managing high potential incidents. Chronic and/or acute stress can be overlooked, if not only in relation to health implications but also the impact on safety.

MATURITY OF HUMAN FACTORS

The level of maturity of human factors varies greatly across the chemical and process industries, not just between specific types of process (e.g., nuclear, offshore, chemicals, pharmaceuticals), but also between specific areas of human factors (e.g., safety culture, human factors in design and fatigue).

In relation to human factors in *design*, industries, such as nuclear and oil and gas, have adopted best practices in Human Factors Engineering (HFE) and Human Performance approaches. Other industries have yet to recognize the relevance of including human factors in detailed design. Where good practices of human factors in design exist they are often limited in scope to specific topic areas such as control rooms, alarm systems, and process control graphical user interfaces. Other major considerations, such as the process plant areas, are ignored. In some areas, HFE is simply seen as applying HFE standards, without defining project-specific user requirements, or testing user performance. There is room for improvement across all industries to ensure that human factors is systematically applied and integral to the system design.

Within operations, there have been several advances through the introduction of programs to enhance organizational safety such as safety culture programs, competence management systems, fatigue risk management systems, stress management, and other areas. Again, not all organizations are taking a comprehensive approach or recognizing the need for human factors in all of these areas. The organizations that have focused on these areas have noticed significant benefits, such as reductions in the occurrence or reoccurrence of incidents (Lardner et al., 2011). While "Human Performance" has started to appear in selected applications, it is not yet well understood or widely applied and is often (in practice) focused mainly at frontline staff rather than the whole system.

There are considerable opportunities for improvements in human factors for the whole industry and the topic may feel overwhelming for those at a lower level of human factors maturity. However, engaging with a specific topic where there is an issue or "unease" may be the best way of moving towards the integration of human factors rather than setting up an elaborate system in advance. The reality is that developing a system is best done by iteration and experience so that there is "real" learning and those involved can see real value. This is reflected in the UK Regulator's "key topics" approach (see Chapter 4: Current Regulatory and Government Focus on Human Factors). Further guidance regarding management frameworks for human factors and improving human factors maturity is presented in Chapter 5, Management Frameworks for Human Factors.

SUMMARY: WHAT PART DOES HUMAN FACTORS PLAY WITHIN THE CHEMICAL AND PROCESS INDUSTRIES?

The simple model of human factors introduced in Chapter 1, What is Human Factors? was used in this chapter to outline the contribution that human factors can make to the chemical and process industries. In discussing each of these human interactions, there has been some reflection on the maturity of human factors management and where opportunities exist to make particular improvements.

In the next chapter, the consequences of failing to consider human factors are illustrated through reviewing past accidents and incidents.

KEY POINTS

- There are similarities in work systems across the chemical and process industries. Effectively they incorporate large-scale processes in which materials undergo chemical reactions and/or physical changes to produce products.
- Humans are an integral part of the work system and perform specific roles within operations, maintenance, and support functions to enable the smooth and efficient running of the process and deal with upset conditions and emergencies.
- There is a significant emphasis on safe and effective control of the process and there is effective use of automation particularly within the process control systems, although there can be drawbacks to the overuse (or inappropriate use) of automation.
- Human factors is relevant throughout the work system, even if processes are automated, to ensure that the design is commensurate with human capabilities and limitations.
- There are specific work equipment, tools, work areas, and environments that are particular to these industries and challenges are evident within the organizational arrangements.
- The maturity of human factors within these industries is variable and many gaps exist offering opportunities to make improvements in health, safety, and system performance.

REFERENCES

Health and Safety Executive (2011). High reliability organisations—a review of the literature. Prepared by the Health and Safety Laboratory.

ISO 11064. Ergonomic design of control centres; Part 1, Principles for the design of control centres, 2001; Part 2, Principles for the arrangement of control suites, 2000; Part 3, Control room layout, 1999; Part 4, Layout and dimensions of workstations, 2013; Part 5, Displays and controls, 2008; Part 6, Environmental requirements for control rooms, 2005; Part 7, Principles for the evaluation of control centres, 2006.

Lardner, R., McCormick, P., & Novatsis, E., 2011. Testing the validity and reliability of a safety culture model using process and occupational safety performance data. Paper presented at Hazards XXII – Process Safety and Environmental Protection Conference, 11–14 April, 2011. Liverpool, UK.

FURTHER READING

Energy Institute: www.energyinst.org/technical/human-and-organisational-factor.
Health and Safety Executive website: www.hse.gov.uk/humanfactors/humanfactors/index.htm.
Step Change in Safety: www.stepchangeinsafety.net/.

Review of common human vulnerabilities and contributions to past accidents

J. Edmonds

LIST OF ABBREVIATIONS

ESD	Emergency Shutdown Systems
F&G	Fire and Gas
HSE	Health and Safety Executive
ICC	Isolation Control Certificate
ISD	Inherently Safer Design
LOTO	Lock Out Tag Out
MIC	Methyl Isocyanate
PTSD	Post-Traumatic Stress Disorder
PTW	Permit to Work
PVC	Polyvinyl Chloride
SMS	Safety Management System
TMI-2	Three Mile Island Unit 2
ULD	Upper Limb Disorder
UPS	Uninterruptible Power Supply

Over the last two decades, the chemical and process industries have achieved significant improvements in system, and health and safety performance, largely due to improving engineering and safety management practices. More recently, many organizations have seen this improvement plateau, and continue to experience major accidents and health and safety related incidents.

It is unlikely that the application of human factors alone will reduce accidents and incidents to zero, but it can have a significant impact on achieving this goal. This chapter presents a review of typical human vulnerabilities in work systems that have contributed to major accidents, personal injuries and ill health.

In relation to major accidents, the recurring themes are used to highlight the areas in which human factors can be used more effectively to reduce the human vulnerabilities in relation to process safety. Personal injuries relating to common human factors causes are also discussed, both in terms of safety-related injuries and ill health caused by work.

The intention is to outline key vulnerabilities related to human factors which are relevant to the chemical and process industries, thereby presenting the opportunity to use human factors to create safer, healthier, and more efficient work systems in the future.

CURRENT SAFETY MANAGEMENT PRACTICES

Safety Management Systems (SMSs) are routinely adopted within the chemical and process industries to systematically identify hazards, control risks, and provide assurance that the risk controls remain effective. There has been a more even balance on process safety and major accident potential, certainly since the Texas City Refinery accident in 2005, but occupational health and safety remains an important aspect of the SMS.

Many organizations aim to achieve an Inherently Safer Design (ISD) by eliminating hazards completely, or by reducing their magnitude sufficiently to minimize the need for complex risk reduction measures. This includes reducing the amount of the hazard, substituting it for a less hazardous substance or material, attenuating the hazard condition or simplifying the process to reduce operating errors. If elimination is not possible, organizations aim to prevent the hazard from being realized or causing an undesirable event.

Prevention and control measures can include passive, active, and operational measures:

- Passive engineered measures are inherent in the design, such as natural ventilation, blast relief panels, welded pipeline connections, drainage, and containment systems;
- Active engineered measures render the plant safe if the event is instigated, such as Emergency Shutdown Systems (ESD); Fire and Gas (F&G) protection system; active fire protection systems; quench systems; and automated cooling systems;
- Operational measures act to reduce the likelihood of the event, such as maintenance, inspection and testing regimes, condition monitoring, manual ESD, ignition source control (eliminating sources of ignition in hazardous areas), equipment tagging, Permit to Work (PTW) systems, Lock Out Tag Out (LOTO) systems, and Isolation Control Certificates (ICCs) to prevent erroneous movement of control elements.

There are also mitigation measures to protect plant and structures where the failure of these may lead to critical escalation, such as the provision of Uninterruptible Power Supplies (UPS) to enable safe control and shutdown in the event of a power failure; fire protection limits to enable sufficient time for safe evacuation; comprehensive design; and planning for effective emergency response.

Learning from incidents, both internally and industry-wide, has also been an important factor in the prevention of repeat incidents and the continuous improvement in safety.

So there have been great improvements in the chemical and process industries through having comprehensive SMSs, implementing effective safety engineering practices, and from incident learning.

THE PLATEAU IN SAFETY PERFORMANCE

Processes within these industries operate within complex sociotechnical systems made up of technical and human elements, which are integral to each other. This has an impact on the performance of the whole work system (as defined in Chapter 1: What is Human Factors?), and means that any major advances in the technical aspects of improving system safety will be limited if the social or human elements are insufficiently considered.

There was a series of notable major accidents between the mid-1970s and the late 1980s, which started to shift attention towards organizational and social factors during incident investigations in addition to the technical factors (some examples are given in Box 3.1 below). One common and significant causal factor identified was the contribution of behavior. People were not doing "what they were meant to do," either intentionally or unintentionally, and consequently, the complex safety systems were undermined.

Around the same time, safety-critical industries were also becoming aware of a plateau in their safety performance. Significant improvements in safety had been made by continuously developing engineering and procedural systems to support safety management. However, further improvements of this nature were not resulting in parallel increases in safety performance, effectively creating a plateau. It was recognized that a focus on cultural and behavioral factors would be required to achieve further improvement.

BOX 3.1 CASE STUDY EXAMPLES SHOWING A SHIFT IN ATTENTION TO SOCIAL FACTORS

Flixborough, 1974—A chemical plant close to the village of Flixborough, England, exploded killing 28 people and seriously injuring 36 others. Prior to the explosion, a reactor was shut down as it was leaking cyclohexane. A bypass assembly was installed to enable production to continue. The bypass assembly subsequently ruptured, resulting in an explosion and initiation of numerous fires on site. Those responsible for designing the plant modification had insufficient engineering competence and there was an emphasis on prompt restarting of production to the detriment of considerations for safety. There was also a lack of engineering management for the modification design, construction, testing, and installation.

Chernobyl, 1986—The Chernobyl Nuclear Power Plant in Ukraine exploded and the resulting fire released large quantities of radioactive particles into the atmosphere. The explosion killed 31 people and longer term effects, like cancer, are predicted to be responsible for around 4000 further deaths. A design flaw in the reactor was being investigated to find a solution. The reactor was being run under test conditions outside the normal safe working envelope. A series of poor decisions were made, such as the disabling of safety systems, operator's knowledge of the design flaw was lacking, and it was assumed that operating instructions were safe.

Herald of Free Enterprise, 1987—The Herald of Free Enterprise ferry capsized soon after leaving the Belgian port of Zeebrugge, leading to a loss of life of 193. The bow door was left open by the assistant bosun, and not identified by the captain prior to entering rough seas. The decks became flooded causing the vessel to capsize. The contributory factors included: the absence of an indicator that the bow door was open, the top heavy design of the vessel, and lack of bulkheads to restrict water ingress. The assistant bosun fell asleep on duty due to fatigue from working an excessively long shift. There were cultural issues meaning that no-one else took responsibility for closing the doors or notifying the captain. There was also pressure to remain on schedule and at a higher level in the organization there was a clear deficiency in safety leadership.

Piper Alpha, 1988—The North Sea oil production platform, Piper Alpha, exploded causing major fires, leading to a total destruction of the platform and the loss of 167 lives. A safety valve had been removed for maintenance, and the initial leak occurred in that area of pipework when it was pressurized during start-up. There were a number of human failures, particularly in relation to the critical communications concerning the status of the plant. There was a crucial failure to notify other teams that the valve had been removed for maintenance. Once the explosion had occurred, the accident escalated due to the erroneous decision to continue production from other platforms in the field. The loss of life was exacerbated by failings in the emergency response.

LEARNING FROM HUMAN FAILURES IN MAJOR ACCIDENTS

If 80% of accidents include human failures, addressing the causes of human failure is essential to achieving further improvements in safety performance. This is the subject of Section II: Managing Human Failure where human failures are discussed in more detail, including the mechanisms by which human factors can be incorporated within risk management systems.

Much can be learned from the human failings in past major accidents. A short review is compiled below to identify some of the key recurring themes. Most of the reviews are taken from the chemical and process industries. Some transport industry incidents have also been included, as they reflect relevant human factors failings. Past accidents provide an opportunity to understand how human failures can be avoided, and safety performance improved.

Bhopal, 1984—A gas release of methyl isocyanate (MIC) occurred at the Union Carbide pesticide plant in India causing 3787 deaths due to the initial gas release, affecting a further 100,000–200,000 people in the surrounding areas. Large amounts of water entered the MIC tank causing an exothermic reaction which raised the pressure beyond the design limit. This forced emergency venting from the MIC holding tank, releasing a large volume of toxic gases into the atmosphere. It was found that the site could have used less hazardous chemicals, and could have stored the chemicals in smaller quantities rather than in large tanks. It was likely that poor maintenance had contributed to the accident through pipework corrosion and multiple failures of the safety systems. Some safety systems had actually been switched off to save money, including the MIC tank refrigeration system which, had it been operating, could have prevented the disaster. There were many other contributory factors, including understaffing, poor introduction of changes, communication and language issues, and a culture of violations and poor safety leadership.

Kegworth, 1989—A British Midland aircraft crashed on to the embankment of the M1 motorway near Kegworth, England, whilst attempting an emergency landing at East Midlands Airport. There were 47 fatalities and 74 serious injuries. Earlier in the flight, the pilots had shut down one of the engines due to a malfunction, indicated by vibrations and smoke entering the cockpit through the vents. Although it was the left-hand engine that had failed, the pilots throttled back the right-hand engine and initially this seemed to cure the problem. The erroneous diagnosis was not challenged by passengers or cabin crew who had seen unburnt fuel igniting in the jet exhaust of the left-hand engine. The pilots were interrupted by air traffic control transmissions during their review of their actions whilst diverting to an alternative airport for landing. There had been changes in the design of the aircraft from the previous version of the Boeing 737, including the operation of the auto-throttle but also the design of a vibration meter which would have indicated that the wrong engine had been shut down. The pilots had received only basic non-simulator training. The investigation revealed that the engine failure was caused by a broken fan blade which had been insufficiently tested during the design stage.

Three Mile Island, 1979—One of the two nuclear reactors at Three Mile Island, Pennsylvania, had a partial nuclear meltdown resulting in the release of radioactive gases and iodine into the environment presenting health risks to a wide area of local inhabitants. Three Mile Island Unit 2 (TMI-2) was too badly damaged and contaminated to resume operations. Mechanical failures occurred, including a pilot-operated relief valve which stuck open allowing reactor coolant to escape. Plant operators failed to diagnose the loss of coolant for several hours due to an ambiguous control room indicator on the user interface. It indicated the stuck valve as closed (the light only indicated the power status of the solenoid providing a false indication, on this occasion, of a closed valve). This caused a lot of confusion and the diagnosis was further hampered by the alarm "flood" that followed. Unfolding events and time pressure created "tunnel vision" in the operators, and they were unable to step back from their assumptions (based on the instrument readings) to correctly diagnose the issue. An operator consequently overrode the automatic emergency cooling system believing there was too much coolant water in the reactor, which was causing a steam pressure release. The oncoming shift team diagnosed the real issue but was too late to prevent the accident. The investigation revealed a lack of training on the need to use a downstream temperature indicator to confirm the real status of the pilot-operated relief valve, and not just to rely on the ambiguous control room indicator. The importance of checking the temperature was further perceptually downgraded by being out of primary view and positioned on the back control desk.

Buncefield, 2005—A series of explosions occurred at the Hertfordshire oil storage facility in England, which was audible 125 miles away in Belgium. No-one was killed, but 43 people were injured, hundreds of people had to be evacuated and significant blast damage occurred in the vicinity. Had the accident occurred in normal office hours, the injury rate could have been significantly worse. Storage tank 912 was being filled with petrol. The tank level was routinely monitored using a level gauge and had an independent high-level trip to shut-off the inflow above a specific set point. On the morning of the accident, the manual level gauge was stuck, tank 912 overfilled and the high-level trip did not operate. The petrol escaped undetected through vents in the top of the tank, formed a vapor cloud, and later ignited and exploded. The ensuing fires lasted for 5 days. The level gauge had stuck intermittently over the previous 4 months and the trip switch was not locked in position to enable the check lever to work. The bund failed allowing petrol to overflow, and the drainage system also failed, later causing firefighting foam to enter the groundwater supplies. Both of these containment systems were inadequately designed and maintained. The investigation also revealed deficient management safety checks. The operators were given insufficient pipeline information to properly manage the storage of incoming fuel and to identify unfolding abnormalities and emergency situations. The control room was poorly designed and laid out. Production was also found to be prioritized over safety, and throughput had been steadily increasing, adding to the pressure and complexity of the operators' tasks. Chronic staff shortages, poor shift and work planning combined to cause excessive fatigue. The procedures were inadequate, especially in relation to ullage management, hazards (such as overfilling tanks), and control room activities. Training was unstructured, with little instruction on upsets and emergencies, and the competence management system was not well developed. In addition, the shift handover was unstructured and informal. There were clearly multiple human factors failures.

Texas City Refinery, 2005—A hydrocarbon vapor cloud was released, which ignited and exploded, killing 15 workers and injuring more than 180 others. The accident occurred during the start-up on the isomerization unit following an outage. The raffinate splitter tower was overfilled, a routine deviation from procedure to avoid causing damage to the equipment, but in doing so the operators left the level control valve in manual which meant that the tower was open to the risk of overfilling. There were equipment issues, including a defective high level alarm, an inoperative pressure control valve, and an uncalibrated level transmitter in the tower. This was made worse by the board operator not having a clear indication of fluid flows in and out of the tower or the actual level in the tower, and field operators also had no indication because the sight glass was dirty. The pressure relief valves opened and flammable liquid was released from the blowdown stack which did not have a flare system. The subsequent investigation also found a lack of supervision on site at the time, inadequate training and procedures, poor shift handover and excessive worker fatigue (some operators had worked up to 30 days in a row on 12-hour shifts). There was also poor management of change including unassessed staffing cuts and a consequent excessive workload, particularly for the board operator who was managing other process units as well as the start-up. At a higher level there was evidence of poor safety culture and leadership.

Formosa, Illinois 2005—An explosion occurred at the polyvinyl chloride (PVC) production unit killing five workers and seriously injuring two others. The incident occurred during the cleaning of Reactor 306 whilst Reactor 310 was midway through a PVC reaction, containing highly explosive chemicals under heat and pressure.

During the cleaning process, the operator was required to empty the cleaning fluid out of Reactor 306 by opening the bottom valve to drain. The operator walked from the top floor of the reactor building to the lower floor, and turned right instead of left at the bottom of the stairwell. He then opened the bottom valve on a working Reactor (310) instead. This reactor did not drain due to a safety interlock installed to prevent the bottom valve being opened whilst under pressure. However, the operator then bypassed the interlock by attaching the emergency air hose, and the reactor contents were released. Rather than evacuate immediately the supervisor and operators tried to deal with the ongoing spillage and five were killed when it ignited. During the investigation, it was revealed that similar errors had occurred in the past and that there had been a failure to properly assess the risk of human error in the design. The emergency and evacuation procedures and training were also identified as inadequate.

SUMMARY OF KEY HUMAN FACTORS THEMES WITHIN MAJOR ACCIDENTS

The UK Health and Safety Executive (HSE) has defined a list of the "top 10" key human factors topics relevant to the high hazard industries. These are briefly introduced (Table 3.1), but covered in more detail within specific book chapters, as indicated. The "top 10" issues have been used in this chapter to map and summarize the recurring themes that are relevant to the 10 major accidents summarized in this chapter.

A map of the "top 10" human factors issues relevant to the 10 past accidents reviewed in this chapter is presented in Table 3.2 to illustrate how these themes have recurred from one major accident to another.

Table 3.1 HSE's Top 10 Human Factors Issues

Topic	Explanation	Chapter Reference
Human failures (this is divided into two 'top' ten issues to cover human failure generally, and specifically in relation to maintenance, inspection and testing)	This relates to when humans "do the wrong thing," in this case committing an unsafe act. Failures are typically classified as human errors (unintentional behaviors) and violations (intentional behaviors). Human failures are caused by a multitude of factors and occur at many levels of an organization but can directly or indirectly contribute to major accidents. They are most often identified reactively through incident investigations, but can also be proactively identified through structured risk assessment processes. Optimizing performance influencing factors can reduce the likelihood of most of these failures.	2

(Continued)

Table 3.1 HSE's Top 10 Human Factors Issues (Continued)

Topic	Explanation	Chapter Reference
Human factors in design	This relates to the design of tasks, tools and equipment, workspaces, and environments in which work is performed. Deficiencies of the human interface can make human failure more likely, for example, poor display screen design can lead to reduced situational awareness and contribute to a subsequent incident.	3
Procedures	Most high hazard organizations have a suite of operating and maintenance procedures. However, deficiencies in document development, organization, management, review, and presentation can have a significant impact on the resulting behavior and performance of the personnel using them. Equally, if they are not valued and trained for, they will likely not be used.	17
Organizational culture	Groups of people form cultures where they share attitudes, beliefs, and ways of behaving. If safety is not prioritized as part of the organizational culture, then unsafe behaviors are more likely to occur.	18
Organizational change	This relates to managing the organizational aspects of change, such as human factors considerations during restructures, redundancies, mergers and acquisitions, rather than just the more traditional technical, engineering, and plant aspects of change.	19
Staffing	Staffing is about having the optimal number and type of people (staff and contractors) to consistently perform at the required standard in all operational scenarios. It includes consideration of optimal (and peak) workload to ensure that the staffing profile and the work tasks are suitably configured to safely manage all operational scenarios. As frontline leaders, supervisors have a key role in enabling safe operation.	20.1 and 20.3
Training and competence	Competence is about the ability to meet role responsibilities and consistently perform to a specified standard. Meeting the competency standard requires training and development of the required knowledge, skills and attitudes/behaviors. The standard should align with the hazards under control and enable a sufficient understanding of them and their associated control measures.	20.2
Safety critical communication	This relates to the transmission and receipt of safety critical information throughout the organization, but is particularly relevant for shift handover and permit to work systems.	21
Fatigue and shift work	Performance is adversely affected when humans are fatigued. This can be caused by poor sleep quality, insufficient amount of sleep, or excessive wakefulness. Poor design or management of shift patterns and other factors can lead to fatigue.	22

Table 3.2 Relevance of the Top Human Factors Issues for the 10 Major Accidents Reviewed

Major Accident	Human Failure (Including Maintenance)	Human Factors Interface in Design	Procedures	Organizational Culture	Organizational Change	Staffing	Training/ Competence	Safety Critical Communications	Fatigue/ Shift Work
Oil and gas									
Piper Alpha, 1988	X		X	X			X	X	
Buncefield, 2005	X	X	X	X	X	X	X	X	X
BP Texas City, 2005	X	X	X	X	X	X	X	X	X
Nuclear									
Three Mile Island, 1979	X	X	X				X		
Chernobyl, 1986	X		X		X		X		
Chemicals and plastics									
Flixborough, 1975	X						X		
Bhopal, 1984	X		X	X	X	X	X	X	
Formosa, 2005	X	X	X				X		
Transport									
Herald of Free Enterprise, 1987	X	X		X					X
Kegworth, 1989	X	X			X		X	X	

It can be seen that most major accidents have suffered multiple human factors failures, some more than others. Implementing human factors improvements across these key areas of weaknesses presents an opportunity to make a vital improvement in safety performance.

FINAL REMARKS REGARDING HUMAN PERFORMANCE

There can be a tendency to over-focus on human variability as a "bad thing." It is important to understand that it is the very performance variability that people bring that is essential to normal day-to-day success in operations, and can save a critical situation from becoming a disaster (a case in point being the air crash on the Hudson River in 2009, discussed further in Chapter 23: Managing Performance under Pressure). So there is a need to understand, that if on most days most things go right, why this is the case, as well as studying the exceptions. The answer is often that people are either working around problems or that those systems concerned are in fact well-designed for people. By finding out what people are working around, it is possible to re-design and re-engineer accordingly and improve system resilience. That means not just investigating comprehensively but learning from operating experience and monitoring, auditing and review, in fact, from all sources.

The "role" of human factors is to understand how humans can contribute to "optimal" performance of the system, assure safe human performance, and thereby enable a higher safety standard.

PERSONAL INJURY

The focus of this chapter thus far has been on the effect of poor human factors on human performance, and its contribution to major accidents. It is important to recognize that poor human factors can, and does, also lead to personal injuries, fatalities, and longer term health effects.

MUSCULOSKELETAL DISORDERS

One of the most common categories of injury from inadequate work place design is musculoskeletal disorders which accounts for between 30% and 46% of all work-related injuries in the European Union (European Agency for Safety and Health at Work, 2000). This rate is reasonably consistent across the developed world.

The areas of the body that can be affected include: bones, joints, muscles, connective tissues (tendons and ligaments), other soft tissues, nerves, and circulatory system. The injury reflects underlying changes in the tissues culminating in a musculoskeletal disease or disorder. Many injuries of the back are muscular strains and small tears of the supporting soft tissues. However, back injuries are typically cumulative and develop over months or years of exposure to the risk factors. More serious

Table 3.3 Categories of ULDs

Type of ULD	Example Disorders
Tendon disorders	Tenosynovitis: inflammation of the tendon sheath
	Lateral epicondylitis: also known as tennis elbow
	Medial epicondylitis: also known as golfers elbow
Nerve disorders	Carpal tunnel syndrome: affects the wrist area known as the carpal tunnel and symptoms include pain and numbness
Neurovascular disorders	Thoracic outlet syndrome: caused by compression of nerves and blood vessels
	Vibration syndrome (e.g., vibration white finger): affects the circulation, nerves, muscles, and joints

injuries may result in damage to the vertebrae, or more commonly the intervertebral discs, causing injuries, such as a prolapsed disc. This can also lead to a protrusion on to the nerves and cause other types of disorder, such as sciatica. The lumbar region is a vulnerable part of the back because it bears the weight of the upper body and can be subject to large forces (compression, shear, and torsional forces) from twisting, bending, and stooping. There is a greater likelihood of injury if the person is also bearing additional loads, such as during manual handling operations, but even static nonneutral seated postures can cause these types of injury. Risk factors relate to the characteristics of the load (force), type and frequency of the activity (posture, duration, and repetition), and other factors, such as vibration and personal variables (such as abdominal girth, age, gender, strength, and mobility).

Upper limb disorders (ULDs) are a specific category of musculoskeletal disorders relating to the neck, parts of the back, shoulders, wrists, arms, fingers, thumbs, and hands. There are several different ULD conditions, generally classified as shown in Table 3.3.

The risk factors are summarized as follows:

- Repetition, essentially over-using the same muscle groups;
- Duration, that is, long periods of undertaking the activity, which can either be a single period or cumulative periods;
- Posture, typically extreme body positions where the body parts are deviated away from a neutral posture, such as a fully prone or flexed wrist;
- Force, requiring extreme effort, such as working with weight or overcoming friction;
- Environmental stressors, such as vibration, cold (which reduces blood flow to the extremities) and light (which can cause people to adopt adverse working postures);
- Psychosocial, such as excessive work pace, lack of control over work;
- Individual factors, including body size, health and physical condition, age, ability, attitude.

Lower limb disorders are less frequent and tend to cause general discomfort rather than a specific diagnosis, although pain may be referred from the back, such as in the case of sciatica. Specific work related activities include:

- Kneeling which can lead to bursitis (housemaid's knee/carpet layer's knee);
- Prolonged standing which can lead to joint problems, fatigue, circulatory problems, heart problems, and varicose veins.

PSYCHOLOGICAL RELATED ILL HEALTH

Psychological ill health is generally the next most frequent work-related injury. Stress can lead to psychological disorders, both acute, such as post-traumatic stress disorder (PTSD), and chronic, such as depression and anxiety disorders. Stress can also contribute to physical ill health, such as cardiovascular disease and gastrointestinal disorders. This topic is discussed in more detail in Chapter 23, Managing Performance under Pressure.

As soon as humans sleep less than 6 hours a night, the immune system and gastrointestinal performance can be affected. In the longer term, fatigue can increase the incidence of cardiovascular disease and colon cancer. From the psychological perspective, fatigue can contribute to anxiety and depressive disorders, and the early onset of dementia and Alzheimer's disease. Fatigue is discussed in more detail in Chapter 22, Managing Fatigue.

OTHER TYPES OF INJURY

Humans can also be injured through exposure to:

- Biological, chemical, and radiation hazards, ranging in its consequence from dermatitis to various forms of cancer and death;
- Mechanical hazards, for example, collision with moving objects or parts, projectiles, and sharp objects;
- Altitude or different atmospheric pressures, such as working at high altitude or subsea working which can affect body functions and performance;
- Outdoor climate and extremes of temperature which can lead to heat or cold stress and eventually death;
- Vibration and motion, causing shock, hand arm vibration disorders, motion sickness, musculoskeletal injury, and other difficulties, such as breathing and vision;
- Noise, causing noise-induced hearing loss or noise-induced stress.

While these consequences may not directly result in major accident events, they are important to keep in focus to maintain a healthy and safe workforce.

SUMMARY: REVIEW OF COMMON HUMAN VULNERABILITIES AND CONTRIBUTIONS TO PAST ACCIDENTS

Given that 80% of accidents include human failures, there is a great opportunity to make a significant positive improvement to safety performance. Human factors interventions can counter plateaus in health and safety performance, as well as preventing larger-scale events. Whether the focus is process safety or personal safety, human factors can greatly contribute to the robustness of an organization's safety management system, to improve overall safety performance.

In the next chapter, the current government and regulatory focus is described, and management frameworks for implementing human factors are presented in Chapter 5, Management Frameworks for Human Factors.

KEY POINTS

- The chemical and process industries have achieved significant improvements in system, and health and safety performance, largely due to improved engineering and safety management practices.
- Despite significant improvements, complex safety systems have been undermined by insufficient attention to the organizational and social factors aspects of the sociotechnical system.
- Eight case studies from these industries and two examples from the transport industry have been used to illustrate some of the common human factors themes which have contributed to these prominent major accidents.
- 80% of accidents include human failures. Further improvement in safety performance needs to include assessment of the human factors contribution, but there is also a need to learn from the successes of human performance.
- Poor human factors can also lead to personal injuries, fatalities, and ill health. The most common of these are musculoskeletal injury and psychological related ill health, but injuries are also caused by hazards inherent within the work environment.

REFERENCE

European Agency for Safety and Health at Work, 2000. Work-related musculoskeletal disorders: prevention report. http://osha.europa.eu/en/publications/reports/204/view.

FURTHER READING

BBC Training DVD: Taking Liberties: Fatal Error (Kegworth Air Disaster).
Health and Safety Executive account: www.hse.gov.uk/comah/buncefield/buncefield-report.pdf (Buncefield Storage facility).

Occupational Hygiene Training Association, 2009. Ergonomics essentials, student manual. Available from www.ohlearning.com/training/training-materials/w506-ergonomics-essentials.aspx.

Ship Disasters: www.ship-disasters.com (Herald of Free Enterprise).

Step Change in Safety: www.stepchangeinsafety.net (Piper Alpha Offshore Platform).

US Chemical Safety Board website: www.csb.gov (Texas City Refinery and Formosa PVC Plant).

US Nuclear Regulatory Commission: http://www.nrc.gov (Three Mile Island Nuclear Plant).

Wikipedia: www.en.wikipedia.org (All major accidents).

World Nuclear Association: www.world-nuclear.org (Chernobyl Nuclear Power Plant).

Current regulatory and government focus on human factors

J. Wilkinson

LIST OF ABBREVIATIONS

ABSG	American Bureau of Shipping Group
ACOP	Approved Code of Practice
ALARP	As Low As Reasonably Practicable
API	American Petroleum Institute
BOEM	Bureau of Ocean Energy Management
BOEMRE	Bureau of Ocean Energy Management, Regulation and Enforcement
BSEE	Bureau of Safety and Environmental Enforcement
CFSSWG	Chemical Facility Safety and Security Working Group
CIMAH	Control of Industrial Major Accident Hazards
COMAH	Control of Major Accident Hazard
CSB	(US) Chemical Safety Board
CTSB	Canada Transportation Safety Board
EO	(US Presidential) Executive Order
EU	European Union
HASAW	Health and Safety at Work (Act)
HOF	Human and Organizational Factors
HSE	Health and Safety Executive
HSEM	Health, Safety and Environment Management System
MAH	Major Accident Hazard
MAHB	Major Accident Hazards Bureau
MHF	Major Hazard Facility
MMS	Minerals Management Service
NOPSEMA	National Offshore Petroleum Safety and Environmental Management Authority
NTSB	National Transportation Safety Board
OIW (DCR)	Offshore Installations and Wells (Design and Construction Regulations)
OSC	Oil Spill Commission
OSD	Offshore Safety Directive
OSHA	Occupational Safety and Health Administration
PFEER	Prevention of Fire and Explosion and Emergency Response
PSM	Process Safety Management
PTW	Permit to Work

Human Factors in the Chemical and Process Industries.

RMP	Risk Management Practice
RP	Recommended Practice
SEMS	Safety and Environment Management Systems
TEAM	Technology, Engineering and Management
SMS	Safety Management System
UK	United Kingdom
US	United States
WHS	Work Health and Safety

There are two main types of regulatory regime: prescriptive or goal-setting. These are also known as prescriptive or performance based in the United States, for example, and as risk management or risk compliance in Australia. A third type might be distinguished as "laissez faire" or "minimalist" but this is usually still accompanied by some basic governmental or more local prescription.

The advantages of a goal-setting regime include placing the primary responsibility for managing the risks on the risk owner. This enables newer technologies and industries that may not yet have standards, guidance or other structures, allowing faster development and change. A prescriptive approach such as regulating solely through specific sets of regulations tends to be focused on the status quo and industry looks to the regulator to lead. The use of the term "standards" in this chapter includes regulations for the regulator (see Health and Safety Executive (HSE) 2015, para. 10) and industry standards, which are industry led.

Arguably, the effectiveness of either regime ultimately relies more on the resourcing of the regulatory body or bodies concerned and less on the detail of what is required. If resourcing enables competent people to visit, inspect, and audit sites, to assess safety reports or cases (or other key documentation) against appropriate and transparent standards and guidance, then there is good chance of establishing, maintaining, and improving health, safety, and environmental performance. Andrew Hopkins calls this *"having a highly 'engaged regulator'"* (Hopkins, 2010 p.24) where the regulator checks directly and at a detailed level for compliance but:

> *"These [risk management and risk compliance] are not mutually exclusive approaches: they are complimentary. The issue is therefore not to decide between the two; it is to get the balance right."*

Ibid., p.4

Even the most thorough goal-setting safety case regime will not work if the reality "on the ground" is not tested rigorously against what is claimed.

The focus in this chapter is on Europe, North America, and Australia in the chemical and processing industries, and on their arrangements for human and organizational factors (HOFs). These countries are largely the models for what other countries have adopted and so make good examples to explain and compare. Two good examples of goal-setting regimes are the European Union (EU) including the United Kingdom (UK) and the Australian model. The UK regime is discussed first with reference to the EU. The focus is on major hazards/process safety in this discussion to reflect the higher hazard nature of the chemical and allied processing industries.

The regulatory arrangements for personal or occupational health and safety are similar and also within the remit of human factors. The US approach is a more prescriptive one and is covered in the second part of the chapter on North America.

THE UK AND SEVESO DIRECTIVES

The general background to safety regulation in the UK is that regulations are made under an "umbrella" (or enabling) act, the Health and Safety at Work (HASAW) etc. Act (1974). This includes the As Low As Reasonably Practicable (ALARP) principle, that is, that risks be controlled to this level. The regulations themselves are underpinned with Approved Codes of Practice (ACOPs) and guidance. ACOPs may refer to standards or preferred/recommended methods to achieve compliance with the regulations. Industry standards may provide a route to compliance or at least a part-route.

ONSHORE

The United Kingdom adopted the Seveso II EU Directive (Council Directive 96/82/EC) as the Control of Major Accident Hazard (COMAH) Regulations 1999 (now modified under Seveso III). Its scope was limited to onshore operations. The Directive was triggered by an industrial accident in 1976 in a small chemical manufacturing plant close to Milan in Italy. Six tons of chemicals were released from a batch reaction including a highly toxic form of dioxin. The intent of the Directive was and is to avoid such disasters in the future. The key difference between this and the previous UK regulations, The Control of Industrial Major Accident Hazards (CIMAH), was that CIMAH was not sufficiently focused on process or Major Accident Hazard (MAH) safety (the terms are broadly equivalent but MAH is preferred in this chapter for reasons given in the discussion of process safety management (PSM) below). Inspecting under the CIMAH regime was difficult given that inspection and assessment were largely viewed through the existing occupational health and safety lens of the HSE (the UK regulatory body). Inspectors at this stage did not consistently have appropriate backgrounds or experience in the absence of a full supporting framework, including training, guidance, benchmarks, or standards.

Preparing for COMAH caused the HSE to reflect on what a more MAH focused framework should look like. Extensive work was done to produce suitable guidance, training, educational materials, and arrangements; both internally for inspectors and specialists, and externally for industry. This was also the case throughout the EU to a greater or lesser degree based on resources and the constraints of existing regulatory arrangements.

OFFSHORE

The offshore UK inspection regime continued to be managed under the "umbrella" or enabling HASAW Act with a more limited set of offshore-specific regulations to support it. At a high level, it was goal-setting (under HASAW) but at regulation level, it was more prescriptive. This combination was not always easy to manage to achieve a proper MAH focus. The recently adopted (2015) EU Offshore Safety Directive (Directive 2013/30/EU) (OSD, 2015) is a wider EU acknowledgment of practical application, in responding to the 2010 Macondo incident (for external waters only). This is reflected in the new offshore guidance:

> *"[The Directive] provides an extra level of regulatory control. This is in addition to regulations such as the Offshore Installations, Prevention of Fire and Explosion and Emergency Response (PFEER) regulations 1995 and the Offshore Installations and Wells (Design and Construction) OIW (DCR) regulations 1996, justified by the major accident potential of the offshore activities within scope."*
>
> **HSE (2015a), para. 9**

The Directive and associated UK implementing regulations require a safety case. The previous regime also included a safety case requirement under The Offshore Installations (Safety Case) Regulations from 1992 and these were revised in 2005.

FACTORING THE HUMAN INTO SEVESO

Neither the Seveso Directive nor COMAH mention human factors at all. Therefore, individual countries needed to integrate human factors into their regulations and guidance when they adopted the Directive and in turn integrate this into standards. OSD (2015) does not mention human factors either, but the associated draft guidance now does (HSE, 2015a). The revised COMAH guidance (HSE, 2015b) now also uses the term HOF relatively liberally. For both Directives however, the requirement is for a systematic approach to risk management to be adopted and demonstrated. This cannot be done without due consideration and integration of HOF.

A regulator still needs competent specialist expertise at a practitioner level to operationalize and support this for effective inspection and assessment, that is, to provide full regulatory engagement. In the UK, this was recognized prior to the implementation of COMAH. From 1999, the HSE developed a small team of HOF discipline specialists to operationalize HOF under the "Top 10" (or Key Topics) approach (see Chapter 3: Review of Common Human Vulnerabilities and Contributions to Past Accidents) and to directly support, develop, and work with inspectors. An essential part of deploying such specialists was to ensure they gained the appropriate industry domain experience (e.g., inspecting with an MAH focus) in the relevant industry processes and activities. Technical and professional expertise was recognized as not being sufficient. Reliance on these alone introduces risks to regulator credibility, particularly from either under or over-specifying standards and guidance so that they are unrealistic or inappropriate.

SAFETY CASES AND SAFETY REPORTS

Safety cases were first developed in the nuclear sector and are usually referred to as being part of a "permissioning" or "licensing" regime. They provide an extra layer of control above sets of regulations. Under Seveso and COMAH, this is subtly different and a safety *report* is required. While this requires a demonstration of safe operation, it does not strictly require regulatory formal permission or a license. In the UK offshore example, the permissioning is managed through licensing and assessment. Whether it is a safety case or safety report, a case for safety has to be made and in practice there is not much difference in what is required. Under the Directives "safety case" or "safety report" is defined only by what is expected to be in it, as set out in the Directives' annexes.

The fundamental requirement for an effective safety case or report is that it should reflect reality, identifying the major hazards, and the associated control measures. It should also be a working document; for example, it should be useable and used by the organization concerned. One UK definition is:

> *"safety case is a document that gives confidence to operators, owners, workers and the competent authority that the duty holder has the ability and means to manage and control major accident hazards effectively. Safety cases are intended to be "living" documents, kept up to date and revised as necessary during the operational life of the installation."*

> **HSE (2015a), paras. 9 and 16**

The sense in which "living document" is most useful is where it is actively used as a means of communicating hazards and control measures to make sure they are understood, checked, and continually improved. If the development of the safety case or report does identify significant gaps or necessary improvements then they should incorporate, or at least refer to an appropriate action plan.

As Hopkins points out: *"A safety case is a case - an argument made to the regulator. Companies must demonstrate to the regulator the processes they have gone through to identify hazards, the methodology they have used to assess risks and the reasoning that has led them to choose one control rather than another."* (Hopkins, 2012).

This underlines the importance of using the safety case to show how and why the process or activity is safely managed. It also provides a convincing and transparent narrative of this to the regulator and the organization and acknowledges work left to do.

A typical flaw with safety cases and reports is that the sites and processes they describe can appear to be run entirely without people. The various hazard and risk analyses are presented without real consideration of the human aspects. In this sense they are "dead" documents and not living. However, if HOF is well-integrated within the safety case, the role and importance of people in safe operation can also be clearly shown. For example, the optimization of human performance and the management of human failure can be as rigorously demonstrated as the more technical and engineering measures.

IMPLEMENTING HOF IN THE UK AND EU

Probably the biggest determinant of the uptake of HOF by a regulator or an industry is the presence of HOF specialists to inspect, assess arrangements, and provide support. While there was initially limited direct HOF support for UK offshore inspectors, the onshore HOF team provided training for all inspectors and specialists on and offshore from 2001. The web-based guidance was a shared resource. Recognizing the gap, the UK offshore regime started its own recruitment of HOF specialists from around 2008. Having HOF specialists regularly inspecting offshore assets has significantly increased the amount of industry and regulator HOF activity and focus.

The difference in the original onshore and offshore approaches can still be seen in the HSE web-based guidance for safety case assessors and authors. For the HOF elements, the offshore guidance is higher level and focused on more general documents such as research. By contrast, the onshore COMAH guidance and support is more detailed and practical. In practice the safety case assessors and authors often use both.

More widely, the offshore regime relied (and still does) almost completely on the existing HSE HOF web-based guidance and inspection tools developed for onshore MAH industries. This makes sense because people remain human whether they work onshore or offshore. Where the offshore industry has developed its own guidance such as through the industry organization, Step Change in Safety, this is generally along similar HOF key topic lines as for onshore. In either case (onshore or offshore), the appropriate integration of HOF into the requirements and guidance and the safety report/case is essential to make it meaningful and effective.

HOF REGULATION IN THE EU: SOME VARIETIES

Not all regulators have the same level of maturity with regard to HOF as the UK HSE, even within the EU, but there are examples of good practice. The UK has learnt from Australia (Esso Longford, 1998) and the US, in particular from the way that the US Chemical Safety Board (CSB) integrated HOF into its timely and thorough investigations, such as for BP Texas City (CSB, 2007). A key issue underlying EU variances (other than resourcing where the UK and for example Germany, are well ahead) is the basic regulatory approach and the inspector selection processes. Also, as is also the case in Australia and North America, the degree of federalism in a country can affect how consistently it is able to establish and implement MAH arrangements. The way in which regulation is organized, such as being part of a wider labor department or verification through third parties, can affect how Seveso is implemented and extended or vary the inspectors' agenda significantly.

INSPECTION OR AUDITING: WHAT IS THE DIFFERENCE?

The two main types of regulatory approach are inspection and auditing. The difference between inspection and audit is worth exploring, and not just for regulators. An inspection program should check the reality against what the safety case or report says. The HSE defines inspection as:

'...the process carried out by HSE warranted inspectors which involves assessing relevant documents held by the duty holder, interviewing people and observing site conditions, standards and practices where work activities are carried out under the duty holders control. Its purpose is to secure compliance with legal *requirements for which HSE is the enforcing authority and to promote improving standards of health and safety in organisations.'*

HSE (2015c)

When the HSE talks about its own audits (as forms of inspection) it is clear these are viewed as different from inspection. For example, an audit is used to verify certain aspects of a safety case or report.

Inspection implies that a basic HOF approach is taken. For example, walking and talking through selected tasks and activities on-plant with the personnel that carry them out. This is essential in order to gain a realistic picture of what is going on as opposed to just verifying documentary arrangements against the safety report as an audit might do. Simply verifying the organization's own audit arrangements and outcomes does not in itself provide a sufficiently real picture. There is an element of triangulation in inspection implied by the HSE definition, for example, converging on the reality through a mix of document review, interviews, and walk or talk-throughs of selected activities against an appropriate performance standard such as a procedure. Also bringing inspection and assessment experience (including incident learning) to bear on what is looked for and what questions are asked. This is a larger, more structured and purposeful approach than auditing. More succinctly it could be summarized as: "Excellent documentation thank you but is your operation any safer as a result? How do you know? Now let's go take a look."

Auditing varies but is usually defined for example as: "... *a systematic and independent examination of data, statements, records, operations and performances (financial or otherwise) of an enterprise for a stated purpose.*" (Wikipedia, 2015— other sources are similar). This does not include active observation and clearly reflects the financial sector derivation of auditing as a process.

Auditing tends to be adopted by some EU regulators because of custom and practice and also to manage lower resource levels or third party use, such as the use of technical bodies or other organizations. Audits range from one-off annual activities or may be part of more integrated and wider inspection program. Whilst the audit may be rigorous, it may still be limited and not always supported by a reality check in the workplace. Although most auditors involve or at least talk to employees as a matter of principle, this is not the same as checking workplace conditions directly. Equally, inspectors tend to be recruited from technical or engineering backgrounds. Therefore, dealing with Safety Management Systems (SMSs) and even more so with HOF issues, can be problematic without effective HOF training and support. Unfortunately, HOF issues are not generally covered in any depth in the engineering and technical fields. This was identified as a continuing issue at a recent EU Major Accident Hazards Bureau (MAHB) EU Inspectors' Forum (MAHB, 2014).

If a regulator is auditing a safety case or report through a third party or its own inspectors, there is a problem with ensuring that the workplace and process reality matches what documents and interviews or discussion show. In a prescriptive regime the main tool is likely to be audit, as demonstrated by the US regulator responsible at the time of Macondo. Inspection in principle identifies workplace conditions and performance standards that are poor or drifting away from good practice. So the degree of regulator engagement is in large part determined by whether they primarily use an inspection or audit approach. In practice an effective regulator will use both appropriately. Without inspection the compliance checking is likely to be less rigorous and less likely to reflect reality.

CONCLUSIONS ON THE UK AND SEVESO PICTURE

The discussion shows that given these variances, even within one regulated entity like the EU, the picture is more complex than just the "safety case/report versus prescriptive regimes" difference. In general, a useful conclusion is that a fully-engaged regulatory inspection program that uses a wide range of tools, including audit, is likely to be better at establishing the MAH safety performance and is therefore likely to be more effective. This is easier done under a suitable over-arching requirement such as ALARP.

Specialist HOF capability is also required that is integral with the inspection or audit program, and not just in a supporting or research role. HOF is a discipline in its own right in the same way as, for example, control and instrumentation, mechanical engineering, and process safety. It has accepted professional institutions and similar bodies worldwide to attest to this. If it was "all common-sense" then everyone would all be doing it and they clearly are not.

AUSTRALIA

Although some readers might assume that Australia just followed the EU/UK lead, there is in fact a richer picture. The seminal accident that turned the focus to MAH/process safety was the Esso Longford disaster in 1998. The ensuing investigation and public enquiry helped drive forward the State of Victoria's regulatory regime and had impacts more widely across Australia and beyond. Andrew Hopkins book "Lessons from Longford" (Hopkins, 2000) made the lessons on MAH/process safety more explicit and clearer and attracted a global audience. Even though the Seveso Directive coincided with this activity, it took the Longford incident and a subsequent HSE investigation into a series of incidents at the BP Grangemouth refinery in 2000 (HSE, 2003), to really turn the focus of regulators towards this specific issue.

Countries like Australia have multiple regulators. Specifically, Australia is a federation of six states (and three self-governing territories), which have individual constitutions, parliaments, governments, and laws. The result is that companies often have more than one regulator to satisfy. This can create tensions and difficulties for the operators in achieving consistent compliance. In practice, the Australian onshore regulators have different levels of focus on MAH safety and HOF, and this can be seen in their varying website guidance. However this is changing. Safe Work Australia was established as an independent Australian Government agency in late 2009 as a tripartite body representing the states and territories, employers, and workers. It develops policy, law, regulations, codes of practice, and other material but it is not itself a regulator. A model Work Health and Safety (WHS) Act and Regulations

(Safe Work Australia, 2011) were adopted in 2011 and have to be individually implemented by the states and territories (two have yet to do so). These include Major Accident Facility (MAF) requirements including safety cases. Human factors are not specially mentioned in the regulations but they are in the guidance. A major review planned for 2017 should indicate how effective all these developments are and how engaged individual regulators are in practice.

Following the Montara offshore incident in 2009 (and reinforced by the subsequent Macondo disaster in 2010), the Australian Commission of Inquiry recognized similar issues for offshore regulation, and recommended:

> *"...that a single, independent regulatory body be responsible for safety as a primary objective, in addition to well integrity and environmental approvals."*
> **National Offshore Petroleum Safety and Environmental**
> **Management Authority (NOPSEMA, 2015)**

The resulting new Australian offshore regulator NOPSEMA operates a safety case regime (adopted in 2009) with direct inspection and verification of claimed control measures. It has recently appointed a HOF specialist and is focusing more on these topics. This is evidenced by the development of a suite of online HOF guidance on their website where the influence of the UK model can clearly be seen. This is work in progress in their drive to "...*standardise Australia's offshore petroleum regulation to a best practice model.*" (*Ibid.*). However, where HOF is currently a focus for Australian companies, this is reportedly more so because they drive it themselves rather than because the regulators are pushing them to do so. The operators recognize the safety and business benefits. That said, there are promising signs of wider activity and engagement even despite the recent oil and gas downturn.

CONCLUSIONS FOR AUSTRALIA

Since 2009 Safe Work Australia has provided an independent drive onshore to implement national health and safety policy and strategy, and has provided an enabling Act and regulatory framework for individual states and territories to adopt. This includes a safety case regime for major hazard facilities (MHFs). The Montara incident prompted swift action in Australia to implement a strict safety case regime for offshore under one regulator. NOPSEMA is enabled to engage fully in checking compliance directly as well as approving cases. There are promising signs of increased HOF awareness and activity.

NORTH AMERICA

In discussing the arrangements in place for Canada and the US, most of the focus in this chapter is on the US.

CANADA

In Canada, while there generally appears to be an overarching legal permissioning requirement for offshore oil and gas operations, there is nothing apparent for onshore major hazards, except in the nuclear sector. A Canadian study by Queen's University of Kingston Technology, Engineering and Management (TEAM), Department of Chemical Engineering on MAH control states that:

> *"Despite the developed state of industry in Canada, there are serious concerns about the state of process safety in this nation."*

> **TEAM (2009)**

It concludes that:

> *"…the state of major industrial accident prevention in Canada was generally insufficient."*

> **Ibid.**

The report also made recommendations for improvement in central oversight and control. This has not yet happened.

The main regulatory focus in Canada is still heavily biased towards occupational health and safety and even within this focus HOF is largely dealt with by reference to lower consequence ergonomic issues. The federal picture appears complex with 14 jurisdictions. There are 1 federal, 10 provincial, and 3 territorial jurisdictions with each having its own occupational health and safety legislation. The Canada Transportation Safety Board (CTSB) has a major accident focus but only for the investigation of transportation incidents in the marine, pipeline, rail, and air sectors. It operates similarly to the US National Transportation Safety Board (NTSB) with which it has close ties. The CTSB was involved in the investigations of the recent and high-profile Canadian railcar and pipeline accidents.

THE US

The US CSB was set up in 1998 as an independent federal agency to investigate major industrial chemical accidents. The US has increased its regulatory focus on major hazards onshore since the BP Texas City disaster following a number of CSB investigation recommendations. This has substantially increased the number of inspectors available for major hazard inspection work in the federal regulator, the Occupational Safety and Health Administration (OSHA). However, given the size of the country and the extent of the major hazard industries this is still a relatively small number. Subsequent incidents, such as the fertilizer explosion in West, Texas, in 2013 (discussed further below), demonstrate the need for an appropriately resourced OSHA a strong PSM standard. (The West incident and the OSHA PSM are discussed further below.)

For offshore installations, the key driver was the Macondo Well disaster in 2010. There was a relatively quick initial response to this in replacing the existing prescriptive and under-resourced regulator, the Minerals Management Service (MMS) with a new one, the Bureau of Ocean Energy Management, Regulation and Enforcement

(BOEMRE). In 2011 this was in turn reorganized into the Bureau of Ocean Energy Management (BOEM) and the Bureau of Safety and Environmental Enforcement (BSEE), separating its main functions more appropriately for its dual role.

The immediate regulatory response in 2010 was the introduction of the Safety and Environment Management Systems (SEMS) regulations for offshore (for outer continental shelf) operators. In effect this is an SMS performance-based regime.

> *"...A SEMS program is a comprehensive system to reduce human error and organizational failure."*
>
> **BSEE (2014), p.1**

This incorporated and enhanced a previous voluntary program to adopt the American Petroleum Institute's (API) recommended practice (RP) 75 which is effectively its standard for SEMS. While this has been developed since 2010, for example to include a workforce participation requirement, SEMS has been and still is subject to criticism. It currently requires the regulated operators to carry out their own SEMS audits and report on them to the regulator, so there is no direct inspection or auditing by the latter. Later the SEMS II amendments stipulated that the auditing be conducted by third parties. The focus is on documents and procedures and although their implementation is also mentioned, this is more as a review exercise of what the operators report they have done themselves (*ibid.*, p.1). In Hopkins' words the regulator is not fully engaged in checking compliance directly at the sharp end.

The SEMS regulations are intended to:

> *"...represent a shift toward more of a performance-based regulatory model versus the traditional focus on strict compliance...."*
>
> **Ibid., p.1**

This goes beyond compliance and the previous prescriptive regime under MMS. The latest evaluation of these reports by BSEE concludes that, although there was 96% compliance in implementing SEMS, this did not tell the whole story. BSEE states that:

> *"...the general finding is that the current status of SEMS implementation is geared toward compliance."*
>
> **Ibid. (2014), p.4**

Operators in general did not provide evidence that they were implementing SEMS as an effective management tool and BSEE states that *"The system maturity and level of SEMS awareness and understanding amongst operators varied significantly"* (*ibid.*, p.2).

More mature operators were able to simply map across their existing Health, Safety and Environment Management System (HSEM) arrangements to SEMS and integrate missing areas. Other less mature operators often treated this as a compliance and checklist exercise rather than putting effort into developing an effective SEMS program. Some implemented the requirements without integrating them with existing arrangements. This had predictable results in that it did not work (*ibid.*, pp.2–3).

STANDARDS

Some of the original MMS regulatory issues were to do with the quality of the API RP75 standard on which the offshore industry based its arrangements (where it did). This standard was integrated into SEMS within a fairly standard international HSEM framework. The problem with the standard and others like it is that the industry strongly influences what goes into it. In practice, this can lead to defensive maneuvering to avoid the perceived threat of additional cost and "burden" from more regulation rather than the sharing and integrating of good practice and development. For example, post Texas City and the ensuing (2007) CSB recommendation to modify the existing API on fatigue management, the CSB declared the proposed revision as still "unacceptable" during a formal board resolution in 2015 (CSB, 2015). In the same way API75 provides a limited basis for regulation intended to focus on MAH or process safety. BSEE are aware of this and working on it as described above.

There is a similar issue with PSM standards generally. OSHA's current PSM standard fails to integrate a full HOF approach and the standard currently applies only to the petroleum industries, not to oil and gas. OSHA invited public comments for revision in 2013 following the Post-West Executive Order (EO) but no revision has yet taken place. The CSB have urged (June 2015) OSHA to extend significantly the scope of its proposed revision including more explicit requirements for HOF (CSB, 2015). BSEE has also recently solicited a report from the American Bureau of Shipping Group (ABSG) consulting on the extent that PSM principles and requirements are integrated (or at least present) in SEMS. The report concludes that there is significant work to do, including the need to:

> *"Adopt a comprehensive human factors standard and require qualitative or quantitative Human Reliability Analysis for each process or operation."*
>
> **ABSG (2015) p.ii**

It is the quality and scope of such a standards development and actual uptake that will determine what is done by the industry and the regulator. It also means using and integrating existing HOF standards and principles in the proposed adoption of such a HOF standard. Finally, a fully engaged regulatory approach will also be required to check compliance directly and at a sufficiently detailed level.

A SAFETY CASE REGIME FOR THE US?

The Presidential Commission of inquiry; the Oil Spill Commission (OSC) into Macondo recommended that "...*the US should develop a pro-active, risk-based performance approach specific to individual facilities, operations and environments, similar to the 'safety case' approach in the North Sea.*" (OSC, 2011, p.252; quoted in Hopkins, 2012).

Hopkins succinctly illustrates the problems for the US, both onshore and offshore in moving to a safety case regime (Hopkins, 2012). It identifies five key elements of such a regime:

- A risk or hazard management framework;
- A requirement to make the case to the regulator;
- A competent and independent regulator;
- Workforce involvement;
- A general duty of care imposed on the operator.

Hopkins argues that only the first two elements can currently be met in the US. The CSB is (as at January 2016) completing its final reports (volumes 3–4) on Macondo which will include recommendations on moving towards a safety case regime. This follows their exhaustive final analysis of the causes of the disaster, of the multiple investigation reports, and of progress made (or not made) since it occurred.

Hopkins points out that:

> *"One of the misconceptions in the US about safety case regulation is that it involves the abandonment of prescription. That is not so. A safety case requires that technical standards be specified and regulators can then enforce those standards."*
>
> **Ibid., p.5**

This underlines the point already made, that whether a regime is a safety case one or a prescriptive one, both rely on the quality of standards (regulations, industry, and organization specific), as well as good supporting guidance. In practice the regimes will always overlap significantly in this way because good practice comes primarily from industry and not from the regulator. This is what should be collected and shared to continually improve standards. In another paper (Hopkins, 2010) Hopkins contrasts the twin approaches of risk management (implied in a safety case regime) and rule compliance (prescription—and compliance implies that there are rule-makers and enforcers as well of course). He makes a case for a balance so that prescriptive technical rules in standards and regulations are also necessary to ensure risk compliance whether or not this is within a risk management framework such as a safety case regime. He also argues that an over-arching requirement such as the ALARP principle is also necessary to enable regulators to insist on consistency in applying standards and regulations whether the regime is prescriptive or goal-setting. Companies cannot then apply lower standards in one country than in another even if standards or regulations vary. They must comply with ALARP or the equivalent. This also helps standards' development to become more consistent.

Safety case regimes are not perfect solutions. For example, Hopkins cites the investigation into the UK loss of a British Air Force Nimrod aircraft in Afghanistan in 2006. He argues that the regulator in that case was not sufficiently competent or independent in assessing or checking the safety case and accepted it despite obvious defects. Equally, as the UK found out through offshore and later onshore experience, after the Buncefield oil storage depot explosion and fires in 2005, a safety case or report regime can get stuck in an assessment rut and not be able to do enough to check actual performance in a timely way in the workplace. At the Buncefield depot, the effort put in to extract a workable safety report from the operator was in the end at the expense of timely inspection effort where arguably some of the precursors to the accident could have been spotted. But caution is also necessary here, as hindsight can easily place unreasonable expectations on those who were there at the time.

An internal HSE review identified the need to revise the safety report assessment arrangements and the post-Buncefield 2007 (and still current) revision has new requirements including an emphasis on the need for timeliness in moving from assessment to verification on site, and on proportionality (HSE, 2007). Earlier, the offshore sector's experience of the repeated cycle of revised submission led to their arrangements being improved in the mid-2000s to speed up the process for the operators and the assessors and to introduce more proportionality into this for subsequent safety case submissions.

DEVELOPMENTS FOLLOWING THE WEST FERTILIZER COMPANY EXPLOSION

A Presidential EO (The White House, 2013) was prompted by the West Fertilizer Company explosion in West, Texas, in 2013. It is aimed at "*Improving Chemical Facility Safety and Security*" and tasks a number of US regulatory executive departments and agencies with identifying additional measures to achieve this improvement. OSHA facilitated a stakeholder meeting (Expert Forum on the Use of Performance-Based Regulatory Models in the U.S. Oil and Gas Industry) in 2012 though not yet with conclusive outcomes. A cross-agency Chemical Facility Safety and Security Working Group (CFSSWG) was established to identify and report with recommendations for action (CFSSWG, 2014):

> "*...to identify ways to improve operational coordination with state, local, tribal and territorial partners; to enhance Federal agency coordination and information sharing; to modernise policies, regulations and standards to enhance safety and security in chemical facilities and to work with stakeholders to identify best practices to reduce safety and security risks in the production and storage of potentially harmful chemicals.*"

Ibid., p.iii

In their report, "safety case" is mentioned only briefly and in terms of:

> "*Evaluating whether EPA, OSHA, and PHMSA should implement a "safety case" regulatory model to lower risks as much as is reasonably practicable in complex industrial processes.*"

Ibid., p.84

The industry and other responses to this proposal are cited as: "*Overwhelmingly, nearly all comments received regarding the adoption of the safety case regulatory model were negative.*" (*ibid.*, p.85) and there is little development of this idea in the report. Even with further endorsement from the forthcoming CSB Macondo reports for offshore activities the concept is likely to at least take significant time and energy to introduce in the US. However, there is discussion about establishing and sharing best practice, and improving standards with stakeholder involvement.

The priority actions from the CFSSWG are being implemented by the agencies concerned. There is a strong focus on improving existing PSM and Risk Management Practice (RMP) approaches both for safety and for the environment. It is possible (albeit a little cynical) that the floating of the safety case idea is an implied threat in case stakeholders fail to engage in establishing and sharing best practices, and in improving standards. HOF integration is mentioned specifically for improving and modernizing process safety (*ibid.*, p.83):

> *"Determining how EPA, OSHA, and PHMSA could better account for human factors in areas such as process safety, management of change, facility operating procedures, incident investigation, training, PHA and other elements."*

This is promising and ties in well with the BSEE/ABSG review discussed above for offshore activities. Although as presented it still appears to stop short of a full HOF agenda, largely because it is based on existing PSM approaches. However, any such improvements would likely have to bring those other areas into play. The question of resourcing the regulators better is mentioned briefly but mainly in relation to emergency planning. There appears to be little appetite for significant expansion and the main focus is on other stakeholders becoming more engaged. The question of improving regulatory engagement (for improved compliance) is not explicitly discussed.

CONCLUSIONS FOR NORTH AMERICA

In Canada there is some awareness that a major hazard regime of some kind is required in the face of a succession of serious accidents but so far no specific activity. In the US, Macondo has prompted a move to an SMS (SEMS) performance-based regime for offshore oil and gas which has not so far been fully effective. BSEE assesses the companies' own SEMS' audits. It is therefore not fully engaged in the sense that it does not (yet anyway) directly inspect the MAH arrangements and control measures on the offshore installations. The West accident has prompted activity from the President downwards towards a performance-based framework for onshore hazardous chemical facilities. So far this has not resulted in concrete change to the regime or to the current more narrowly-applied and incomplete PSM standard. Problems with standards' making, consistency, and compliance are part of a wider problem in the US which has not yet been addressed.

WHAT DOES "GOOD" LOOK LIKE FOR REGULATION?

It would be easy to say that every regulator should operate a safety case or safety report goal-setting regime, but it is unlikely that such an approach would suit every country's situation onshore and offshore at every time. It is also not appropriate just to say that more (industrially) mature countries should progress to the latter because this is easily interpreted as condescending or patronizing and would not necessarily be welcomed or productive and indeed, may not be true. Hopkins points out that historically it is not necessarily correct to see a linear progression from risk compliance towards risk management and that there may be a necessary corrective shift back towards compliance currently (Hopkins, 2010, p.4).

A better argument might be to focus on standardizing good practice and guidance globally for all areas including key HOF topics. Current PSM standards do not do this well enough on their own. As an example of this on a global scale, the International Civil Aviation Organization (ICAO), a specialized agency of the United Nations (and not a trade association or regulatory body):

> *"works with the Convention's 191 Member States and industry groups to reach consensus on international civil aviation Standards and Recommended Practices (SARPs) and policies in support of a safe, efficient, secure, economically sustainable and environmentally responsible civil aviation sector."*

ICAO (2016)

The US EO and the CFSSWG action plan support this view in focusing pragmatically on best practice and standards and there is evidence of a desire to integrate HOF. This is also the case for offshore (ABSG, 2015). This should be supported by making sure that the business case for implementing and continually improving it is also well communicated, and the safety benefits recognized. New and emerging technologies or other developments will take time to evolve into standards so there will always need to be an over-arching goal-setting requirement as well. The good practice comes from *practice* and cannot be developed in a vacuum. Whatever inspection or auditing program is used, it also needs to be effective in checking the workplace and process reality at a sufficiently detailed level. This includes appropriate HOF specialist input and adequate resourcing for the regulator concerned.

Generally, the MAH industries still sometimes struggle to share fully and effectively their own good practices and learnings from incidents and from other industries. If there are "live" and practical standards reflecting good practice coupled with practical guidance, then a prescriptive regime can still achieve a lot, providing industry drives the standards and does not wait for the regulator's lead. In the EU the regulators are often closely involved in standards' making but this is not the case in the US. Ideally the standards' making would be driven by a body like the ICAO which is sufficiently independent of industry and the regulator. The ICAO is more than just a standards organization.

In making standards and managing their hazards, industries need to avoid using incident and investigation learnings as a kind of "screening out" or defensive checklist otherwise it is not possible to learn more effectively and promote the right kind of "chronic unease" that MAH control requires. Essentially, organizations need to remain curious about both what is working in their own systems and what is not. The key is to understand what is done well (and why) and also not taking it for granted that a lack of incidents or apparent issues in one area means all is well. Regulators also have a key role in promoting both the sharing and the learning, including their own.

SUMMARY: CURRENT REGULATORY AND GOVERNMENT FOCUS ON HUMAN FACTORS

This chapter discussed the regulatory and government focus on human factors in different parts of the developed world. A distinction was made between two main types of regulatory regime: prescriptive and goal-setting and the relative advantages and disadvantages of each approach. The next chapter discusses management frameworks for human factors and possible approaches that can be taken. It is suggested that risk owners are best placed for establishing the management framework.

KEY POINTS

- Two main regulatory models exist, goal-setting (as exemplified by a safety case or report regime) and prescriptive through specific sets of regulations.
- Comparing examples of both regimes shows the seminal role of major accidents in driving countries towards a more goal-setting regime.
- There is no "one size fits all" regulatory approach. The two main types will suit different countries at different times and in practice it is the balance between the two that is more important.
- To achieve a good balance the regulator needs to be fully engaged, that is, checking compliance at a sufficiently detailed level, and directly. This implies a sufficient level of resourcing for the regulator and effective selection, recruitment, and training arrangements for inspectors. An over-arching framework or requirement such as ALARP is also essential for consistency.
- A focus on inspection, including auditing but not dominated by it, is a significant contributor to regulatory engagement and effectiveness. The distinction between inspecting and auditing is not always well understood but a proper inspection focus on the reality on MAH sites is essential.
- While a safety case or report regime may seem to be the "gold-plated" option, it is not necessarily the only way of improving safety and has its own potential problems. Without a full HOF integration (both required by the regulator and delivered by the safety case or report owner) these can be weak documents.
- The deployment of practical and credible direct HOF support for inspection is identified as a key driver for regulatory and industry improvement. If the business benefits from HOF integration are also well articulated, then industry is more likely to want to establish and share good practice.
- It is argued that a proper focus for both types of regimes is in developing good practice standards and guidance and that while there is a goal-setting role for

the regulator with this as well, the good practice can ultimately only come from industry sharing and from technological advances. An independent body can more effectively lead standards' development and improve consistency and compliance.

- Current US proposals support this view in trying to develop robust standards for MAH or PSM arrangements and for fuller HOF integration within these. The adoption of a safety case regime for the US is currently being vigorously debated for on- and offshore.

REFERENCES

ABSG, 2015. Process Safety Assessment Final Report Submitted to the Bureau of Safety and Environmental Enforcement (BSEE). ABSG Consulting Inc., Arlington, VA, (e-report) Available from: <http://www.bsee.gov/uploadedFiles/BSEE/Technology_and_Research/Technology_Assessment_Programs/Reports/700-799/732aa.pdf> (accessed 25.02.2016.).

BSEE, 2014. SEMS Program Summary—First Audit Cycle (2011–2013). BSEE Issue Memorandum, 23 July 2014. Available from: <http://www.bsee.gov/uploadedFiles/BSEE/Regulations_and_Guidance/Safety_and_Environmental_Management_Systems_-_SEMS/SEMS%20Program%20Summary%208132014.pdf> (accessed 25.02. 2016.).

CFSSWG, 2014. Executive Order 13650 Actions to Improve Chemical Safety and Security—a Shared Commitment. Chemical Facility Safety and Security Working Group. May 2014. Available from: <https://www.osha.gov/chemicalexecutiveorder/final_chemical_eo_status_report.pdf> (accessed 25.02.2016.).

CSB, 2007. Investigation Report BP Texas City, Texas, 23 March 2005, Report No. 2005-04-I-TX Refinery Explosion and Fire (15 Killed, 180 Injured), US Chemical Safety Board, Washington, DC. Available from: <http://www.csb.gov/assets/1/19/csbfinalreportbp.pdf> (accessed 25.02.2016.).

CSB, 2015. Rulemaking Watch: OSHA's Process Safety Management Standard and EPA's Risk Management Program. Available from: <http://www.csb.gov/mobile/recommendations/regulatory-watch-process-safety-management/> (accessed 25.02.2016.).

Hopkins, A., 2000. Lessons from Longford. CCH Publishing Australia Ltd, Sydney, ISBN:9781864684223.

Hopkins, 2010. Risk Management and Rule Compliance: Decision Making in Hazardous Industries. Working Paper 72. (e-paper) National Research Centre for OHS Regulation, ANU, Canberra. Available from: <http://www.processsafety.com.au/free-downloads/working-papers/> (accessed 25.02.2016.).

Hopkins, 2012. Explaining "Safety Case." Working Paper 87. (e-paper) National Research Centre for OHS Regulation, ANU, Canberra. Available from: <http://www.csb.gov/assets/1/7/WorkingPaper_87.pdf> (accessed 25.02.2016.).

HSE, 2003. Major Incident Investigation Report BP Grangemouth Scotland: 29 May–10 June 2000. HMSO. Available from: <http://www.hse.gov.uk/Comah/bpgrange/index.htm> (accessed 25.02.2016.).

HSE, 2007. Statement by the Competent Authority on the Safety Report Assessment Manual. Version 2 May 2007. Available from: <http://www.hse.gov.uk/comah/sram/sramchanges.pdf> (accessed 25.02.2016.).

HSE, 2015a. The Offshore Installations (Offshore Safety Directive) (Safety Case etc.) Regulations 2015 Guidance on Regulations. L154 Second Draft July 2015. Available from: <http://www.hse.gov.uk/pubns/priced/l154.pdf> (accessed 25.02.2016.).

HSE, 2015b. A Guide to the Control of Major Accident Hazards Regulations (COMAH) 2015. L111 3rd Edition. ISBN:9780717666058. Available from: <http://www.hse.gov.uk/pubns/books/l111.htm> (accessed 25.02.2016.).

HSE, 2015c. (Undated webpage). Available from: <http://www.hse.gov.uk/foi/internalops/og/ogprocedures/inspection/> (accessed 25.02.2016.).

ICAO, 2016. Home Web Pages. Available from: <http://www.icao.int/about-icao/Pages/default.aspx> (accessed 25.02.2016.).

MAHB, 2014. Safety Management Systems in Multinational Companies. (e-guidance) Mutual Joint Visit Workshop for Seveso Inspectors: 17–19 September 2014, Arona, Italy. Available from: <https://minerva.jrc.ec.europa.eu/en/content/f30d9006-41d0-46d1-bf43-e033d2f5a9cd/publications> (accessed 25.02.2016.).

NOPSEMA, 2015. Home Web Pages. Available from: <http://www.nopsema.gov.au/about/history-of-nopsema/> (accessed 25.02.2016.).

OSC, 2011. Oil Spill Commission, National Commission on the BP Deepwater Horizon Oil Spill and Offshore Drilling, Deepwater: The Gulf Oil Disaster and the Future of Offshore Drilling, Report to the President. ISBN:9780160873713. Available from: <http://www.gpo.gov/fdsys/pkg/GPO-OILCOMMISSION/pdf/GPO-OILCOMMISSION.pdf>.

Safe Work Australia, 2011. Model Work Health and Safety Act 2011. Available from: <http://www.safeworkaustralia.gov.au/sites/swa/model-whs-laws/model-whs-act/pages/model-whs-act>.

TEAM, 2009. The Control of Major Accident Hazards in Canada. Technology, Engineering and Management (TEAM), Department of Chemical Engineering at Queen's University (Kingston, Ontario), 2009. Canadian Society for Chemical Engineering. Available from: <http://team.appsci.queensu.ca/documents/CSChE_ControlMajorAccidentHazardaCanada_Report.v4.pdf> (accessed 25.02.2016.).

The White House, 2013. Improving Chemical Facility Safety and Security. Executive Order (EO) 13650—1 August 2013. Available from: <https://www.whitehouse.gov/the-press-office/2013/08/01/executive-order-improving-chemical-facility-safety-and-security> (accessed 25.02.2016.).

Wikipedia, 2015. Webpage available from: <https://en.wikipedia.org/wiki/Audit> (accessed 25.02.2016.).

Management frameworks for human factors

5

J. Edmonds

LIST OF ABBREVIATIONS

EHS	Environment, Health, and Safety
HSE	Health and Safety Executive
HFI	Human Factors Integration
HFIP	Human Factors Integration Plan
MANPRINT	MANpower and PeRsonnel INTegration
PIFs	Performance Influencing Factors
SMS	Safety Management System

Human factors is an integral part of achieving successful organizational, and health and safety performance and a comprehensive approach to managing human factors is required to realize these benefits. The approach requires consideration of the depth of the different topic areas, whilst recognizing the breadth of coverage and the need to approach the subject from a holistic perspective. A management framework will help an organization to achieve this level of coverage whilst enabling the human factors effort to achieve the intended outcome. There is no single definitive management framework, and it is preferable for each organization to develop its own to ensure that it is relevant and integrated with existing practices and infrastructure. Having a management framework does not mean that it needs to be complex or unwieldy. In fact the simpler the system and the more integrated it is with the organization's existing infrastructure, the more likely it is to be successful.

It is quite likely that an organization already addresses some aspects of human factors, but it may not be recognized as falling within the human factors remit. For example, there may be a process to consider control room design, or measure safety culture for example. While some human factors issues may already be receiving attention, they may require a deeper or broader focus to attain a higher standard that is sufficiently comprehensive. It is likely that some organizations will greatly benefit from developing a management framework to acknowledge what they already do, and to highlight gaps and improvements that need to be addressed.

The most obvious "home" for human factors within an organization is for it to be integral to health and safety management. However, it also needs to be integral to engineering management and considered within operational management. Some organizations tackle specific issues within the human resources function, such as organizational culture, work hours and fatigue, staffing, workload, and management of change. So wherever it actually resides, it does need to encompass each of these aspects.

There are models of existing practice that can be applied or used as a starting point, and this chapter focuses on how to integrate human factors within an organization, particularly in relation to where to start.

BASIC PREMISE OF A HUMAN FACTORS MANAGEMENT FRAMEWORK

Any effective management framework should provide a structure to apply the human factors effort. There will likely be numerous activities with a range of goals, but they will ideally be undertaken within a single framework, so that the effort is coordinated. There are some fundamental aspects that the organization will need to understand:

- What it is trying to achieve in relation to human factors (*Goals*)
- How it will achieve the goals (*Structure*)
- How well the organization is performing in relation to human factors (*Measurement*, *Action*, and *Review*)

This simple premise applies to an organization that is just starting to consider human factors or one that is already on its journey.

An illustration of a basic human factors management framework is presented in Fig. 5.1. It is then discussed in more detail in the following subsections.

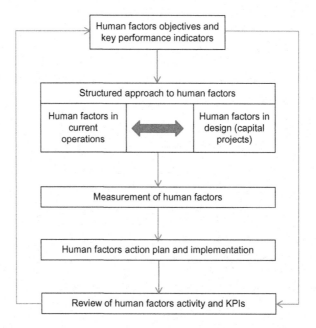

FIGURE 5.1

Basic management framework for human factors.

GOALS

The first aspect is to understand what the organization needs to achieve with regard to human factors. The high-level objectives need to include specifics about achieving exemplary health and safety performance and effective human performance. This may include a reduction in human contributions to incidents, lower design modification costs, fewer production interruptions, or whatever is relevant to the organization. In particular, the organization needs to ensure that human factors contributions to major accidents is understood and well managed.

In conjunction with goal-setting, the organization may benefit from defining the measures to assess subsequent performance against the primary objectives.

STRUCTURE

The next step is to outline how the organization will meet its objectives. What does it look like? How can performance be improved? This can be defined using a basic structure of the human factors relevant to the organization. Given that different topic areas within human factors are intrinsically linked (e.g., changing rosters to manage fatigue, relies on effective management of organizational change), it is important to ensure broad coverage and not just a specific slim view of the subject, even if some aspects will not be tackled for some time. Different approaches for a human factors structure are presented later in the chapter.

In Chapter 1, What is Human Factors? a simplistic view of a work system life-cycle was presented, and this can relate to any type of facilities, such as an offshore installation or chemical processing plant. It was stated that a system will go through three distinct phases:

- *Design*: where it is conceptualized, designed, developed, and built;
- *Operation*: where it performs the function it was designed to perform;
- *Decommissioning*: where it is taken back to its starting point where it no longer exists as a system.

The lines between the phases are blurred. For example, design can occur during operations in the form of midlife updates, operations may continue whilst certain processes are commissioned or decommissioned, and the decommissioning phase may include design and the performance of decommissioning functional activities (or decommissioning operations).

The broad topic of human factors can be grouped within a single structure, but it can be useful to make a distinction between human factors in *design* and human factors in *operations* because they are quite different in how, and by whom, they might be managed (decommissioning would fall into both aspects). If it is divided into two aspects, it is useful for the frameworks to be linked and cross-refer to each other.

MEASUREMENT, ACTION, AND REVIEW

There needs to be a means to measure current and future progress against the objectives for the different areas of coverage and this is likely to require some form of gap analysis. The gap analysis can be used to prioritize the organization's effort and formulate an action plan to structure its implementation. It is recognized that it will take time to develop a management framework for human factors and to fine tune it, which is why it is necessary to prioritize efforts so that the action plan is practical and achievable.

The action plan should ensure that sufficient resources are made available. This includes relevant knowledge and expertise, a clear outline of the activities to be undertaken and the dates by which they should be achieved. The basic action plan should be supported by raising awareness and engagement in the organization and gaining an understanding of workforce opinions. As with any new program, endorsement from senior level management will be a key determinant of success.

Any action plan must be reviewed to ensure that actions have been implemented and are effective. There is also a need to periodically review the human factors effort to ensure that it is enabling progress to be made. This may be done by identifying a set of Key Performance Indicators (KPIs) relevant to the organization. As with any safety metric, this should consider both "leading" and "lagging" indicators. Leading indicators provide an opportunity to proactively confirm that controls are functioning as required, such as the completion of safety critical task analyses, procedure reviews, or competence assessments being current and in date. Lagging indicators are used reactively to monitor performance, such as failure data and incident data, in this case the percentage of incidents with a human factors contribution. Human factors KPIs can sit alongside other organizational KPIs, particularly process safety KPIs given that human factors is a key element in the control of major accidents.

CURRENT MODELS FOR HUMAN FACTORS MANAGEMENT FRAMEWORKS

This section provides a review of some of the approaches which can be used to structure the human factors framework. Four types are discussed: topic-based, risk-based, maturity-based, and a specific framework for human factors in design. There are overlaps between the approaches, but they allow illustration of the breadth of coverage that will ideally be considered.

TOPIC-BASED APPROACHES

A topic-based approach can be used to define the content of a human factors management framework. As already discussed, the UK Health and Safety Executive (HSE) defines a "top 10" key human factors topics relevant to the high hazard industries. This has been derived on the basis of the HSEs experience of common themes that have repeatedly been identified during inspections, and identified as key issues from research and consultation with industry. The top 10 is further subdivided and is not intended as a fully exhaustive list of human factors issues nor presented in any priority order. The top 10 are described in Table 5.1. They have also been presented in

Table 5.1 HSE's Top 10 Human Factors for the High Hazard Industries

Top 10 Human Factors Issue	Subtopic	Description	Relevant Chapter for More Detail
1. Managing human failures	1.1. Human error	Structured inclusion of influences on human failure (violations and errors) in design and risk assessment	Chapter 7
	1.2. Incident investigation	Structured assessment of human behavior within incident investigation	Chapter 8
2. Procedures		User-friendly procedures to support error-free performance	Chapter 17
3. Training and competence		Staff and contractors perform activities to a recognized standard on a regular basis	Chapter 20.2
4. Staffing	4.1. Staffing levels	The right level of skilled people to perform tasks	Chapter 20.1
	4.2. Workload	Manageable workload, especially during upsets and emergencies	Chapter 20.1
	4.3. Supervision	Critical role in achieving safe behavior	Chapter 20.3
	4.4. Contractors	Competent contractors integrated into the work force	Chapter 20
5. Organizational change		Human aspects of organizational change, risk-assessed and controlled	Chapter 19
6. Safety critical communications	6.1. Shift handover	Effective processes for shift and task handover	Chapter 21
	6.2. Permit to work	Effective permit-to-work systems	Chapter 21
7. Human factors in design	7.1. Control rooms	Ergonomic design of control rooms	Chapter 11/12
	7.2. Human–Computer Interface	Ergonomic design of control systems	Chapter 13
	7.3. Alarm management	Ergonomic design of alarm systems	Chapter 13
	7.4. Lighting, thermal comfort, noise and vibration	Ergonomic design of the work environment	Chapter 16
8. Fatigue and shift work		Fatigue risk management to prevent/mitigate fatigue, and reduce error	Chapter 22
9. Organizational culture	9.1. Behavioral safety	Programs target critical behaviors, and include process and occupational safety	Chapter 18
	9.2. Learning Organizations	Chronic unease exists, always looking for system causes of failures, and opportunities to learn or improve	Chapter 18
10. Maintenance, inspection, and testing	10.1. Maintenance error	Structured process to minimize maintenance errors	Chapter 7
	10.2. Intelligent customers	The capability of the organization to have a clear understanding and knowledge of the product or service being supplied	Chapter 18/20

earlier chapters as they are so effective for summarizing the primary and recurring deficiencies in the high hazard industries.

The HSE provides guidance on the HSE website: https://www.hse.gov.uk, and all of these topics are covered within this book as indicated in Table 5.1. The top 10 is underpinned by the HSE's inspector's toolkit which effectively defines the key human factors topics in a lot more detail (HSE, 2005). This can be used in support of performing an audit and review of the organizations performance in respect of human factors.

The Energy Institute provides guidance on the top 10 issues which can also be used to structure the content of a management framework, available from www.energyinst.org. A human factors toolkit is available from Step Change in Safety: https://www.stepchangeinsafety.net/safety-resources/human-factors.

Guidance from other industries can be reviewed to aid the development of a management framework, such as the rail industry (see Rail Safety and Standards Board (RSSB), 2008).

Whichever structure is used, there will be a need to undertake more detailed assessment of each of the areas of human factors. The chapters within this book will aid the reader to understand the scope, method, and outcomes of more detailed assessment.

RISK-BASED APPROACH

The HSE presents a "human factors roadmap for managing major accident hazards" (https://www.hse.gov.uk), which links across to specific topics defined in the HSE inspector's toolkit. The roadmap begins with the milestone of identifying safety critical tasks related to the major accident hazards on site. This includes tasks where humans may initiate a threat, undermine a threat barrier or fail to mitigate a major incident.

Tasks identified as safety critical are analyzed in more detail (task and human error analysis). During this analysis, Performance Influencing Factors (PIFs) are identified. These are factors that make human failure more likely and can include factors that are internal to the person (such as being fatigued or stressed) or external (such as poor lighting, shift systems, human interface, or procedures). PIFs are discussed in more detail in Section II, Managing Human Failure and specific topics in chapters throughout the book, as indicated in Table 5.1.

For identified human failures, the first consideration is to engineer them out. This may mean automation of a task. Caution is required when engineering out the human, as this can lead to hidden vulnerabilities or a less effective overall system performance. It may place the human vulnerability elsewhere, such as towards the maintenance teams, who can also fail. Therefore, in this example, maintenance performance also needs to be analyzed.

There is a need for "human" controls that are lower down in the hierarchy of control than the "engineered" controls, such as the development of robust procedures, effective supervision, training, and competence of staff. These topics are covered within this book.

Ultimately, the aim of the HSE's roadmap is to provide assurance for human performance in the context of managing major accident hazards.

HUMAN FACTORS MATURITY® MODEL

The Keil Centre developed a model specifically for measuring an organization in respect of its Human Factors Maturity®. The model presents an opportunity for organizations to measure how mature they are in managing human factors topics, and pinpoint where to focus their future efforts.

In an offshore technology report (HSE, 2002), a five-level framework for assessing human factors capability was advocated for defining different levels of capability based on following best practice, planning, procedures, processes, and feedback. At lower levels of capability, human factors provision is typified by an ad-hoc approach that is not planned and has no set policies or procedures. At higher levels, best practice approaches are applied in a planned and systematic way and are monitored for their impact on performance.

The Human Factors Maturity® Model (Mitchell, 2005) is a measurement tool to define what each human factor topic "looks like" at each level using a five-level model. Each level of the model (1–5) represents a level of maturity from emerging to leading (Fig. 5.2).

The model incorporates specific human factors topics with an associated methodology for measuring Human Factors Maturity®. It provides a measurement of the human factors topics with guidance on the strengths and areas to develop at an organizational level.

The initial set of human factors topics was based on the HSE's top 10, but these have been, and can be, modified according to organizations' needs, including additional topics that can be substituted in, thereby making the tool flexible to the needs of the organization. An example of the topics is shown in Fig. 5.3.

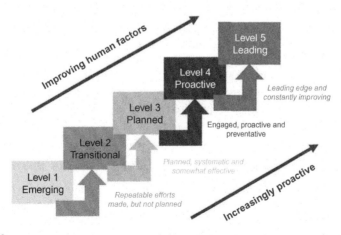

FIGURE 5.2

Human factors maturity® model.

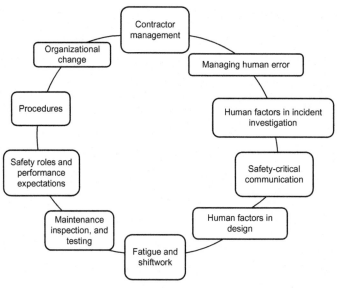

FIGURE 5.3

Example human factors topics.

The measurement is undertaken in a workshop context with company representatives who know about the human factors practices within the company. The model uses a "card-sort" methodology where the company representatives are asked to review five statements (written on the cards) and to select the statement that best reflects their organization's level of Human Factors Maturity® for each topic. Each statement reflects a level of Human Factors Maturity® ranging from level 1 ("emerging") to level 5 ("leading").

A facilitated discussion is then undertaken to identify why the group chose the level of maturity they did, how to improve, and to highlight priorities.

The Human Factors Maturity® workshop does not need to be replicated across the site as the measurement is less about perception and more about what is physically in place in terms of procedures and best practice.

An action plan is subsequently prepared to help the company develop its capability in the prioritized human factors topics.

MANAGEMENT FRAMEWORK FOR HUMAN FACTORS IN DESIGN

There has been significant development of management frameworks for human factors within design (more commonly referred to as Human Factors Integration (HFI)). This is discussed in more detail in Chapter 10, Human Factors Integration within Design/Engineering Programs.Chapter 10

Table 5.2 HFI Domain Areas

Domain	Description
Manpower	Determination of staffing requirements (numbers of people required to operate, maintain, and support the system). It also covers analysis related to identifying whether personnel can cope with the workload.
Personnel	Identification of skills required for the operation, maintenance, and support. It includes the identification of skills, knowledge, and attitudes within the current "resource pool" and matches this against requirements.
Training	Analysis and planning to meet the training requirements for all end users of the system.
Human engineering	The identification of user tasks, functional allocation between human and machine, specification of human machine interfaces, work area, and environmental requirements. It is concerned with the specification, design, and assessment of the human interfaces within the system.
Health hazards	The identification, analysis, and reduction of health and safety hazards created and imposed by the operation of the system. This includes environmental factors such as noise, climate, vibration, lighting, chemical/substance hazards, radiation, and other inherent hazards.
System safety	The identification, analysis, and reduction of safety hazards created and imposed by the system with focus on reducing human failure.
Accommodation and habitability Survivability	The accommodation of the user within their working environment, ensuring that it is sufficiently "habitable." The mitigation against threats which may actually occur ensuring that personnel survive during an adverse event, such as crash protection, and escape and rescue facilities.

Its early development was undertaken in the defense sector in the 1980s with what was then called MANpower and PeRsonnel INTegration (MANPRINT), and has since been further developed and adopted for different industries. Within the framework the HFI domain areas are applied during the engineering development of a new system (see Table 5.2).

The latest, most relevant guidance for the chemical and process industries has been produced by the international Oil and Gas Producers (OGP, 2011). It sets out a framework for integrating human factors engineering within capital projects and defines the scope as: people; work; equipment; work organization; and environment. Although, it has been developed within the oil and gas industry, it provides guidance which is relevant to other chemical and process industries.

Table 5.3 Typical Roles and Responsibilities for Human Factors

Typical Roles	Typical Responsibilities	Typical Organizational Function
Senior Management/Board	Sets the requirement for, and is accountable for the implementation of human factors within the organization	Senior Management/Board
Environment, Health and Safety (EHS) Manager	Accountable for implementing the organizations' strategy for human factors	EHS function (the EHS function may include EHS for operations and engineering)
	Accountable for the implementation of human factors standards.	
Company Human Factors Specialist(s) – may be internal or external to the organization	Provides technical knowledge and advice to the organization for human factors best practice.	May reside within EHS, Engineering, Operations or its own function
	Develops and maintains technical human factors standards.	
	Implements the human factors standards with support from other suitably trained personnel for operational issues.	
	May act as the human factors design authority to provide assurance for engineering projects, but reside external to the project team.	
	May provide human factors technical expertise for specific engineering projects.	
Head of Engineering Projects	Sets the requirement for, and is accountable for the implementation of human factors across the organizations' engineering projects.	Engineering function
Project manager or project EHS manager or project engineering manager and team	Accountable for the implementation of human factors within a specific engineering project.	Engineering function
	May undertake a coordination role in support of the human factors technical specialist to aid the implementation of human factors within the project.	
Project and Discipline Engineers	Responsible for implementing human factors requirements/standards and reporting/supporting the resolution of human factors issues for specific engineering projects.	Engineering function
Operations/Production	Responsible for providing operations expertise within engineering projects, specifically supporting human factors analyses, design reviews and issue resolution and user testing. Involvement in procedure development.	Operations function
	Responsible for providing operations expertise and opinions in support of human factors analyses, assessments, and implementation programs during operations. For example, safety critical task analysis, task analysis, workload reviews, safety culture assessments, procedure reviews, and stress risk assessments.	

It is best practice for an engineering project to have a project-specific Human Factors Integration Plan (HFIP) which specifies how human factors will be implemented, including the organization of the effort, the integration across the project, and the specific work breakdown structure.

PRACTICAL ADVICE FOR IMPLEMENTING A MANAGEMENT FRAMEWORK

ORGANIZATIONAL POSITION OF HUMAN FACTORS

It was mentioned earlier in the chapter that human factors is relevant to different areas of the organization, such as health and safety, operations, and engineering development. A typical "home" for human factors is within the health and safety domain. However, some organizations have a specific human factors function so it sits alongside health and safety, engineering, and production rather than being within one of these functions. As long as there is integration and cross referencing so that human factors areas are not missed, it is more dependent on what would be most suitable for the organization.

ROLES AND RESPONSIBILITIES FOR HUMAN FACTORS

There needs to be recognition that human factors is not a "one person job," just the same as with health and safety. There are responsibilities that need to pervade the organization. A generic (but not exhaustive) description is presented in Table 5.3.

SUMMARY: MANAGEMENT FRAMEWORKS FOR HUMAN FACTORS

A basic management framework for human factors is illustrated in Fig. 5.1. The framework is intended to help organizations to consider its own strategy for human factors. This is a reasonable place to start.

The next step is to take a baseline human factors measurement of the organization, possibly using one of the approaches described within this chapter to identify priorities for action and then to implement that action plan. As the organization matures, consideration may be given to the development of its own standards and procedures to support the implementation of human factors throughout the organization. This helps to embed human factors and reduce duplication of effort throughout the organization.

The important point is to actually get started in a structured manner and to recognize that human factors is not just a short-term initiative.

The next chapters in the book provide more detail on all of the topics introduced in the introductory chapters.

KEY POINTS

- Human factors is integral to the organization's operation, engineering development, and health and safety management. A human factors management framework needs to incorporate each of these elements.
- Successful implementation is best achieved by creating a simple system which is integrated within the organization's existing infrastructure.
- The basic requirements are to define the objectives (and KPIs) for the human factors effort, a comprehensive structure, and means of measuring performance so that the implementation can be prioritized and effectively planned.
- There are approaches available to aid the structuring, measurement, and action planning of the human factors effort. These have been described as topic-based, risk-based, and maturity-based methods.
- There may also be a need to define a management system specifically for human factors in design which links with an overarching human factors management framework.
- There are different roles and responsibilities to consider within the organization, which will include EHS, engineering, and operations.

REFERENCES

Health and Safety Executive, 2002. Framework for Assessing Human Factors Capability. HSE Offshore Technology Report: 2002/016. HSE Books.

Health and Safety Executive, 2005. Inspectors Toolkit: Human Factors in the Management of Major Accident Hazards. HSE.

International Oil and Gas Producers, 2011. Human Factors Engineering in Projects, Report 454.

Mitchell, J. 2005. Five Steps to Maturity. SHP, February 2015.

Rail Safety and Standards Board, 2008. Understanding Human Factors. A Guide for the Railway Industry.

FURTHER READING

Health and Safety Executive website: www.hse.gov.uk/humanfactors/humanfactors/index.htm

Energy Institute: www.energyinst.org/technical/human-and-organisational-factor

Step Change in Safety: www.stepchangeinsafety.net/

Managing human failure

Human factors in risk management

R. Scaife

LIST OF ABBREVIATIONS

ALARP	As Low As Reasonably Practicable
COMAH	Control of Major Accident Hazard
CREAM	Cognitive Reliability and Error Analysis Method
DCS	Distributed Control System
EPC	Error Producing Condition
FMEA	Failure Modes Effects Analysis
GEMS	Generic Error Modeling System
HAZID	Hazard Identification
HAZOP	Hazard and Operability
HEART	Human Error Assessment and Reduction Technique
HEP	Human Error Probability
HRA	Human Reliability Analysis (or Assessment)
HSE	Health and Safety Executive
HTA	Hierarchical Task Analysis
ISS	International Space Station
MTBF	Mean Time Between Failures
NARA	Nuclear Action Reliability Assessment
PIF	Performance Influencing Factor
PSF	Performance Shaping Factors
SHERPA	Systematic Human Error Reduction and Prediction Approach
SIL	Safety Integrity Level
TRACEr	Technique for Retrospective Analysis of Cognitive Error
TTA	Tabular Task Analysis

Managing risk is a core requirement in every industry sector. It is essential to ensure operational risks are defined, have been assessed, and controls implemented to meet the ALARP principle: "As Low As Reasonably Practicable."

When it comes to risk, most hazards and threats will include some aspect of human performance. This may include the potential for an operator or maintainer to make an error when carrying out a task, or a poor design decision regarding a piece of equipment that will subsequently impact on human performance.

When it comes to managing the risks associated with human performance, many organizations still use lessons learned from incidents. Whilst it is good that these lessons are being learned and applied, there is an opportunity to proactively identify hazards and threats by incorporating human factors into risk assessment processes. By doing so organizations have an opportunity to prioritize and manage human factors risks proactively thereby reducing the likelihood of human failures causing incidents. It is also more cost-effective to identify potential threats and hazards and manage them than to wait for an incident to highlight the problem.

With the benefit of hindsight there are many examples of how human performance at various levels in an organization has introduced risk. A well-known example of this is the Kegworth air disaster in 1989. In this incident a British Midland Boeing 737-400 aircraft crashed on the embankment of the M1 motorway just short of East Midlands airport in the United Kingdom. The two-engined jet aircraft had suffered a failure in one of its engines but could have flown to its diversion airport successfully on the remaining engine. However, the cockpit crew mistakenly decided to shut down the healthy engine. As they attempted to increase thrust to the damaged engine for landing maneuvers, it failed catastrophically with no opportunity to restart the healthy engine before the aircraft crashed. The cockpit crew certainly made an error in diagnosing which engine was faulty, but there were also fundamental design flaws which led the aircraft to fail in the first instance. The key question is "could the risk of these human failures have been predicted?"

Several factors to do with the design of the aircraft systems could have indicated an increased risk of human failure. For example, when the pilots were interviewed following the incident, neither of them recalled having seen any abnormal indication on the engine instrument displays associated with the faulty engine. The displays were new digital displays that had replaced the analogue displays used in the previous model of the aircraft. One of the key symptoms of the failed engine was a high level of vibration caused by a fractured fan blade in the engine. However, the displays that showed engine vibration were small and did not highlight the abnormal situation using warning lights or other attention-getting signals. If the potential for misreading the instruments had been identified prior to the aircraft entering service, the likelihood of human error during abnormal situations, such as engine failure, could have been reduced through redesign.

One piece of information used by the captain and first officer to diagnose which engine had failed was their knowledge that the air in the cockpit in a Boeing 737 aircraft was supplied via the right-hand engine. When the left-hand engine failed, they both smelled smoke in the cockpit and in their minds; this helped confirm that their decision to shut down the right-hand engine was correct. They did not realize that when the Boeing 737-400 was designed, the source of the cockpit air was changed to the left-hand engine. The designers did not consider the fact that the smell of smoke would be used to help diagnose an engine failure. However, involvement of the end users, in the design may have identified the potential for human error, and managed it accordingly.

The human may not always be the source of hazard or threat that needs to be managed. The Swiss Cheese Model (Reason, 1990) describes how people can also be the defense against errors or other system failures. In some cases, people can be better at managing risk than an automated system or software program. Although automated systems are used to produce consistent results for a given action, they cannot do anything that they are not programmed to do. Designers cannot foresee all future possibilities. Automated systems are based on logical arguments, such as "IF X HAPPENS THEN DO Y." This provides a consistency of performance that is not typical of a human. However, the human still has an advantage. They are not reliant on a program and series of logic statements to perform. They can use judgment, knowledge, and problem-solving to react to novel situations. A computer program designed to initiate a certain action when it detects a value of X will be unable to do anything when presented with a value of Y. The human, if properly trained, competent and alert can react.

Examples of human performance successfully overcoming unexpected events happen every day. A dramatic example occurred in April 1970 when an oxygen tank on board the Apollo 13 spacecraft ruptured en route to an attempted moon landing. There was no set procedure defining the best way to recover from such a situation, and there was no way to use the command module to return to earth. The decision was made to attempt a return to earth by piloting the spacecraft from the lunar module, but in order to do so the lunar module would need to be made habitable for the duration of the return journey. The crew used their knowledge and training, together with ground support, to modify lithium hydroxide canisters from the command module to fit the lunar module. This was used to remove carbon dioxide from the cockpit thereby providing life support for a safe return to earth. The canisters used in the two modules were of different design, so the modification was required to make the canisters fit the lunar module. After Apollo 13, components used in spaceflight were designed to be compatible with each other so that such modifications would no longer be required.

Using human factors knowledge to manage risk is based on two key principles:

1. how potential human failures could contribute to risk;
2. how to mitigate such risk, including how people can be best utilized to reduce risk.

As shown in Fig. 6.1, these principles can be applied either at a high level to understand the potential human performance threats at a process or system level, or at the task level to target specific modes of human failure.

The remainder of this chapter discusses ways in which human factors can be applied to the management of risk using these principles. Many organizations have started with task-level human factors risk assessments because they have identified a specific issue they want to address. This type of risk assessment will therefore be discussed first. High-level risk assessment will be discussed later in this chapter, including consideration of how human factors can be integrated with existing safety engineering processes.

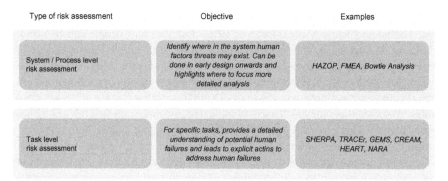

Type of risk assessment	Objective	Examples
System / Process level risk assessment	Identify where in the system human factors threats may exist. Can be done in early design onwards and highlights where to focus more detailed analysis	HAZOP, FMEA, Bowtie Analysis
Task level risk assessment	For specific tasks, provides a detailed understanding of potential human failures and leads to explicit actins to address human failures	SHERPA, TRACEr, GEMS, CREAM, HEART, NARA

FIGURE 6.1

Overview of human factors in risk assessment.

HUMAN RELIABILITY ANALYSIS

Human Reliability Analysis (HRA), sometimes referred to as Human Reliability Assessment, is a generic term applied to any systematic method used to quantify the impact of human factors on organizational risk.

That is, HRA is used to:

- predict the errors people could make when completing a specific task;
- quantify the impact of these errors;
- identify ways to reduce the likelihood or consequence of errors in the future.

Any such analysis needs to consider a number of factors: the characteristics of the person(s) completing the task; the task itself and the steps involved; and the conditions under which the task is completed.

Kirwan (2005) describes a broad methodology for HRA that defines best practices when using any HRA tool. The 10 step methodology describes the following stages:

1. Problem Definition—precise definition of the conditions of the task, goals, and potential for the human to contribute to failure in achieving the goals.
2. Task Analysis—description of the task in detail, including the equipment used, procedures, and communications required at each step.
3. Error Identification—identification of the credible errors at each step of the task and how they can be recovered from.
4. Representation—determination of the impact of the errors on the system.
5. Screening—definition of the level of detail that the analysis will go to.
6. Quantification—the likelihood of the error and the likelihood of recovery to quantify the overall impact in the system once recovery mechanisms are in place.

7. Impact Assessment—determination of the human reliability impact on the system and whether additional control measures are required to manage the risk of error.

8. Error Reduction—recommended actions to reduce errors, improve detection and recovery, or improve human performance.

9. Quality Assurance—checks that the proposed error reduction measures are effective in practice.

10. Documentation—record of the results of the analysis so they are understandable by the end user and actions can be tracked and audited.

Not all tools and techniques for HRA follow all of these steps, and they do not necessarily occur in the same order as Kirwan suggests. However, the general principles for HRA will help ensure that any analysis is a success and the results are usable for practical improvements.

Specific tools and techniques for performing HRA fall into two broad categories:

- Qualitative approaches that seek to describe the potential error, the consequences and potential error reduction measures; and
- Quantitative approaches that attempt to estimate the probability of each error type in numerical terms, before identifying potential error reduction measures.

Qualitative and quantitative approaches can be used in isolation or combination. Both approaches have benefits and drawbacks, and in most cases one approach is chosen over the other depending on the project. Typical considerations on the use of the two broad approaches are the availability of subject-matter experts, time constraints, and the number of tasks that are likely to require analysis.

A description of qualitative and quantitative approaches follows, including a sample of relevant tools within each category, illustrated with industry examples.

QUALITATIVE APPROACHES

In the non-nuclear major hazard industries, qualitative approaches are usually the preferred option. These approaches normally begin by categorizing potential credible errors using a set of guidewords or a taxonomy of error types. The aim is to describe what might happen and how the potential errors could be reduced, either by enhancing detection and recovery or by preventing the error through design or other interventions. Although numerical probabilities are not usually assigned to the potential errors identified, it is not uncommon for a qualitative technique to assess the likelihood in terms of a qualitative frequency scale such as high to low or frequent to incredible.

Qualitative approaches include tools specifically designed for the assessment of potential human errors. Some well-known examples of these tools include:

- Systematic Human Error Reduction and Prediction Approach (SHERPA),
- Technique for Retrospective Analysis of Cognitive Error (TRACEr), and
- Generic Error Modeling System (GEMS).

Qualitative approaches also include modifications of more general safety assessment methodologies used within industry, to focus more explicitly on human error. For example, Hazard and Operability (HAZOP) is a qualitative approach used in safety analysis in a wide range of settings. By using a set of guidewords designed to investigate the potential for human failure, HAZOP can support HRA.

Qualitative approaches tend to focus on specific tasks identified as requiring assurance that human performance will be within acceptable limits. There is typically a need to demonstrate that potential for human error has been suitably controlled by the design of the system or some other means. This requires a systematic and reliable method prior to detailed analysis.

IDENTIFICATION OF SAFETY-CRITICAL TASKS

There are different ways to identify which tasks require the level of analysis described above. Organizations with safety-critical operations, for example, may have undertaken other forms of analysis that can be used as the basis for identifying tasks requiring HRA. These may have been compiled from existing sources such as the Control of Major Accident Hazard (COMAH) report, Bow Tie diagrams, Hazard Identification (HAZID) studies, HAZOP studies, Safety Integrity Level (SIL) assessments, incident data analysis, amongst others. By amalgamating the evidence provided by these sources of information it is possible to collate a list of activities to consider for HRA. As part of this process, there is still a need to establish the extent to which human performance is important in the successful control of risk for these activities. Some safety-critical functions have little or no intervention by an operator, others rely significantly on the operator to control or monitor a process.

At the end of this stage the safety-critical functions that have some element of human input will have been identified. The next step is to determine how critical successful human performance will be in each case.

SAFETY-CRITICAL TASK SCREENING

Critical tasks have different levels of human input. To avoid the analysis process becoming too laborious, it is necessary to screen the list of tasks and establish an order of priority for further analysis.

The UK Health and Safety Executive (HSE) produced guidance for the identification of safety-critical tasks (HSE, 1999). It uses a set of five screening questions to determine how vulnerable the task is to human failure. As shown in Table 6.1, each of these questions is answered using a three-point scale from high to low, with high attracting a score of 3 and low attracting a score of 1. If the question is not applicable a score of zero is assigned. It should be noted that this screening process was designed for use in the oil and gas sector, and as such the methodology and some of the generic tasks may need to be modified for use in other industries.

The results for all five questions are summed, resulting in a possible score between 0 and 15. A series of criticality bands are assigned, linked to the total score,

Table 6.1 Screening Tool to Determine Task Vulnerability to Human Failure

	High	Medium	Low	N/A
1. The intrinsic hazards associated with the task (in terms of substances, energies, or conditions)	3	2	1	0
2. The extent to which ignition sources are introduced by the performance of the task	3	2	1	0
3. The requirement to override safety protection systems as part of the task	3	2	1	0
4. The extent to which incorrect performance of the task could lead to damage to the system	3	2	1	0
5. The extent to which the task requires changes to the configuration of the system	3	2	1	0

N/A, not applicable

allowing tasks to be defined as High (9–15), Medium (5–8), or Low (1–4) criticality, and prioritized for further analysis.

TASK ANALYSIS

Qualitative approaches to HRA typically examine a task in detail, and identify credible human errors or deviations that could occur for each task step. It is important to analyze the actual task steps, as opposed to what people "should" do, as the aim is to predict what could potentially go wrong when the task is performed. This sounds obvious, but unless this premise is fully understood, the analysis could fall foul of the old computing adage—"Garbage in, Garbage out." Fundamentally, task analysis is about understanding the task as it is actually done and at the level of detail appropriate for the task type and level of hazard. An illustration of this is provided in Box 6.1.

For the purposes of developing the task analysis, it is important to use multiple sources, including documented procedures and walk-through or talk-through by an experienced operator to determine what they actually do when completing the task. Observation of the task can also be used to supplement other sources of information.

Task analysis can take many forms (see Kirwan and Ainsworth, 1992). The two most common are Hierarchical Task Analysis (HTA) and Tabular Task Analysis (TTA). HTA allows the basic task structure and purpose to be understood. TTA is commonly used in qualitative HRA approaches as it provides the ability to list the tasks and subtasks alongside descriptions of the equipment, communications, conditions, and performance standards required to perform the task properly. An HTA can readily be transferred into TTA format to allow further detail to be recorded. It is also possible to record the nature of the task (such as action, communication, judgment, information retrieval) alongside the task step to assist in identifying appropriate error types.

BOX 6.1 IMPORTANCE OF UNDERSTANDING ACTUAL TASK STEPS FOR EFFECTIVE ANALYSIS

An organization performed an assessment of its safety-critical tasks and developed a list of more than 100 tasks that they wanted to analyze using qualitative HRA. They knew they would need to employ an expert to do this, and thought they could save time by defining the tasks themselves. They took the procedures and wrote out every step in the procedure into a spreadsheet, which they called their task analysis. When it came to performing the HRA, it was found that many of the steps in the task analysis were not actually done anymore by the operator, as an automated system had been introduced, and the procedure had not been updated. There were also a number of task steps the operator now did that were not mentioned in the procedure as a result of the introduction of automation. Therefore, the entire task analysis had to be rewritten during the HRA, a time-consuming and expensive process.

BOX 6.2 KEEP IT SIMPLE: DEFINING TASK STEPS ACCURATELY

If the task according to the procedure is "Open and lock open valves A to H" then this may on first reading appear to be a relatively simple step to analyze for potential errors. However, it introduces the possibility of errors in the opening of valves, the sequence of opening them, and also errors to do with locking them open afterwards. This one simple statement may lead to prolonged discussion about the step. Breaking the step down by consulting with experienced operators helps to simplify the analysis of potential errors. For example, it may be that all of the valves are first opened and then locked open. Alternatively, each valve may be opened and then immediately locked open before moving onto the next valve. It may be that plant modifications mean that the valves are no longer operated in alphabetical order, and so the task steps included in the procedure need to be amended to ensure that the analysis is accurate. In this example, the procedure seemed to define a single task step that would be relatively simple to analyze. However, further detail means that the analysis can be more thorough and therefore identify more potential sources of error by defining what the operators actually do when completing the task.

The initial task analysis will require validation. Often separate subject-matter expert(s) walk through the steps to confirm they are adequate descriptions of what is done, in the correct order, and sufficiently detailed.

Before proceeding to HRA, it is important to ensure each step only defines a single activity (Box 6.2).

WORKSHOP-BASED ANALYSIS

HRA involves operators or maintainers (depending on the task being analyzed) participating in a workshop to perform the assessment, alongside other suitably qualified participants such as technical safety, engineering, and human factors specialists. Participants require sufficient knowledge of the task to make accurate judgments on the potential for human failure.

The HRA workshop is an opportunity to assess the conditions under which the task is performed. These conditions are described throughout this book as Error Producing Conditions (EPCs), Performance Shaping Factors (PSFs), or Performance

Influencing Factors (PIFs) depending on the error analysis model being used. A checklist of factors, based upon research or existing human error analysis tools, is used. At this stage, consider whether the task will always be conducted in the same set of conditions such as in the control room only or in multiple locations by multiple personnel. For example, part of the task being completed in the control room with assistance from a plant-side operator. In the former case it may be possible to assess the conditions once, whereas in the second case these may need to be assessed for each location or on a task-by-task basis.

Depending on the specific approach used to perform the HRA, each step of the task is examined using a series of guidewords to identify one or more types of potential human failure.

For each potential human failure, the consequences of the error are identified as a description of what could actually happen. This is more useful than just using a severity scale rating, as it helps to determine the most appropriate types of error reduction measure.

It is also necessary to define current methods to prevent, detect, or recover the error (i.e., what are the existing defenses). The group will then explore additional defenses to further prevent, detect, or recover the error. The level of detail required for this stage depends largely on the defined consequences of the identified potential error and the strength of the existing defenses. If the defenses are already strong, or the consequences result in little more than having to redo the task, then the level of effort expended should be low. If the consequences are high or the existing defenses are weak, greater effort is required to find solutions to reduce the likelihood of error.

It is worth emphasizing that error recovery is an important point to consider during the analysis. Defenses that help error detection do not necessarily aid recovery from the error, and in fact, if the error is detected and the design does not support recovery from the error, the potential problems that could occur when operating or maintaining the system have not been resolved. This is illustrated in Box 6.3.

Facilitators of HRA workshops should draw on participants' experience and knowledge to focus on effective recovery. It is also important for facilitators to challenge participants if their proposed defenses are unlikely to be effective in prevention, detection, or recovery of errors.

Following any HRA workshop, the facilitator must analyze and report results. The report should summarize recommended actions to reduce the likelihood or impact of potential errors, EPCs, and the requirement for additional defenses. The resulting recommendations are based on a description of the error, the existing defenses, and the recommended additional defenses.

BOX 6.3 IDENTIFYING ERROR RECOVERY PROCESSES

Consider the potential error of misreading the temperature of a vessel on a Distributed Control System (DCS) in a control room. If the proposed defense against the error is a high temperature alarm to aid detection, this may do little (depending on the set point) to aid recovery. However, if the defense is a rate of change temperature alarm then the operator would be afforded more time to recover from their error.

QUANTITATIVE APPROACHES

In safety engineering, quantification of failure rates is common. There are existing standards that define how such exercises should be undertaken for both software and hardware systems. For example, in the US Military, MIL-STD-882E defines the Department of Defense's standard practice for system safety and provides detailed guidance on how the safety of any system should be assessed through all stages of the lifecycle of the system from development to disposal. The functional safety of programmable electronic systems is covered by IEC 61508, an international standard that provides guidance on the assessment of SILs and the required reliability of software used at each SIL level. For example, at the highest SIL software should have a probability of failure of between 10^{-5} and 10^{-4}.

Standards to quantify failure rates are useful for software, hardware, and safety engineers to assess systems in terms of their integrity and likelihood of failure to work out whether or not a management system will be sufficiently robust. For example, safety engineers measure or predict the Mean Time Between Failures (MTBF) of hardware and can provide probabilities (albeit based upon a number of assumptions about how the system will be operated) to describe the reliability of the system. If these figures do not meet the required standard for safety-critical systems, alternative designs or actions are required to improve reliability.

In addition to hardware and software elements, it can be argued that a crucial part of a safety-critical system is the people who operate and maintain it. Therefore, to fully assess the reliability of the system, it is necessary to calculate the probability of human failure as well.

This feels like a daunting prospect when the figures being produced by software safety specialists tell us that the probability of failure of a piece of software in a safety-critical system is 10^{-5} and the hardware engineer is telling us that the infrastructure has a MTBF of 10,000 hours.

What kind of values could be expected for the probability of failure of the human operator performing part of a safety-critical task? Would it be approaching 10^{-3} or 10^{-5}? It is probably not surprising to learn that this would depend on the characteristics of the task, and the conditions under which it is performed (Box 6.4).

Methods for arriving at these probability values vary, but the fundamental principle of determining human error probability (HEP) is to measure how often the task

---

BOX 6.4 EXAMPLE OF NASA QUANTIFICATION OF HUMAN ERROR RATES FOR MISSION CONTROL FLIGHT CONTROLLERS

A study in 2010 performed by the National Aeronautics and Space Administration (NASA) suggested that with well-trained Flight Controllers at Mission Control, the likelihood of errors in sending commands to the International Space Station (ISS) ranged from approximately 0.1 to around 10^{-4} (Chandler et al., 2010). These errors included selecting the wrong procedure to use, or sending the wrong command to the ISS, and were affected by working conditions such as fatigue, time pressure, and cognitive overload.

is completed over a given period of time, and divide this by the number of times an error is made when performing the task. Predicting the probability of failure requires the effect of the conditions under which the task is completed to be incorporated in recognition that tasks are not completed in a vacuum.

In some industries the frequency of the task and the error rate can be measured directly and therefore the calculation can be based upon real data. The aforementioned NASA study used data logs that automatically recorded error rates every time a specific task was completed. In other cases, however, these data are not available and have to be estimated.

There are several quantitative HRA tools. Some popular examples of these are Cognitive Reliability and Error Analysis Method (CREAM) (Hollnagel, 1998), Human Error Assessment and Reduction Technique (HEART) (Williams, 1988), and Nuclear Action Reliability Assessment (NARA) (Kirwan et al., 2005). More information on the selection of tools can be found in Bell and Holroyd (2009).

HEART illustrates the process of conducting quantitative analysis as it is one of the first generation of HRA techniques, and several tools that have been developed since have used HEART in one way or another as their basis.

The objective of HEART was to develop a tool that was relatively simple to use and would be applicable to a range of industrial settings rather than being focused on one sector. As a result, rather than providing the user with an extensive list of specific types of task, it uses a set of nine generic tasks. The nine tasks were developed based on 40 years of psychological studies into how humans fail when performing tasks. This provides a generic HEP associated with each generic task type. Theoretically, this is the probability of failure when completing the task under perfect conditions for human performance.

HEART takes into account that perfect conditions are unlikely, and a set of 38 EPCs is presented for selection. Based on a meta-analysis of human factors research, the developers of HEART selected EPCs that consistently reduce human performance, and therefore influence the likelihood of error to the same extent whether they occur alone or in combination. If multiple EPCs are identified for a given task, then each one acts as a multiplier of the error probability.

So, for example, according to HEART high levels of emotional stress increase the probability of error by a factor of 1.3. HEART takes into account the fact that just because a condition is present does not mean that it exerts all of this effect on the person performing the task. Operator inexperience may be a factor (it has a weighting of ×3 in HEART) but the person performing the task may have other experience which is transferrable to the task and therefore inexperience may not be attributed the full weighting of 3. This requires judgment on the part of the analyst, and Williams acknowledges that this is the most difficult aspect of the analysis. A background in human factors is likely to be required to perform these judgments consistently.

Once these judgments on the strength of effect have been made, the analysis is completed by multiplying the generic probability of failure of the task by the proportional weightings of the EPCs using the HEART formula. This provides an assessed HEP for the task under the prevailing or expected conditions for the person completing the task.

Following the analysis, recommendations are made for reducing the influence of the EPCs on the task. When this has been done, the analyst can perform "What If" analysis by reapplying the formula assuming that the recommendations have been implemented to estimate the potential change in reliability. This provides an indication of the potential benefit of implementing the recommendations.

COMPARING QUALITATIVE AND QUANTITATIVE APPROACHES

Qualitative and quantitative techniques are similar in a number of ways. They are systematic ways of conducting an analysis and they both require judgments to be made. Qualitative analysis is typically done by a facilitated group of task experts, whereas quantitative analysis is normally completed by a lone analyst. Both team-based and individual-based analysis approaches have pros and cons. Combining the two approaches can be beneficial.

It is possible that subject-matter experts without human factors knowledge can make unrealistic judgments about the credibility of errors unless they are carefully guided by a suitably experienced facilitator with specialist human factors knowledge. Likewise, an experienced human factors expert does not necessarily know the task in depth, and might be subject to bias when calculating the strength of EPCs effect. Using subject-matter experts for the task being analyzed can help the analyst, and to identify suitable measures to reduce the influence of the EPCs.

Quantitative analysis techniques produce an actual number. In safety engineering, if the probability of failure of a safety-critical computer program had been calculated at 10^{-5} then this figure would likely be used in the safety assurance for the system. If a quantitative HRA resulted in the probability of human failure being reported as 0.08, it can be compelling to use the numbers in a similar way to present a stronger argument for assurance of the human element, or indeed to present a case for introducing greater levels of automation in the task. However, although both are presented as a probability value, the HEP has been derived using a high degree of judgment in terms of the effect of the EPCs, and should be treated as a starting point for further analysis rather than a stopping point. To defend against the potential misuse or misinterpretation of human error probabilities, it can be useful to translate them into qualitative probability descriptors as used in MIL-STD-882E (Frequent, Probable, Occasional, Remote, Improbable, and Eliminated). This is a simple task requiring reference to a table of probability values with associated qualitative descriptors and does not add significant time or effort to the analysis. Generally, the probabilities that quantified methods produce have a likelihood of error themselves of about one order of magnitude and so the qualitative descriptions reflect this better. With the above taken into account, quantitative analysis is effective in assessing a specific safety-critical task, for example, to determine whether further controls may be justified. It allows "What If" analysis to be performed to estimate the impact of new controls or the mitigation of EPCs. See Box 6.5 for an example.

BOX 6.5 USING QUANTITATIVE ANALYSIS TO ASSESS NEW CONTROL MEASURES

An organization performed a quantitative HRA on a filling task which had been the subject of a near-miss with the potential for an explosion on their site. Their analysis found the following:

Task	Probability of Failure
Filling vessel	0.003

EPC	Cumulative Effect on probability of failure
Unfamiliarity with a situation which is potentially important but which only occurs infrequently or which is novel	0.012
Low signal-to-noise ratio	0.046
Little or no independent checking or testing of the outcome	0.102
Inconsistency of meaning of displays	0.107

The organization was able to assess the additional control measures identified from the analysis in terms of potential impact. For example, improving the signal-to-noise ratio (making different vessels more distinctive and segregating different groups of vessels) could reduce the likelihood of error by approximately four times.

SYNERGIES WITH SAFETY ENGINEERING

Human factors practitioners often work as part of a multi-disciplinary engineering, safety or human resource team. Therefore, identifying synergies between human factors and other disciplines can have mutual benefits and help to prevent duplication of effort. Human factors approaches have developed from gaining experience of the other disciplines, and safety engineering is one example. On major design projects, safety engineers and human factors engineers tend to be involved in similar analyses with different foci at similar stages of the project lifecycle, including the early (e.g., concept) stages. Over the years, some safety engineering tools have been modified to incorporate human factors, or to refocus more explicitly on human performance. Some examples of these are outlined below.

HAZARD AND OPERABILITY STUDIES

HAZOP studies are a good source of information on the potential human contribution to risk, and how to manage the risk. Even traditional HAZOPs begin to uncover human factors issues associated with the system, but this tends to be by exception.

A traditional HAZOP study takes a process component-by-component, and applies a small number of guidewords to each component in turn to identify credible deviations from process design intent. For example, a guideword "More" applied to a vessel may result in the credible deviation of "too high a level in the vessel" with potential causes of the deviation. This may result in a cause of human error by an operator, or a maintainer. As a result of the standard HAZOP process human error is only identified as a credible deviation from the process design intent if the application of the guidewords to the process components happens to highlight the possibility (i.e., by exception). Even then, the cause of the deviation tends to be recorded in simplistic terms (e.g., "human error") rather than focusing in on how the error could occur. A more systematic approach to considering the potential errors in detail can be taken during a HAZOP study, but this relies on the workshop participants having some knowledge of human factors.

Various sets of human factors guidewords have been developed over time to refocus HAZOP on the operator's performance more systematically. Some of these have simply taken the types of guidewords used in qualitative HRA techniques such as SHERPA and used these in place of the usual HAZOP guidewords. This has the disadvantage of resulting in up to 20 guidewords, which can make it difficult to maintain the pace of the workshop, making it less effective. It is possible that a HAZOP analysis will highlight a specific task that requires more in-depth analysis so using too many guidewords can result in duplication of effort.

Other human factors guideword sets are more focused and cover the main error-based deviations such as omissions, wrong order, too early/late, and instead of (see, e.g., Ellis and Holt, 2009). These would be applied in the same way as a traditional HAZOP, but with guidance from the HAZOP Chair regarding how this could apply to the human.

Other approaches to incorporating human factors into HAZOP have focused on the development of prompts for the discussion of how the identified deviation could occur (see, e.g., OGP Report 454). Some of the less-developed approaches simply add the term "human factors" to a set of prompts, whilst others go into more depth in terms of the types of errors that could result in the identified deviation.

FAILURE MODES AND EFFECTS ANALYSIS

Failure Modes Effects Analysis (FMEA) is a well-established safety engineering methodology first applied in the aviation industry in the 1960s and adopted by a range of other industries. In the FMEA approach, the analysts (usually a multidisciplinary team) determine the potential failure modes that could occur with the system. Each failure mode is rated on 3 scales; severity of effect (1–10), likelihood of failure (1–10), and ability to detect the failure before it happens (1–10). The three scores are multiplied together to obtain an overall score between 1 and 1000. A criterion is agreed before the analysis on actions required to control the failure effects.

Traditionally, FMEA has focused on the failures in general (e.g., embrittlement, electrical short circuit, corrosion, etc.) rather than specifically analyzing the potential human failures involved (e.g., error in visual inspection, failure to follow procedure, etc.). Like HAZOP, however, when it is possible that a failure could be initiated by human factors issues, they are typically identified by exception rather than being identified systematically.

Village et al. (2011) describe the development of an approach referred to as HF-FMEA, which is focused on human failures and how they could cause the system to fail, how they could result in harm to personnel. Modifications were made to the severity, occurrence, and detection scales to focus more on the risk of injury and how this would be controlled by the task. The researchers state that this is a work in progress and in need of further development, but it has potential for another aspect of risk that could be analyzed and managed through the integration of human factors. There are parallels with the approach used in the screening of safety-critical tasks prior to HRA, and it may be that the rating scales used could be applied to that process to make it more generally applicable outside of high hazard industries.

BOW TIES

Bow tie diagrams provide an intuitive means of describing a hazard, its associated critical event (such as loss of containment), its causes, and consequences. They also provide opportunity for the analyst to introduce controls to prevent the critical event, and to mitigate the consequences of it. They are known as bow tie diagrams because the critical event is shown as a single entity in the center of the diagram with the causes (i.e., threats) fanning out from the left of the critical event and the consequences fanning out from the right, as illustrated in Fig. 6.2.

In between the causes and the critical event, the preventive barriers can be shown (along with potential causes or failures of these barriers). To the right of the critical event, mitigation barriers can be included to prevent the consequences of the hazard, and any potential failures of these barriers can also be recorded.

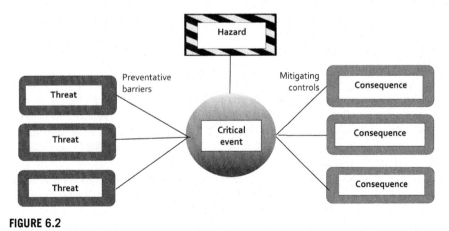

FIGURE 6.2

Bow tie diagram.

> **BOX 6.6 CASE STUDY: HUMAN FACTORS CONSIDERATIONS IN PREVENTION AND MITIGATION BARRIERS**
>
> The critical event in examining the BP Texas City refinery fire and explosion in March 2005 was the overfilling of the raffinate splitter tower. One of the causes was the operators routinely failing to follow the procedure and filling the tower too high. There was a barrier in place in the form of a second high-level alarm. However, this barrier failed because the maintenance crew did not check the operation of sensors and safety devices before start-up. Once the overfill had taken place, one consequence was the release of hydrocarbons from the blowdown drum. The defense in place was a high-level alarm on the blowdown drum, but this too was not checked prior to start-up and was unserviceable. Based upon this analysis, additional barriers could be introduced to prevent overfilling and to mitigate the effect of overfilling. For example, a design solution could be introduced to prevent a level above 6.5 feet being reached. If this proved not to be feasible, an administrative barrier could be introduced to train or educate operators in the danger of filling the vessel above the level specified by the procedure. Designing the DCS system to calculate the volume of liquid in the vessel based upon flows into and out of the vessel would provide an additional indication of an impending overfill in the event that the level sensors failed.

Human factors can be incorporated into bow tie diagrams to determine the threats, barriers, and controls from a human performance perspective. However, this can generally only be done at a higher level of detail and further task-specific analysis is likely to be needed as well. Bow ties can rapidly become unmanageable if they try to consider too much detail. However, Box 6.6 provides an example of how bow tie human factors considerations may have provided insight to prevent a disaster in the 2005 BP Texas City refinery fire and explosion.

OTHER FORMS OF RISK MANAGEMENT

Human failure risks are not limited to the operation, development, and maintenance of systems. Failures in the operation or maintenance of a system are often referred to as active failures; they happen when the task is being carried out. Failures in the design or development of systems are often referred to as latent failures; they happen further back in the timeline and represent a potential failure of the system at a later time.

Some latent failures can extend upwards into the organization as well, and can cause equally concerning situations to develop. These are often referred to collectively as "organizational factors," often seen as "the wallpaper" in an organization, that is, "the way things are round here" (Wilkinson and Rycraft 2014) and can be harder to identify and tackle.

If these risks can be identified early and mitigations put in place to control them or reduce their impact on people working in the organization, then potential severe impacts can be avoided. Often it is difficult to identify these risks from within the organization because those involved are also part of the organization. Having a structured method of assessing the risks helps, as it provides/allows those examining change to do so systematically.

BOX 6.7 CASE STUDY: ASSESSING THE RISK OF ORGANIZATIONAL CHANGE

In the BP Texas City disaster, latent human failures were created at the time of the acquisition of Amoco by BP. The organizational changes had significant impacts on events.

Cost reduction postacquisition included losing a board operator. Even during normal operations one board operator found it difficult to control three process units. During plant upset and emergency conditions this became impossible to manage and information went undetected and undiagnosed. It was also likely that the staff reductions resulted in more opportunities for overtime. The operator was likely to have been severely fatigued on the day of the accident as he had worked 12-hour shifts for 30 days. The downsizing did affect not only operations but also maintenance. The failure to check the safety devices and sensors prior to start-up may well have been due to a shortage of maintenance technicians.

The personal injury rates were linked to the refinery bonus scheme and process safety had been overlooked for a long time. The warning signs of a potentially catastrophic event went undetected. The bonus scheme encouraged attention on personal safety to the detriment of process safety.

One example of an underlying "organizational factor" that causes risk is organizational change. Following on from our previous case study, Box 6.7 demonstrates how failure to consider organizational change contributed to the 2005 BP Texas City incident.

Managing organizational change is discussed in more detail in Chapter 19 Managing Organizational Change. Other organizational risk management issues, such as fatigue risk management, competency management, and staffing are discussed in detail in Section IV, Understanding and Improving Organizational Performance.

SUMMARY: HUMAN FACTORS IN RISK MANAGEMENT

Many organizations rely on incidents to gain insight into human factors issues. If this is done well it can make a significant difference in reducing future human failure. However, incorporating human factors into the risk management process provides opportunity for organizations to proactively understand, prioritize, and manage human factors before an incident occurs.

The following chapter provides a detailed discussion of human failure, taking account of errors (unintentional unsafe behaviors), and violations (intentional unsafe behaviors).

KEY POINTS

- There are several human factors techniques to proactively identify potential human failures and determine the most effective control measures. These include qualitative and quantitative techniques. Safety-critical task analysis is typically undertaken to prioritize which tasks are selected for HRA.

- There is synergy between human factors risk management and safety engineering. This enables safety analysis work to be refocused if required to examine some of the human factors issues that may otherwise go unnoticed.
- Organizational risk management, which if not done or if done without due care and attention, can introduce significant risks to the organization. If these risks are not explicitly identified via structured assessments, they can lay hidden for long periods of time before their effects are discovered. Such risk assessments are not limited to assessments of organizational changes, but could also include fatigue risk assessments, workload analysis, and ergonomic assessment of new designs amongst others.

REFERENCES

Bell, J. & Holroyd, J., 2009, Review of human reliability assessment methods, health and safety laboratory. Research report RR679.

Chandler, F., Addison Heard, I, Presley, M., Burg, A., Midden, E. & Mongan, P., 2010, NASA human error analysis. National Aeronautics and Space Administration.

Ellis, G. R. and Holt, A., 2009, A practical application of "Human-HAZOP" for critical procedures. Hazards XXI Symposium Series No. 155, IChemE.

Health and Safety Executive (HSE), 1999, Offshore technology report. Human Factors Assessment of Safety Critical Tasks, OTO 1999 092.

Hollnagel, E., 1998. Cognitive Reliability and Error Analysis Method: CREAM. Elsevier, Oxford.

Kirwan, B., 1994. A Guide to Practical Human Reliability Assessment. Taylor & Francis, London.

Kirwan, B., 2005. Human Reliability Assessment. In: Wilson, J., Corlett, N. (Eds.), 2005, Evaluation of Human Work, third ed. Taylor & Francis, Boca Raton, FL.

Kirwan, B., Ainsworth, L.K. (Eds.), 1992. A Guide to Task Analysis Taylor & Francis, London.

Kirwan, B., Gibson, H., Kennedy, R., Edmunds, J., Cooksley, G., Umbers, I., 2005. Nuclear action reliability assessment (NARA): a databased HRA tool. Safety Reliability 25 (2), 38–45.

OGP Report No. 454, 2011. Human factors engineering in projects, August 2011.

Reason, J.T., 1990. Human Error. Cambridge University Press, Cambridge.

United States Department of Defense (USDoD), 1974, Procedures for performing a failure mode, effects and criticality analysis (1974/80/84). Washington, DC.

USDoD, 1974, Procedures for performing a failure mode, effects and criticality analysis. US MIL-STD-1629(ships), Washington, DC.

USDoD, 1980, Procedures for performing a failure mode, effects and criticality analysis. US MIL-STD-1629A/Notice 2, Washington, DC.

USDoD, 1984, Procedures for performing a failure mode, effects and criticality analysis. US MIL-STD-1629A, Washington, DC.

Village, J., Annett, T., Lin, E., Greig, M., Neumann, W.P. 2011. Adapting the failure modes effect analysis (FMEA) for detection of human factors concerns. Proceedings of the 42nd Annual Conference of the Association of Canadian Ergonomists, London, Ontario, 17–20 October.

Wilkinson, J., & Rycraft, H., 2014, Improving organisational learning: why don't we learn effectively from incidents and other sources? Proceedings of the Institution of Chemical Engineers, Hazards 24 Conference, Edinburgh.

Williams, J.C., 1988. HEART—A proposed method for assessing and reducing human error Proceedings of the 9th Advances in Reliability Technology Symposium. University of Bradford, Bradford.

Reducing human failure

R. Scaife

LIST OF ABBREVIATIONS

ABC	Antecedents, Behaviors, Consequences
ATM	Air Traffic Management
DCS	Distributed Control System
EPC	Error Producing Condition
ESSAI	Enhanced Safety through Situation Awareness Integration
HEA	Human Error Analysis
HRA	Human Reliability Analysis
IVMS	In Vehicle Monitoring System
MGB	Model of Goal-Directed Behavior
NASA	National Aeronautics and Space Administration
PBC	Perceived Behavioral Controls
PIF	Performance Influencing Factor
PPE	Personal Protective Equipment
PSF	Performance Shaping Factor
RHT	Risk Homeostasis Theory
STAR	Stop, Think, Act, Review
SA	Situational Awareness
SHEL	Software, Hardware, Environment, Liveware
SRK	Skills, Rules, Knowledge
TEAVAM	Team Error And Violations Analysis Method
TPB	Theory of Planned Behavior
TRACEr	Technique for Retrospective Analysis of Cognitive Error

This chapter describes ways to reduce human failure. As it is such a large topic, it is introduced in this overview and explained in more detail in Chapter 7.1, Human Error and Chapter 7.2, Intentional Noncompliance.

In understanding the concept of human failure and the specific conditions under which it can occur, human factors specialists are concerned with ways of reducing the likelihood that humans may fail when completing a particular task. As all readers will be acutely aware, human beings are random, nondeterministic systems. When subjected to a given stimulus, the response can vary and therefore it is difficult to empirically prove whether or not a human will perform as expected on a given task.

With deterministic systems, such as conventional software code, it can be tested by changing input variables. The output will be known for a given input, and if it fails to produce the correct output it is possible to correct it. There is no such luxury when it comes to people. However, there has been much research into how people may fail, and the conditions under which this is more likely.

If a car started behaving erratically, the driver would not start randomly applying different fixes until something worked. The driver would work out the problem (or get a mechanic to do this for them) and apply the fix that is most likely to reduce the recurrence of the problem. The same should be done with human behavior, but the human is seen as complex and nondeterministic and therefore this can put people off the idea of working out how they failed, and how to reduce the failure rate.

This chapter provides an overview of human failure and is followed by two subchapters covering the two key human failure types: unintentional human errors (see chapter: Human Error) and intentional noncompliance with rules and procedures (see chapter: Intentional Noncompliance). The overview provides an introduction to the broad concepts of error (or unintentional action) and violation (or intentional noncompliance) and the ways in which people may fail to perform as expected.

OVERVIEW OF HUMAN ERROR

One type of human failure is an unintentional human error. The person does not set out to do something wrong. Their behavior is unexpected. For example, a driver attempts to turn at a junction in front of on-coming traffic but misjudges the distance and speed of the on-coming vehicle. The driver does not deliberately misjudge the situation and has not decided to put their vehicle in harm's way, nor violated any traffic laws. They simply made an error of judgment.

Ever since psychology emerged as a scientific discipline, researchers have tried to understand human error and to develop methods to prevent it. Many models of human error exist. Three of the most renowned authors of human error models are Reason (1990), Rasmussen (1983), and Wickens (1992). Most models describe different types of human error and, depending on the model used, help to explain how the error might occur and relevant control measures to reduce the likelihood of the error occurring. These models will be discussed in more detail in Chapter 7.1, Human Error, where the tools and methods are introduced. These tools use theoretical models but define practical analysis techniques to help understand why errors occur and to develop actions to help manage human error.

A large body of research has also developed on specific conditions which increase the likelihood of human error (Kirwan, 1994; Shorrock and Kirwan, 1999; Williams, 1986). These conditions are variously referred to as Performance Shaping Factors (PSFs), Performance Influencing Factors (PIFs), and Error Producing Conditions (EPCs). Regardless of name, understanding the conditions, and how they influence human behavior, provides an opportunity to manage human performance. This chapter refers to them as PSFs.

It is important to draw the distinction between conditions that affect human performance and root causes of human error. The analyst may believe that they have found a way of preventing a particular error when all they have done is increase the reliability of human performance and not tackled the cause. This can be a difficult point to grasp, as PSFs such as fatigue, noisy work environments, or high workload can be seen as root causes of error. In reality however, what they do is interfere with the reliability of performing the task making error more likely. How the error happens is an entirely different mechanism rooted in psychology and the brain's higher cognitive functions. Consider the example where the driver misjudged the speed and distance of an oncoming vehicle when turning across its path. An analysis of the human error may uncover that the misjudgment was caused by a failure to integrate the speed of the vehicle and its distance and whether it was safe to turn. It may also reveal it was dark, the driver was tired, and the traffic was particularly heavy. The light conditions, fatigue, and traffic density are examples of PSFs. They will have affected the driver's performance but the root cause of the error remains the failure to integrate several pieces of information to make the judgment. The failure to integrate information could have happened whether the PSFs were present or not, so the PSFs are not root causes, they just made the likelihood of degraded performance greater.

Chapter 7.1, Human Error, goes into detail on how to reduce the occurrence of unintentional errors. Several of the models developed to help understand human error are discussed, as well as the influence of PSFs on human performance. The chapter discusses various ways to manage human error including preventing future errors. It is also important to identify error prevention mechanisms in the context of the task so that different errors in another part of the operation are avoided. Automating an error-prone activity may eliminate opportunity for the operator to make an error, but it may complicate the maintenance task and introduce a new type of error that was not possible prior to automation. The chapter also discusses how it may be more practical to make it more likely that errors will be detected early so they can be corrected, or to make the method of recovering from errors more efficient.

So far the concept of human error and the conditions that increase its likelihood have been introduced. However, this is only part of the story. The other half concerns intentional noncompliance with rules and procedures.

OVERVIEW OF VIOLATIONS AND NONCOMPLIANCE

It is important to understand what motivates humans to break the rules, which can differ depending on the situation.

The worker who breaks the rules to rescue a coworker does this because he/she feels that the cost of not doing so is significantly higher (the potential injury or death of a friend). These are exceptional circumstances, and thankfully relatively rare events. The motivation is to make things right, to save the situation or a person. All very strong motives. These are referred to as exceptional violations.

Situations where the person does not believe they can stay within the rules in the current circumstances are commonly observed as well. For example, a worker may consider themselves unable to perform a difficult task while wearing the "mandated" gloves. They want to do the job well, and believe they can do the job better without the gloves, even though they know the site rules say gloves must be worn. In the conditions imposed by the task the rules do not help so the worker is motivated to perform well and temporarily ignore the rules. This is referred to as a situational violation.

The third type of violation is where rules are ignored because it has become custom and practice to do so. At the BP Texas City refinery in 2005, the operators believed that under-filling a vessel on start-up could damage the plant. They had adopted a routine practice during start-up of filling the vessel higher than the procedure instruction. On this occasion the routine practice, combined with other circumstances, resulted in the vessel over-flowing into a blowdown drum causing a liquid and vapor release, and subsequent fire and explosion. This type of violation is referred to as a routine violation.

Each of the three scenarios above has at least two things in common; they all involve people intentionally breaking the rules, but those involved did not expect any harm to come to them, the workplace, or other people as a result of their actions. In fact, with the benefit of hindsight, it could realistically be expected that all people involved would have been shocked that anything bad happened as a result of their actions.

Cases do of course exist where the person breaking the rules does so with full knowledge of the damage they will cause. For example, a worker who uploads a computer virus to their work network because they have been disciplined. Such cases are thankfully rare and are examples of sabotage, but how to deal with them is clear. These violations as are labeled as malicious.

It can be difficult to decide how to address exceptional, situational, and routine violations to reduce the chances of similar behaviors occurring in the future. There are differences in motivation. In some cases, people are motivated by personal benefit; in other cases, their motivation is organizational benefit (e.g., helping to keep production on target or avoiding damage to company equipment).

As with human error, the working conditions can influence the likelihood a person will break the rules. For example, if they do not have easy access to equipment, are not well trained, do not understand the risks of straying from the rules, the individual may believe that not following the rules is acceptable. When planning how to reduce the likelihood of violations, it is necessary to consider workplace factors that may have encouraged people to break the rules. Some of these things can be relatively easy to fix. For example, low stock of Personal Protective Equipment (PPE) can quickly be replenished and monitored. Similarly, insufficient signage can be addressed with visible signs to remind people of when and where a rule applies. Other interventions such as building people's risk awareness can take a greater level of effort to overcome, but the solution is still relatively simple.

In addition to work conditions, other factors come into play when people violate procedures and rules. A person not wearing hearing protection in a high-noise area of plant may be doing so because of perceived benefits they gain as a result of not wearing their PPE. Benefits could include being able to communicate with work colleagues better, which will have a positive impact for them in getting the job done. Another benefit could be avoiding the discomfort associated with the current brand of hearing protection. Such factors will have a strong influence on the motivation to break the rules because some benefit is expected for the individual or the team as a whole. The individual may also consider the negative consequences, such as the potential to damage their hearing, or the potential to be disciplined for being caught without the correct PPE. Depending on levels of knowledge and the culture at their site, consequences may motivate people not to violate the procedure (e.g., if they understand and believe they will harm their hearing and/or be disciplined) or they may be seen as inconsequential risks (e.g., the supervisor is never on-site to catch them, and they are only going to be exposed to noise for a few minutes so hearing damage is unlikely).

To reduce the likelihood of intentional human failure, it is necessary to understand the conditions that may make someone believe that breaking the rules is worth doing, and also perceived consequences if they do break the rules. If both of these things are known, it is possible to work on ways of reducing the likelihood of someone choosing a course of action that they know is not within the rule book.

Chapter 7.2, Intentional Noncompliance, goes into greater depth on the subject of intentional noncompliance and how it can be better understood. Various methods of analyzing noncompliance are discussed and compared, and the links between noncompliance and safety culture are examined. The chapter also includes some guidance on the actions that can be taken to reduce noncompliance once the reasons and motivations have been understood.

KEY POINTS

- This chapter overview has highlighted the distinction between human error (an unintentional behavior) and violation or noncompliance (an intentional behavior).
- To reduce human failure, it is necessary to understand why people make errors and why people sometimes choose to ignore the procedures and rules.
- There is a vast amount of knowledge about human failures, and how to help to diagnose and reduce failures.

Human error

7.1

J. Mitchell and R. Scaife

Humans make errors and always will. Research in this area has been prolific, as scientists seek to define and understand human error and the mechanisms that lead humans to make errors in the first place.

In its simplest terms, human error is unintended human behavior. The person has done something they did not set out to do and is usually puzzled by how their unintended behavior came about. To state that someone made an error, it is necessary to demonstrate they unintentionally did something that they did not set out to do. In doing so, it is necessary to rule out the intervention of an external chance event causing the outcome. For example, if the individual unintentionally pressed the wrong button on the control panel, resulting in the valve remaining closed when it was expected to open, this is an error. If the individual pressed the correct button but the valve actuator catastrophically failed (a chance event) then there was no error.

MODELS OF HUMAN ERROR

THE SKILLS, RULES, KNOWLEDGE FRAMEWORK

The field of human error is awash with models that describe human error in different ways. One of the most well known is the Skills, Rules, Knowledge (SRK) framework developed by Jens Rasmussen (1983), which classifies errors under skill-, rule-, or knowledge-based performance. This model describes how different types of tasks require different levels of cognitive processing or conscious effort. Fig. 7.1 illustrates this.

Skill-based performance involves completion of well-practiced physical actions in familiar surroundings. These actions are conducted subconsciously, with little concentration required from the individual. Reactions to cues are automatic, such as driving a car, checking a gauge, or turning a valve. As a result, there is potential for workers to become overfamiliar, overconfident, or even complacent with the task and the inherent risks involved. Errors in this category may involve a slip of action such as inadvertently pressing the wrong button; or a lapse in memory, such as misrecollection of the correct sequence of button-presses. Although not included explicitly in this framework, it could be argued that some errors associated with sensing basic information would fit in this category too, such as failure to see the warning label next to the button that was pressed.

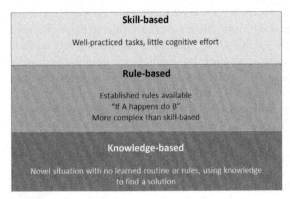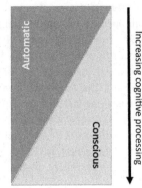

FIGURE 7.1

The SRK framework.

BOX 7.1 EXAMPLE OF RULE-BASED ACTION

IF high-pressure alarm THEN check temperature and level indications on vessel.
   There would then be further rules to govern the action required, for example:
   IF level indication >15 THEN close inlet valve and open relief valve.

Rule-based performance occurs when people are faced with familiar problems and are required to apply memorized or written rules to manage the situation. Errors in this category refer to situations where there is more conscious monitoring and analysis of what is happening, and where the individual is responding based upon their "know-how" of which procedure to apply to a situation. Through training and experience of the system being operated or maintained, an understanding of what should be done in a range of system states is gained. For example, a control room operator would have a rule governing what to do if the high-pressure alarm on the plant being monitored is activated. This rule might require the checking of several other system variables, such as level and temperature before taking an action to control the pressure. In this case, the operator knows what to do and has likely done such things before. As shown in Box 7.1, there is a standard sequence of task steps to follow before acting. Rule-based errors occur when the diagnosis of the situation based upon these rules goes wrong, for example, choosing the wrong rule to apply to the situation being observed.

Knowledge-based performance occurs in a completely conscious state. Types of tasks in this category would be beginners performing a new task, or experienced individuals faced with unplanned and/or an unfamiliar situation. High levels of concentration are required during knowledge-based performance. Effort is required to maintain awareness of the current status of the system and relate this to knowledge of the system. Experience of similar situations may be recalled to assist in diagnosis of the current situation, along with general process knowledge. For example, the control room operator may encounter a novel plant failure never encountered before, so may have no rules for dealing with the situation. The operator would need to rely on their knowledge of how the plant works to control the situation, a mentally demanding situation. If there is a lack of knowledge or expertise, the chosen solution may not work, and this would be an example of a knowledge-based error.

In any one task, workers may well switch between these three levels of processing in order to resolve a situation. For instance, if a worker is turning a valve (skills-based) and a liquid starts rushing out, they would likely apply a rule-based approach to preventing the release, perhaps closing the valve. However, if the valve did not close the worker is then reliant upon knowledge-based performance to rectify the situation.

SLIPS, LAPSES, AND MISTAKES

Another popular model is Reason's model which builds on the SRK framework by introducing the concepts of slips, lapses, and mistakes (Reason, 1990).

In Reason's model the skill-based errors are subdivided into slips of action and lapses of memory, while the rule- and knowledge-based errors are grouped under the heading "mistakes" which in this context means poor decision-making or errors of judgment.

As shown in Fig. 7.2, the SRK framework and the Reason model interact with each other, but slips could occur while undertaking action at either the skill-, the rule-, or the knowledge-based level. Lapses of memory could also occur in selecting the right rule to apply or in planning to take action.

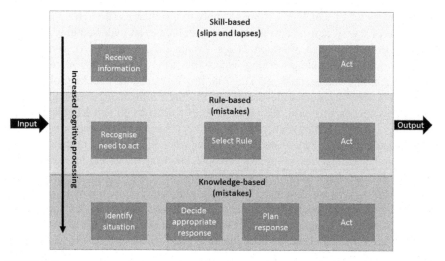

FIGURE 7.2

Reason and SRK.

The SRK framework and Reason's model highlight processes in how people interact within their environment. SRK talks of people receiving, recognizing, and identifying information from the world around them. This is central to information processing. If people fail to sense and perceive information correctly, they may fail to complete the task correctly.

SRK also defines the processes involved in making decisions, associated with selecting the correct way to deal with the situation, or making decisions and plans to deal with a novel scenario. If the wrong decisions or judgments are made, or plans are not fit for purpose, the end result of the task will not be correct, referred to as "mistakes" in Reason's model.

Reason's model identifies the importance of memory in how people process information. Lapses in memory are linked in his model to skill-based activities. Failure to recall information or the misrecollection of information at the deeper levels of mental processing can also result in unsuccessful task completion.

HUMAN INFORMATION PROCESSING MODEL

Another model of human error was developed by Wickens for a different purpose: the design of complex aircraft systems to match human information processing capabilities and reduce the likelihood of errors (Wickens, 1984). The model draws together some of the key points of the Reason and SRK frameworks, but goes further to help the user develop specific recommendations to prevent, detect, or recover from different types of errors (Fig. 7.3).

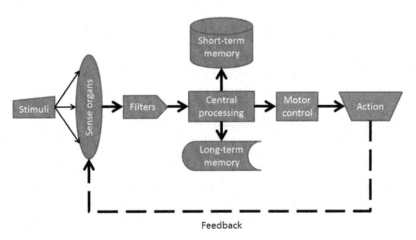

FIGURE 7.3

Human information processing model.

Based on Wickens (1984).

FIGURE 7.4

Optical illusion.

The Wickens model describes how people process information by sensing what is happening around them (using sight, hearing, smell, taste, touch, and balance). The information passes through a number of filters that help make sense of the information. These are not physical filters necessarily, but ways the brain has developed to process complex information in a complex world. For example, the brain attempts to differentiate between an object and the background, filtering out information the observer does not need to see. This results in the ability to be tricked by optical illusions, such as Fig. 7.4, where some people will see a trophy and others will see two faces. It also results in the ability for us to fail to sense or perceive information correctly.

Once the information has been received by the senses and filtered, the brain then processes the information to make decisions and judgments about what to do next. Sometimes the amount of mental effort required is very small (like skill-based activity in the SRK framework) or greater (such as responding to a novel situation). Depending on the depth of processing required, information will be stored and retrieved from both short-term and long-term memory. It may also be that a decision or judgment cannot be made without additional information, some of which will be stored in memory. This assists the decision-making process to check information that has been delivered via the senses with information in memory about the procedures and experiences. This helps the person determine the best course of action.

Finally, the brain exerts motor control over the muscles to initiate a physical action (or speech) to be taken based on the decision or judgment that has been made.

This model is the basis of several human error analysis tools, notably TRACEr (Technique for Retrospective Analysis of Cognitive Error), developed for use in Air Traffic Management (ATM) (Shorrock and Kirwan, 1999).

By determining the point during information processing where the error occurred, the most effective course of action to combat the error in future can be targeted on the

FIGURE 7.5

The three levels of Mica Endsley's model of situation awareness (1995, p.35).

mechanisms that allowed the error to occur. For example, consider the error of filling a petrol car with diesel. A sensory error may occur because the two pump nozzles looked similar. The solution would be different for a decision error where the person decided it was acceptable to use diesel because their knowledge was deficient. More detail on the application of these human factors techniques during incident investigation can be found in Chapter 8, Human Factors in Incident Investigation.

MODEL OF SITUATIONAL AWARENESS

One final concept that helps build an understanding of human error is Situational Awareness (SA). SA is the ability of an individual to perceive and understand the conditions of the environment in which they are working, so that they can accurately predict how changes in the working environment will affect them and those working around them. SA covers the higher-order processing of information to predict the effect of conditions in the immediate future.

A well-known model of SA, first proposed by Endsley (1995), describes the cognitive process in which an individual, in a restricted period of time and space, is able to perceive information (e.g., events, information, people, and actions) from the outside world (Perception) make sense of it (Comprehension) and predict the future status of the situation (Projection). The model shown in Fig. 7.5 illustrates how our level of SA interacts with our decisions and actions. In addition, it highlights that our level of SA is impacted by factors such as stress, workload, complexity, training, and preconceptions.

In most high-hazard organizations, the SA of workers is paramount for safe operation. Therefore, changes to conditions that impact on levels of SA require careful consideration. For instance, decreasing the workload of staff through increased automation a task, which at some levels may be desirable, can considerably degrade levels of SA (Wickens, 2008).

The *perception* level of the model (Level 1) involves attending to external cues. For instance, drivers constantly scan the road for pedestrians, bikes, signage, and other cars and make decisions accordingly to avoid danger. The level of constant scanning often reduces when driving on roads that are familiar or where there are fewer visual cues requiring attention. Endsley (1995) identified four possible errors relating to the perception level:

- data are not available (e.g., no road signs);
- data are difficult to detect/perceive (e.g., insufficient street lighting at night);

- failure to scan or observe data (e.g., familiar with the area so do not read the road signs);
- misperception of data (e.g., misreading the road sign).

The second level of Endsley's model is *comprehension*. This involves making sense of the information that has been perceived. As individuals gain experience in a particular task, they develop "mental models," that is, an understanding of what cues mean, recognizing patterns, and how the cues normally interact. An expert combines several pieces of information quickly, often subconsciously. An experienced operator might quickly identify that an alarm can be ignored in certain conditions based on past experience of that particular combination of events. However, experience leading to quick comprehension and judgment can also lead to bias. When an individual attempts to save mental time and energy, there is a susceptibility to applying previous knowledge of similar situations (mental model) without sufficiently considering the context of the current situation.

Projection (Level 3) involves building on the comprehension of the situation and thinking ahead to consider what might happen next. For instance, an experienced driver observing that a school is nearby and children are congregating on the street may well predict that it is quite likely for children to run on to the road and adjust his/her driving accordingly. In dynamic work environments, it is essential that workers think two or three steps ahead to plan ("what if" scenarios) and execute their work safely.

HUMAN ERRORS IN THE CONTEXT OF A FLAWED SYSTEM

For the majority of human errors, research suggests there is a link with latent organizational weaknesses (US DOE, 2009). Latent organizational weaknesses are weaknesses in managements systems within the company (e.g., procedure or equipment design approvals or resource management) that may not be immediately obvious, but nevertheless could influence how strong the defenses against human error are. For example, a weakness in calculating the required number of personnel to safely perform a task could lead to low staffing levels and therefore the potential for peer checking to suffer, resulting in an increased risk that errors will not be spotted quickly. These weaknesses are often collectively referred to as a "flawed system."

The means by which errors can be understood in the context of a flawed system is shown in Fig. 7.6. The more complex the system, the more susceptible workers are to error. Latent organizational weaknesses such as poorly defined policies and processes can give rise to Performance Shaping Factors (PSFs) (also known as error precursors or Error Producing Conditions, EPCs) and flawed controls. PSFs are discussed later in this chapter, but include factors such as time pressure, unworkable procedures, and poorly designed tasks. Flawed controls relate to the inability of the system to prevent, detect, or recover from error such as an alarm that indicates that there is a problem. The system can be improved by trying to prevent errors occurring in the first place or by identifying and reducing the impact of errors when they do

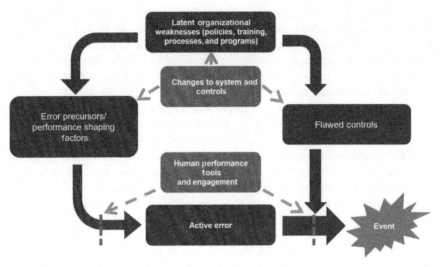

FIGURE 7.6

Framework for an event from a human performance perspective.

occur. An organization can strengthen the system reactively following an incident or proactively through attempting to understand the likely errors that could occur and the current weaknesses in the system.

Errors can be managed in two ways (refer to Fig. 7.6):

1. through making changes to systems and controls;
2. through applying human performance tools to engage the workforce in managing errors.

Current methods to proactively manage human error are discussed later in this chapter.

PREVALENCE OF HUMAN ERROR

It is commonly accepted that approximately 80% of all incidents are attributed primarily to human failure (Perrow, 1984; Reason, 1990), many of which are unintentional errors. Through using some of the models described above, it is possible to gain an understanding of the prevalence of the different types of errors. First, the errors that can occur during skills-, rules-, and knowledge-based performance are considered.

The main errors relating to the skills-based mode are slips ("action" errors) or lapses in attention or concentration ("memory" and "sensory" errors). Research in the nuclear power industry has found that the chances of a skills-based error are less than 1 in 10,000 and only account for 25% of all errors (US DOE, 2009).

The main errors resulting from the rule-based mode are misinterpretations ("decision" errors) and the application of rules incorrectly. This can be due to cognitive biases, where the person restricts the information they use to make a decision. An example of this is the application of "strong" rules that have been useful to the person in the past. Research has identified that the chances of a rule-based error are about 1 in 1000 and account for roughly 60% of all errors (US DOE, 2009).

Knowledge-based errors can stem from "sensory" errors (e.g., misreading a level gauge), "memory" errors (e.g., misrecalling a previous situation), or more frequently "decision" errors (e.g., making an incorrect decision on how to progress) due to a lack of knowledge or incomplete information. Given the nature of knowledge-based performance (i.e., high levels of attention, complex/incomplete information) it is perhaps unsurprising that the performance of workers is severely degraded by high levels of stress (e.g., time pressure, high severity of "wrong" action) and the chances of errors are high (1 in 2 to 1 in 10). Again, in these high-pressure situations cognitive biases can occur, for instance a worker may fixate on one aspect of the problem to the detriment of the bigger picture or only use the information that is most readily available to them, rather than correctly identifying the information required. Research has revealed that in the nuclear industry 15% of all errors are knowledge-based (US DOE, 2009). Due to the high chances of error during knowledge-based performance, the focus of attention for most companies is on moving from performance based on knowledge to performance based on rules.

There has also been research to identify the prevalence of errors within the context of the model of SA. Jones and Endsley (1996) analyzed 143 aviation accidents and found that 78% of accidents were due to problems relating to the "perception" level, most commonly a failure to scan for cues (35% of all SA errors). This was replicated by a similar analysis of 135 incidents conducted in the oil and gas industry (67% of errors occurred at the perception level) by Sneddon et al. (2006). Often, in the heat of the moment and under significant stress, individuals become so fixated on a particular cue or task that they fail to scan to pick up other salient cues (Box 7.2).

Failure to scan for all salient cues is therefore a key concept in the quest to manage error more effectively. This is useful to know as it is a skill that can be developed. For instance, experienced drivers tend to scan ahead and to the side for cues while less-experienced drivers are more likely to observe only what is in front of them. A study by Curry et al. (2010) found that a limited scanning of hazards was the biggest reason (21%) for motor vehicle crashes among teenagers.

Sneddon et al. (2006) found that 55.6% of comprehension errors (11.6% of all errors) involved the use of incorrect mental models. A mental model is an internal representation of how the real world operates. People use their mental models to

---

BOX 7.2 EASTERN AIRLINES CASE STUDY

In 1972 an Eastern Airlines jet crashed in the Florida Everglades as the three crew members fixated on solving the problem of a faulty landing gear light. They failed to observe that the autopilot had accidentally been disengaged and the altitude-hold turned off.

BOX 7.3 AIR FRANCE AIRBUS FLIGHT 447 CASE STUDY

On June 1, 2009, Air France Airbus Flight 447 stalled and crashed into the Atlantic Ocean, killing 228 passengers. One of the key contributing factors linked to this incident was the incorrect decision of one of the copilots to pull back on the joystick and send the plane into a steep climb when conditions were not suitable (the temperature was too high). Furthermore, the copilot subsequently ignored all warning signals, such as the "stall" warning, indicating the incorrect decision, and continued with the climb. Experts suggest that one explanation of this seemingly irrational behavior is that the copilot, in the stress of the situation, applied a familiar "mental model" incorrectly to this novel situation, Take Off, Go Around (TOGA). Writing in *Popular Mechanics* Wise (2011) reports that:

When a plane is taking off or aborting a landing—"going around"—it must gain both speed and altitude as efficiently as possible. At this critical phase of flight, pilots are trained to increase engine speed to the TOGA level and raise the nose to a certain pitch angle. Clearly, Bonin (co-pilot) is trying to achieve the same effect: He wants to increase speed and to climb away from danger. But he is not at sea level; he is in the far thinner air of 37,500 feet where the engines generate less thrust, and the wings generate less lift. Raising the nose to a certain angle of pitch does not result in the same angle of climb, but far less. Indeed, it can—and will—result in a descent.

---

predict what will happen, so if mental models are inaccurate or incorrect then the actions people take can have unanticipated consequences (Box 7.3).

Jones and Endsley (1996) found that only 5% of errors were due to a failure to project ahead, if the situation had been observed and correctly interpreted. Sneddon et al. (2006) found that it accounted for only 13.3% of all SA errors. Again, the most likely cause of a projection error is a poor mental model of how events will unfold.

PERFORMANCE SHAPING FACTORS

PSFs are defined as conditions that are likely to interfere with the successful performance of a task and make error more likely. PSFs do not in themselves cause error, the error is caused by a failure to correctly process the information.

A useful model to assist the understanding of what constitutes a PSF and how it can detrimentally affect performance is the SHEL (Software, Hardware, Environment, Liveware) model (Edwards, 1972, further developed by Hawkins, 1987).

The SHEL model considers the interfaces with the Liveware (the human) as a basis for explaining the topic of human factors. The model depicts the Liveware as having four interfaces with the Hardware, Software, Environment, and other Liveware.

At each of these interfaces, there are opportunities for misalignment between the Liveware in the center and the other elements in the model. When this happens it is likely that performance of the task at hand will be affected negatively.

For example, poor performance could result from:

- Liveware–Hardware interface—factors such as the ergonomic design of the equipment being used or the design of the task itself;

- Liveware–Software interface—software (such as the Distributed Control System (DCS));
- Liveware–Environment interface—factors such as noise, vibration, temperature, or lighting if they are not adequately controlled;
- Liveware–Human interface—factors such as communications between operators, teamwork quality, procedures, training, and handover quality.

In practical terms, the misalignment of these interfaces is likely to result in issues associated with aspects of the task such as:

- the task, including its complexity or the time pressure involved;
- communications, including written and verbal information;
- procedures, including poor or confusing design;
- equipment, including poor ergonomic design (including information displays);
- training for the task, including its quality and how frequently it is provided;
- environment, including adverse weather conditions and temperature;
- personal factors, including situations that distract attention;
- team, including how mature the team is and how well individuals work together.

The literature on Human Error Analysis (HEA) and Human Reliability Analysis (HRA) includes a number of different taxonomies of PSFs. HEA and HRA are often used interchangeably, but there is a difference; HRA is concerned with the predictive analysis of human reliability on a given task under a set of specific conditions; HEA refers to any analysis of human error, which can also be retrospective (such as analyzing errors that have occurred during an incident).

There is no unified model of PSFs, although there are many similarities between the different methods to which they belong. Some methods are designed to be used in specific industry sectors and provide PSF's tailored to the issues in that industry. Others are more generic and require the application of experience by the analyst.

Flin et al. (2008) describe people's capacity for picking up new information and maintaining mental awareness as being likened to the capacity of a jug. The size of the jug depends on individual factors including competency, fatigue, and stress. Individual capacity is represented by the liquid. When the jug is not full, the person has capacity to add more (such as extra tasks or distractions). When the jug is full, there is no capacity to add more unless some of the existing content is poured out or spilled. During safety critical tasks, there should be spare capacity to enable the individual to cope with new information, tasks, and distractions without exceeding their jug capacity. The issue of workload is discussed in more detail in Chapter 20.1, Staffing and Workload.

A study conducted by the Indiana University of Pennsylvania (IUP) asked around 2300 employees to rate the degree to which EPCs were present at their workplace (Watcher and Yorio, 2013). In addition, they were asked to report the number of injuries and near misses they had experienced in the preceding 6 months. The results indicated that the EPCs most frequently encountered were time pressures, multitasking, nonroutine tasks, and distractions. As shown in Table 7.1, of these EPCs, distractions and time pressures had the most significant impact on the rate of near misses, first-aid events, and beyond first-aid events.

Table 7.1 IUP Study Showing the Presence of PSFs and Correlation with Safety Performance

Measure	Mean Response (1 = Strongly Disagree, to 5 = Strongly Agree)	Near Misses Correlation	First Aid Correlation	Beyond First Aid Correlation
At work, there are time pressures. I feel rushed.	3.44	0.19*	0.10*	0.08*
At work, there are mental pressures. I find it difficult to concentrate.	2.94	0.16*	0.08*	0.06*
At work, I conduct many nonroutine tasks.	3.31	0.12*	0.06*	0.04
At work, I conduct many new/unfamiliar tasks.	3.01	0.10*	0.06*	0.05*
At work, I typically have a high workload.	3.51	0.11*	0.02	0.04
At work, I typically multitask—doing many different things at the same time.	3.72	0.03*	0.03	0.01
At work, I receive work guidance that is at times vague or imprecise.	3.06	0.16*	0.08*	0.05*
At work, there are many distractions around me.	3.19	0.23*	0.09*	0.08*

Note: *indicates that the result is statistically significant (p<0.05).

The results are not surprising, especially considering that time pressures, distractions, and interruptions are found frequently in safety critical environments, such as operating theatres (Healy et al., 2006) and frequently identified as contributing factors in incidents (Lardner and Maitland, 2009).

HUMAN LIMITS AND SUSCEPTIBILITY TO BIAS

Although many errors are instigated by a flawed system, some errors occur due to workers' own biases and susceptibilities. For instance, many people overestimate their own ability to perform tasks and underestimate the risks inherent in the tasks they are completing (such as working at height without fall protection). There is real danger when workers fail to appreciate their own personal performance limits, in particular with regard to attention span and working memory capacity. Human working memory is a limited resource, identified as being between 7 (+ or −2) items at any one time (Wickens, 1992) and is a key capability for performing tasks safely. An over-reliance on working memory, without checks, can lead to an increase in the likelihood of errors.

Lardner and Maitland (2009) describe how five isolation incidents occurred on an offshore platform, three involving human error. One of the incidents involved a supervisor being called away during a complex isolation on a turbine generator and on returning continued the isolation on the wrong (adjacent) turbine generator. Lardner and Maitland (2009) found that by addressing the PSFs (distractions, time pressure, labeling, and competence) and introducing a competent second checker, the error rate was reduced by 66%. Resources such as checklists, procedures, and second checking can help reduce the likelihood of errors linked with human limitations.

In his seminal book *Thinking Fast and Slow* Kahneman (2011) explores the concept of "system 1" and "system 2" thinking. "System 1" thoughts are automatic, fast, emotional, and easy, such as reading a poster or adding "2+ 2." "System 2" thoughts require more effort and are deliberate and logical, such as solving a complex problem. The key point is that humans are programmed to avoid the effort and energy of using "system 2" by applying mental shortcuts, known as heuristics. This involves applying common sense, personal experience, stereotyping, and intuition to make educated responses quickly and easily. An example of a shortcut would be choosing something to eat in a restaurant based only on what the waiter has recommended (instead of reading the menu). Another example is liking someone on first meeting them based on him or her reminding us of someone else we like. People make decisions based on the information at hand. The downside of "system 1" thinking is that people do not always possess or use all the information needed to make a "good" decision and occasionally the mental shortcuts are not logical.

"System 1" thinking is highly susceptible to cognitive bias. Examples of cognitive bias include:

- *confirmation bias*, where an individual attends only to information that confirms his or her thinking and ignores evidence to the contrary; or
- *availability bias*, where an individual bases his or her decision on their most vivid and memorable examples rather than considering all relevant examples.

These cognitive biases (of which there are many others) can have some serious consequences (Boxes 7.4 and 7.5).

It is vital, when it comes to critical decisions, that thinking is a conscious, deliberate, and methodical process that is free from bias.

POTENTIAL SOLUTIONS FOR MANAGING HUMAN ERROR

The optimal way to manage human error is through the design of a workplace that decreases or removes the likelihood of the error occurring in the first place and has sufficient controls in place to quickly detect and recover from any errors that do occur. For example, mistakenly putting petrol in a diesel car should not be possible; the nozzle should have been designed not to fit. A comprehensive overview of how this can be achieved is presented in Section III, Human Factors within Design and

BOX 7.4 BP TEXAS CITY CASE STUDY

In March 2005, a catastrophic explosion occurred at the BP Texas City Refinery killing 15 people and injuring 180 others as well as causing considerable damage to the plant and surrounding areas. The explosion occurred during the start-up of an isomerization (ISOM) unit when a raffinate splitter tower was overfilled; pressure relief devices opened, resulting in the release of flammable liquid from a blowdown stack that was not equipped with a flare. One of the key contributing factors to this catastrophic incident was the inability of the operators to diagnose why the raffinate tower was overfilling. In fact, the decisions they made to open valves worsened the impact of the already unfolding event.

> *In each instance, operators focused on reducing pressure: they tried to relieve pressure, but did not effectively question why the pressure spikes were occurring. They were fixated on the symptom of the problem, not the underlying cause and, therefore, did not diagnose the real problem (tower overfill).*
>
> **p.92, CSB REPORT NO. 2005-04-I-TX**.

The decision errors made by the operators can best be understood in the context of the human error framework (Fig. 7.5). There were several PSFs and flawed controls that stemmed from latent organizational weakness. Operators were working in "knowledge-based" mode without the sufficient knowledge or experience to call upon to help them make the right decisions. Operators did not have sufficient procedures, operating limits, training, or sufficient experience that would allow them to understand and handle abnormal start-up conditions. They were fatigued as a result of working thirty 12 hours shifts in a row, causing them to fixate on reducing the pressure, rather than looking at the whole situation. In addition, they were receiving inaccurate information from readings and gauges due to poor maintenance. In the words of the Chemical Safety Board (CSB) (US Chemical Safety and Hazards Investigation Board, 2005) "These safety system deficiencies created a workplace ripe for human error to occur" (p.102, CSB REPORT NO. 2005-04-I-TX).

BOX 7.5 NASA CASE STUDY OF BIAS

On January 27, 1986 engineers at National Aeronautics and Space Administration (NASA) had a critical decision to make on whether to launch the space shuttle Challenger based on safety concerns raised by their contracting engineers (Morton Thiokol) about potential failure of the O-rings in low temperatures. Although the concerns raised were serious and the advice given was clear, NASA (and Thiokol) managers decided to place more weight on the evidence available in favor of launching. This example of confirmation bias, as well as "sunk-cost" bias (where thinking is biased due to the time/money/effort already invested), resulted in a decision to launch, and ultimately the destruction of the Challenger space shuttle due to the failure of the O-rings.

Engineering. The methods that can be used to predict errors that might occur during a critical task are described in Chapter 6, Human Factors and Risk management. However, this is a lengthy process and it is unlikely that organizations will want or need to perform in-depth analysis for every task. Even when tasks have been thoroughly analyzed for potential errors, it is unlikely that management strategies will ever completely eliminate the potential for error. Furthermore, many work activities are unpredictable and evolve so there is a need for workers to be skilled in "thinking on their feet" and managing the evolving situation without error. Lastly, there are simpler and more accessible tools that gain greater involvement of the wider work-force in assessing and managing errors.

Applying the hierarchy of control, the first consideration in addressing error is to determine whether it can feasibly be engineered out of the system. If a task has been demonstrated to be error prone and/or of significant risk, an automated solution to performing the task could be considered. Caution needs to be applied however, as the introduction of automation may introduce an opportunity for error elsewhere in the system, such as in maintenance.

As part of the management of change process, the proposed changes should be assessed to determine which design is more robust; the one with a higher degree of human operation or the one with a higher level of automation. It is useful to regard the degree of automation as a continuum which can also be either fixed or dynamic. This is discussed in Chapter 9, Overview of Human Factors Engineering.

Addressing error requires examination and resolution of the work conditions that negatively affecting performance (the PSFs). Addressing PSFs may require engineering or administrative.

It is also necessary to understand the underlying cause of the error. Understanding the psychological error mechanisms can help to define actions that help to reduce the chances of the same error occurring in future.

It is important to consider what happens if the error does occur and how it can be contained and/or its impact reduced. Often the prevention of error is not possible. Therefore, attention should focus on increasing the detection of errors (possibly through automated systems or peer checking) and recovery from error (such as allowing sufficient time to recover should an error occur).

One of the options for managing error is through educating the workforce. As well as educating participants, training should involve a practical element to support skill development. Gaba et al. (1995) proposed the following techniques could be taught to anesthetists to improve their situational awareness:

- practice in scanning instruments and the conditions in the operating theatre environment;
- greater use of checklists to ensure relevant data are not missed;
- allocating attention more effectively;
- practice in multitasking;
- pattern recognition and matching of cues to disease and fault conditions.

Banbury et al. (2007) have shown that instructional materials, such as— Enhanced Safety through Situation Awareness Integration (ESSAI), for airline pilots have helped to develop situational awareness skills during flight time. Techniques to reduce human error should be linked to practical changes in the workplace (such as new equipment that reduces the amount of information the person is required to digest). Greater use of checklists is another effective method. The airline industry has a long history of using checklists; for instance, completing a "before take-off" check-list. Even though most pilots are familiar with the tasks involved before taking off, the checklist reduces the likelihood of a memory error that would be more likely if distractions are present. Breaking down complex tasks into smaller steps through the use of checklists has been strongly advocated for the healthcare (and other) industry in popular books such as *The Checklist Manifesto* (Gawande, 2009).

Other practical approaches that can be implemented include: minimizing distractions when workers are completing safety critical tasks, reducing multitasking, and reducing time pressures. These points seem obvious, but require effective supervisory leadership and cooperation of the team. All team members need to be fully aware of the negative impact that PSFs can have on error and create an agreement to cooperate with any working practices put in place.

Worker-centric *human performance tools* can be utilized to remind workers to actively think through work tasks, become aware of their surroundings and potential error traps, gain a second opinion, and think logically about next steps. This (ideally) provides an opportunity for workers to reduce the likelihood of errors and be more alert to detecting and recovering situations where they do occur.

Watcher and Yorio (2013) sought the opinions of several high performing organizations in various industries (nuclear, aviation, and power generation) regarding the human performance tools they have successfully used. Some of the tools are practical tools that can be implemented and some are related to shaping the cultural behaviors of workers. The 10 tools they identified include:

1. *Pretask and posttask briefings*—Organizations identified these tools as effective when they fully engaged workers in considering PSFs, critical steps, stop-work criteria, and additional tools that could be applied. In concept, directly before and after tasks is an opportune time to prime workers on warning signs and patterns to identify, which can reduce errors relating to inadequate scanning. Posttask briefs are particularly useful when complications have occurred, and where improvements or any other learnings are realized.
2. *Self-checking/STAR*—This tool is particularly useful for workers who operate in a skill-based or rule-based performance mode. STAR stands for Stop, Think, Act, Review and requires the worker to pause before completing a task and think through the expected outcome of the task, verify the results, and apply a contingency, if necessary.
3. *Jobsite review*—This is intended to act as an opportunity for workers, on arrival at the worksite, to complete a quick review of the workplace and compare with prejob briefing expectations. This can help workers have a more complete "mental model" of the work environment. Many organizations have a specific name for this tool such as "take two" or "take five" and a short document to record observations.
4. *Stop when unsure*—If "stop work" criteria have been defined during the pretask brief stage, then workers are primed to stop work and seek help. This encourages workers not to go beyond their capabilities (or continue in the knowledge-performance mode without the requisite knowledge) and deal with uncertainties by getting help rather than "soldiering" on.

5. *Questioning attitude*—This is a cultural approach, where questioning is an acceptable practice and promoted within the organization. A questioning attitude helps prevent cognitive biases (for one's self and others) by encouraging workers not to accept situations at face value but to gather the facts, think logically, and consider their understanding carefully to ensure they have an accurate mental model. Similar to "stop when unsure," workers should not continue when in doubt.

6. *Identifying critical steps*—this tool helps workers become more aware of their situation and potential errors and become more cautious when they are completing the steps that could cause them and others harm.

7. *Coaching and observation*—This tool involves coaches (managers, colleagues, or designated coaches) providing support to workers and encouraging them to think through potential issues, PSFs, and likely errors. This is similar to an observation conversation except the focus of the conversation is on maintaining SA and managing errors across the whole job (not just the immediate observed behavior).

8. *Three-way communication*—This tool helps ensure that all team members have a shared and accurate "mental model" of the task or any changes that have been introduced. The sender of the message states the message and the receiver acknowledges the message and repeats it back in a paraphrased form. The sender then acknowledges the senders reply.

9. *Concurrent verification/peer checking*—Concurrent verification involves two workers working together to individually confirm the safety of an action or a component before, during, and after an action. This tool is particularly important when the wrong action (such as pressing the wrong button) would cause an immediate harmful event. At each stage the two workers independently verify and agree on the next steps in the process and the verifier observes the performer during execution of the action. Peer checking involves an independent check of a performer's completed work. This is particularly useful when the harmful event is not immediate, but as a result of the correct implementation of a task (such as the isolation of a component). It is vital that checkers avoid assumptions that no mistakes will be made and therefore are not thorough in their approach.

10. *Procedure use, adherence and review*—Procedures should be followed step-by-step, while maintaining situational awareness and a questioning attitude (Are things going to plan? Is this procedure correct?). Workers should be aware that the procedure might be incorrect and in this case the answer is not to work-round the procedure, but to revise and assess the procedure before work starts. This way, workers should be involved in the continual review and improvement of procedures. Procedures should be carefully designed to avoid common flaws (such as vague, poorly laid-out, confusing options).

There are a variety of options available to organizations to manage error more effectively. Organizations need to develop a clear strategy so the workforce is not overwhelmed or unclear about the different things they should be doing. One solution, currently used in the Oil and Gas industry, is the Team Error And Violations Analysis Method (TEAVAM; Mitchell, 2015). The TEAVAM approach involves workers assessing their own tasks, identifying the potential for error, and developing a set of actions that include one-off organizational changes, improvement of controls (to detect, prevent, and manage errors), and the application of a set of human performance tools. TEAVAM uses the "pretask brief" method to focus on when and how to apply the human performance tools. The application of these techniques in a structured way contributes to reducing the likelihood of error.

SUMMARY: HUMAN ERROR

A distinction was made in this chapter between errors and noncompliance on the basis of whether behavior is intentional or not. Different models for describing human error and the influences of PSFs which make error more likely were discussed. Potential solutions for managing human error were presented. The next chapter discusses noncompliance which is related to intentional behaviors.

KEY POINTS

- There are several models that can be used to understand how and why human errors occur. These include the SRK framework, Slips, Lapses, and Mistakes, Human Information Processing Model, and the Model of Situational Awareness.
- According to the SRK framework, errors are most likely during knowledge-based performance, where the situation may be complex and the individual may have an incomplete "mental model."
- Errors related to situational awareness commonly occur due to insufficient "scanning" (individuals fail to take in all the information they need to) or poor comprehension due to applying the wrong "mental model."
- Most human errors are caused by flawed systems.
- PSFs such as poor procedures, inadequate training, and fatigue contribute to the degradation of human performance and an increase in the likelihood of human error.
- Humans have limitations (such as memory) and an error is more likely when working at full capacity and when mental shortcuts are incorrectly applied.
- The first consideration for frequently occurring or high potential errors is to consider engineering the error out of the system. If this is not possible, consideration should be given to early detection of the error and opportunity to recover.
- Organizational practices such as the use of checklists and the application of human performance tools can be used to reduce the likelihood of error.

REFERENCES

Banbury, S., Dudfield, H., Hormann, J., Soll, H., 2007. FASA: development and validation of a novel measure to assess the effectiveness of commercial airline pilot situation awareness training. Int. J. Aviat. Psychol. 17, 131–152.

Curry, A.E., Hafetz, J., Kallan, M.J., Winston, F.K., Durbin, D.R., 2010. Prevalence of teen driver errors leading to serious motor vehicle crashes. Accid. Anal. Prev. http://dx.doi.org/10.1016/j.aap.2010.10.019.

Edwards, E., 1972. Man and machine: Systems for safety. In Proc. of British Airline Pilots Associations Technical Symposium, British Airline Pilots Associations, London, pp. 21–36.

Endsley, M., 1995. Toward a theory of situation awareness in dynamic systems. Hum. Factors 37, 32–64.

Flin, R., O'Connor, P., Crichton, M., 2008. Safety at the Sharp End. A Guide to Non-technical Skills. Ashgate Publishing Limited.

Gaba, D., Howard, S., Small, S., 1995. Situation awareness in anaesthesiology. Hum. Factors 37, 20–31.

Gawande, A., 2009. *The Checklist Manifesto. How to Get Things Right*. Metropolitan Books.

Hawkins, F.H., 1987. Human Factors in Flight, second ed.. Uniepers.

Healy, A., Sevdalis, N., Vincent, C., 2006. Measuring intra-operative interference from distraction and interruption observed in the opening theatre. Ergonomics 49, 589–604.

Jones, D., Endsley, M., 1996. Sources of situation awareness errors in aviation. Aviat. Space Environ. Med. 67, 507–512.

Kahneman, D., 2011. Thinking Fast and Slow. Penguin Books.

Kirwan, B., 1994. A Guide to Practical Human Reliability Assessment. Taylor & Francis.

Lardner, R., Maitland, J., 2009. To err is human. A case study of error prevention in process isolations.

Mitchell, J., 2015. Learning human factors lessons—before you have an incident. Presented at Human Factors Application in Major Hazard Industries Conference. 6–7 October 2015, Aberdeen, UK.

Perrow, C., 1984. Normal Accidents. Living with High-Risk Technologies. Princeton University Press, Princeton, NJ.

Rasmussen, J., 1983. Skills, rules, and knowledge; signals, signs, and symbols, and other distinctions in human performance models. In: IEEE Transactions on Systems, Man and Cybernetics, vol. smc-13 No. 3 May 1983.

Reason, J., 1990. Human Error. Cambridge University Press, Cambridge, UK.

Shorrock, S.T., Kirwan, B., 1999. The development of tracer: a technique for the retrospective analysis of cognitive errors in ATC In: Harris, D. (Ed.), Engineering Psychology and Cognitive Ergonomics, vol. 3 Ashgate, Aldershot, UK.

Sneddon, A., Mearns, K., Flin, R., 2006. Situation awareness in offshore drill crews. Cognition, Technology and Work 8, 255–267.

U.S. Chemical Safety and Hazards Investigation Board, 2005. Investigation report: investigation and fire. *CSB REPORT NO. 2005-04-I-TX*.

U.S. Department Of Energy, 2009. Human Performance Improvement Handbook: Concepts and Principles, vol. 1. Author, Technical Standards Program, Washington, DC, DOE-HDBK-1028-2009.

Watcher, J.K., and Yorio, P.L., 2013. Human tools performance. Engaging workers as the best defense against errors and error precursors.

Wickens, C.D., 1984. *Engineering Psychology and Human Performance.*. Charles E. Merrill Publishing Co, Columbus, OH.

Wickens, C., 2008. Situation awareness. Review of Mica Endsley's 1995 articles on situation awareness theory and measurement. Hum. Factors 50 (3), 397–403.

Wise, J., 2011. What Really Happened Aboard Air France 447? Available from: http://www.popularmechanics.com/flight/a3115/what-really-happened-aboard-air-france-447-6611877/

Intentional noncompliance

7.2

R. Scaife and J. Mitchell

Intentional noncompliance is defined as situations where, for whatever reason, an individual or group of individuals consciously decide to perform a task in a way they know is not correct. For example, they may do something that is not written in the procedure or go against instructions they have been given. In either case, they know they are not doing the task in the way it is meant to be done.

As an external observer looking at an intentional noncompliance, it is often difficult to understand why someone would behave that way, particularly if it results in an incident. The observer may judge the person based on what they would have done in the same situation. This approach can lead to a lack of understanding of why the person did what they did.

To reduce intentional noncompliance in the workplace, it is necessary to understand what was going on from the point of view of the person who engaged in the intentional noncompliance.

This chapter differentiates intentional noncompliances from unintentional human errors (as discussed in Chapter 7.1: Human Error). It also outlines tools and methods (and their theoretical underpinning) that can be used to understand why people commit intentional noncompliances. The link between noncompliance and safety culture is discussed, including individual differences in risk perception and actions that can be taken to improve compliance at work.

RECOGNIZING INTENTIONAL NONCOMPLIANCE

In theory, distinguishing noncompliant behavior from human error is relatively easy. Intentional noncompliance implies that the person deliberately chose not to behave as they should have done, whereas human error implies that the individual unintentionally failed to behave correctly. This seems quite a stark distinction, but differentiating intentional noncompliance from error in reality it is not always as simple. Take for example a driver who drives through a red light. How would it be possible to determine if this was deliberate or an error? In the eyes of the law this would always be treated as an intentional noncompliance and there would be negative consequences for the driver.

An author of this chapter once interviewed a police driver who was asked to inform the parents of an accident involving one of their children. The couple were so upset that on leaving to return to the police station, the police driver was distracted by feelings of sympathy for the parents and drove straight through a red light, narrowly avoiding another accident. In this case the behavior was clearly unintentional, but without detailed knowledge of the state of mind of the driver at the time, would have been difficult to classify.

It is necessary to determine two aspects to classify a behavior as intentionally noncompliant:

- the person knew the correct way to behave;
- there was intentional deviation from the correct way to behave.

Different sources of evidence are required to make a reliable judgment on both of these factors. Interviews with the individual and witnesses may not prove conclusive. This is particularly the case if there is negative consequence associated with being "caught." However, if interview data is corroborated with physical or documentary evidence, it is more likely that conclusions will be more accurate (Box 7.6).

In some cases the person involved will state their intent not to comply with procedures, often with justification of why it is appropriate to deviate. Interviews with witnesses may reveal this kind of information, which is an indication that, if corroborated by other evidence, the behavior was intentional (Box 7.7).

BOX 7.6 BEHAVIORAL PREDICTORS FOR NONCOMPLIANCES

A study into driver behavior revealed several features that point to intentional noncompliance when driving (De Winter et al., 2007). A number of measures were recorded while drivers were performing simulated tasks. The researchers found strong predictors of noncompliant behavior such as pressing the clutch while decelerating and turning, and driving fast when making a turn. Pressing the clutch when decelerating and turning is not undesirable in itself, but the research found that drivers who performed behaviors similar to this were also likely to engage in unacceptable behaviors such as speeding. This implies that such behaviors, if they could be measured as easily in real driving incidents, could provide indicators to support whether or not intentional noncompliance had been present. Although measures such as this may be appealing, they must still be used in conjunction with other sources of evidence to draw reliable conclusions.

BOX 7.7 CASE STUDY: MEASURING NONCOMPLIANCES

An onshore oil and gas company introduced an In Vehicle Monitoring System (IVMS) to monitor and report on safe driver behavior. This included measurable data such as speed, harsh breaking, length of time driving (more than 2 hours), and seatbelt usage. Some of the drivers resented this and saw it as an invasion of their privacy. During a vehicle inspection audit, it was found that some of the drivers, in their time off-site, had visited a spares yard, purchased seatbelt buckles, and inserted them into the vehicle so that the system was fooled into believing that the seatbelts were always buckled. This resulted in them not recording a noncompliance event on the system. The behavior was evidently started by a group of operators who had to regularly jump in and out of the car to open and shut farmers' gates. Doing up their seatbelt, in their view, took too much time for low speed, low risk driving.

A common mistake in establishing the "intentionality" of a behavior is to focus on the outcome of the event, rather than whether the person intended to behave the way that they did. For example, a pedestrian crossing the road when the pedestrian crossing light is red and being hit by an oncoming vehicle. The pedestrian had intended to perform the behavior of crossing the road when the light was red, but they had no intention or expectation of getting hit by a vehicle. The analyst needs to focus on the "intentionality" of the behavior, which was clearly intentional, and not be tempted to say that being hit by a vehicle was unintentional.

TOOLS AND METHODS FOR ANALYZING NONCOMPLIANT BEHAVIORS

Developing effective strategies to modify noncompliant behaviors requires analysis of the factors leading to the behavior.

Analysis techniques used in incident investigation are discussed in detail in Chapter 8, Human Factors in Incident Investigation. In this chapter, a high-level overview of three techniques is presented:

1. Work Compatibility Model;
2. Antecedents/Behaviors/Consequences (ABC) Model;
3. Theory of Planned Behavior (TPB).

There are many theories that are useful to understand different job aspects such as motivation, wellbeing, or strain, which may influence behavior. However, some of these theories fail to consider interactions between different factors or provide an integrated structure to use in "real-world" settings.

WORK COMPATIBILITY MODEL

One model that introduces a systems-based, integrated approach to understanding behavior is the Work Compatibility Model (Genaidy et al., 2007). This model is based on a review of various work assessment methods. It measures the state of the human–work system by examining the balance between cultural, social, and physical work characteristics and the resultant experience of the worker. This experience is mediated by the design of the organization, the job, and the wellbeing of the workforce (Fig. 7.7).

The effect of these factors on the individual is measured in terms of effort, perceived risk and benefit, satisfaction levels, and performance. It is used to understand the direct influences on behaviors such as intentional noncompliance. A number of techniques for measuring this have been proposed by Genaidy and colleagues, however they are complex and impractical for everyday use.

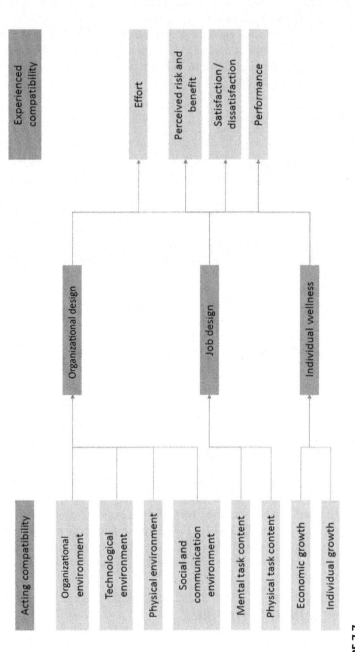

FIGURE 7.7

Work compatibility model (Genaidy et al., 2007).

From Genaidy, A., Salem, S., Karwowski, W., Paez, O. & Tuncel, S. (2007). The work compatibility improvement framework: an integrated perspective of the human-at-work system. Ergonomics Vol. 50, No.1, 15 January 2007 pp. 3–25.

Analyzing why people behave the way they do is complex for a number of reasons. People do not always respond the same way to the same stimulus and are highly susceptible to the context in which they find themselves.

In order to have practical value, any tool or method used to understand behavior needs to be simple yet diagnostic. It must also lend itself to being developed into an "easy to deploy" tool to aid the practitioner in understanding the behavior and formulating actions to help address it.

ABC MODEL

Techniques for understanding "why people behave the way they do" date back to the early 1950s when psychologists like Skinner (1953) proposed that the perceived consequences of a behavior would either encourage or discourage that behavior. As humans learn that a specific response will result in consequences that they find acceptable, they become conditioned to behave in this way.

The analysis of the antecedents or conditions that may trigger the behavior initially and the perceived consequences from the point of view of the person engaging in the behavior is known as "ABC" Analysis.

The presence of certain conditions in the environment where the human is operating can positively encourage the behavior to be initiated. For example, a manager who always wears the correct Personal Protective Equipment (PPE) does so to encourage others to behave in the same way (leading by example). The absence of conditions such as this may conversely fail to encourage the behavior.

ABC analysis has formed the basis of applied behavioral analysis and modification approaches. From the point of view of the individual, what they have learned about the best way to behave under different circumstances has an influence on behavior. As discussed in this chapter, it is necessary to analyze behavior from the point of view of the person performing the task, not from an objective point of view. Factors such as social norms, the example set by colleagues, and experience of the task contribute to the behavior.

The application of ABC analysis to incident investigation is covered in more depth in Chapter 8, Human Factors in Incident Investigation, but it can also be used proactively to promote desirable behavior change. ABC analysis is a simple model of human behavior and presents a systematic tool for the analysis of behaviors. Unlike other models that draw together other influences of behavior in the workplace, ABC analysis lends itself to practical application to understand and modify noncompliant behavior.

THEORY OF PLANNED BEHAVIOR

Another established theory of intentional behavior is the TPB (Ajzen, 1991), illustrated in Fig. 7.8.

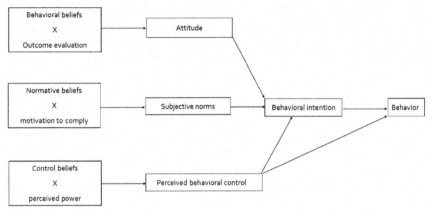

FIGURE 7.8

Adapted from theory of planned behavior (Ajzen, 1991).

Copyright Notice: The theory of planned behavior is in the public domain. No permission is needed to use the theory in research, to construct a TpB questionnaire, or to include an original drawing of the model in a thesis, dissertation, presentation, poster, article, or book.

The TPB has frequently been applied in the health sector to understand and influence factors that lead to intentional behavior, such as smoking. Specifically, these factors include:

- attitudes (i.e., the beliefs about the consequence of the behavior, and the desirability of these beliefs);
- subjective norms (i.e., the individual's perception of the beliefs and behaviors of managers and other colleagues and their personal motivation to comply with these expectations); and
- Perceived Behavioral Controls (PBC) (the individual's perception of the ease or the difficulty of performing the behavior and the power they have to influence the situation).

The impact of these factors on behavior is mediated by the intention of the individual to act, except for PBC which also has a direct effect. For example, an individual may intend to perform a behavior but may believe that he or she does not have the time or the correct equipment available.

The TPB intuitively fits with theories of safety culture. For instance, studies have shown the link between group norms and safety behavior (Zohar, 2000), measures of PBC and self-reported injuries (Huang et al., 2006), and safety attitudes and behavior (Johnson and Hall, 2005). Despite these links, very few studies have applied the TPB to safety outcomes.

Fogarty and Shaw (2010) examined the relationships in this model with the introduction of a new factor: management attitude to safety. They asked 308 members of the Australian Defence Force, the majority of whom were involved with aircraft maintenance to complete a questionnaire that covered a safety version of the TPB. PBC became a scale relating to workplace pressures, and group norms were about group practices in relation to noncompliances. The results identified that management attitudes, as outlined by other researchers (Mearns et al., 2003), have strong correlation with personal safety attitudes, perceived group norms, and work pressures. As Fig. 7.9 illustrates, the relationships in the study worked slightly differently from the TPB, with group norms and "own" attitudes also having a direct relationship with intentional noncompliance (violations). The researchers also found that perceived group norms had a strong relationship with "own" safety attitude and "work pressures," as well as intentions to violate and actual violations.

This research highlights the importance of management attitudes and group norms on intentional noncompliance. One surprising result was that "work pressures" did not have a direct correlation with violations. However, it was highlighted by the authors that it is likely that the scale used was inadequate for capturing the true nature of PBC.

Over time the TPB has been extended into new models, for example, the Model of Goal-Directed Behavior (MGB) (Perugini and Conner, 2000), which incorporates the concept of goals, emotions, and past behaviors. Although it is not as well researched or applied, some studies have illustrated that the MGB delivers superior explanatory performance when compared to TPB (Bagozzi and Lee, 2000). However, as yet, there is little research with regard to its applicability for safety. Furthermore, the complexity of the MGB may detract from the performance benefits gained.

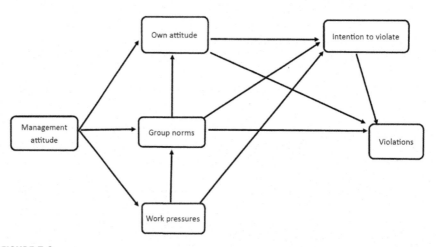

FIGURE 7.9

Safety climate and the theory of planned behavior (Fogarty and Shaw, 2010).

From Fogarty and Shaw - Safety Climate and the Theory of Planned Behaviour, Towards Prediction.
Accident Analysis and Prevention. September; 42(5), 1455–1459.

Intuitively, the importance of past behavior seems to be a glaring omission from the TPB. Conner and Armitage (1998) revealed that past behavior accounted for an additional 13% of variance in behavior. This concept (past behavior or habit) was added to an extended version of the TPB proposed by Maddux (1993), which also highlighted the importance of comparing attitudes toward old (unhelpful) behaviors and new behaviors. These two concepts appear to be useful additions to the model and further research is needed to provide evidence for the use of this extended model in safety.

THE LINKS BETWEEN NONCOMPLIANCE AND SAFETY CULTURE

Incident data from an oil and gas company reviewed for human factors root causes indicated that approximately 50% of incidents involving human failure were due to intentional noncompliance. The remaining 50% involved human error. The organization concerned was at the mid-point of a scale depicting the maturity of safety culture (i.e., the safety culture mature assessment measures indicated that they had a moderately mature safety culture).

An aviation organization presented similar data, but in their case incidents were 90% error and 10% intentional noncompliance. Their safety culture maturity rating was high (i.e., the Safety Culture Maturity® assessment indicated they had a very mature safety culture).

This suggests that as safety culture improves, the incidence of intentional noncompliance reduces. This makes sense because at lower levels of safety culture, there is a lack of widespread acceptance that everyone has a role in improving safety performance. In a more mature safety culture, one would expect high levels of involvement and cooperation in safety performance improvement and stronger shared values and beliefs about how behaviors influence safety.

In addition, there are many studies that link specific aspects of safety culture to safety compliance. For example, Mitchell (2008) found that the trust in managers made a unique and significant contribution in predicting personal safety compliance. Zohar (2002) demonstrated that when supervisors talk more about safety and reinforce these behaviors, safety compliance improves. Cohen and Cleveland (1983) compared 42 heavy industry sites and found that staff are more compliant when they are involved in the decision-making process, have specific responsibilities and authority, and receive prompt feedback on their work. Methods to develop safety culture factors are discussed later in the chapter. Safety culture and behaviors are discussed in more detail in Chapter 18, Safety Culture and Behavior.

RISK HOMEOSTASIS AND SAFETY NONCOMPLIANCE

Another concept that helps explain safety noncompliance is risk homeostasis. Risk Homeostasis Theory (RHT), first proposed by Wilde (2014), suggests that for any task people will accept a level of subjective risk (a "target" level of risk) to their health and safety in order to gain the benefits associated with the risk taken. If people subjectively perceive the level of risk to be less *or* greater than their "target" level, they will adjust their exposure accordingly and not always respond as expected to rules and controls. The theory helps explain why some initiatives may not work as intended. When initiatives are put into place to resolve one safety issue, they may inadvertently create another. For instance, sticking to marked walkways through areas where forklift trucks are operating may feel "too" safe to some people, and they may choose to increase their risk by taking shortcuts or walking to the side of the walkway. However, if the walkways did not exist, the same people may well believe that the level of risk is higher and warily stick to safer routes. The lesson from this theory is not to remove initiatives that reduce risk, but to help consider unforeseen behaviors that might occur when people perceive an activity to be "too safe." In some cases the solution might be to educate or remind people of the risks so that they adjust their own subjective perception of the risk to a more accurate level.

ACTIONS TO REDUCE INTENTIONAL NONCOMPLIANCES

Solutions to combat intentional noncompliance following a specific event or an incident should be determined by a detailed analysis of the behaviors (e.g., using ABC analysis) to understand the root causes of the behaviors.

There are principles that can be applied to reduce intentionally noncompliant behavior. These principles have been found to influence the key concepts discussed in this chapter such as social norms, attitudes, and perceived behavior controls. They can be used proactively to reduce the likelihood of intentional noncompliance or reactively by choosing the appropriate solution based on analysis.

- *Make it easy for people to comply*:
  - *Engineer compliance by design*—equipment, procedures, and associated training can make compliance difficult if they are badly designed or ineffective. Consider the ergonomic design of equipment (including PPE) and procedures and the effectiveness of training to deliver the required knowledge and skills to perform the task in a compliant manner. Also consider the effort required to be compliant compared to the effort required to take a shortcut. If there is an issue with risk perception, this can lead workers into taking the easiest option. Engineering the compliant way of performing the task to be easier will help (Box 7.8).
  - *Improve the rules or procedures*—rules or procedures that are impractical, outdated, or difficult to follow are likely to impede compliance. Habitual intentional noncompliance with impractical rules and procedures can result in workers placing less value in following rules and procedures in general.
  - *Physically engineer out intentional noncompliance*—some behaviors happen because there is an opportunity to perform a task in a way not prescribed by

---

BOX 7.8 CASE STUDY: UNINTENDED USE OF DESIGN

On an offshore installation, a number of pipes had been positioned horizontally at approximately knee height underneath a set of valves that were known to be difficult to reach without a temporary platform. As a result, workers used the pipes as a step to reach the valves when required. An alternative design (such as positioning the pipework at chest height or routed away from the area) would have helped discourage the undesirable behavior instead of installing a temporary platform.

---

    procedure. For example, boxed in pipework at or below knee height provides a flat surface that could be used as a step to reach a high valve rather than installing the required access platform. Removing the boxing or making it impossible to stand on could help reduce intentional noncompliance in conjunction with a rule of using access platforms for the task.

- *Tell people what their colleagues are doing (clarify the social norms)*:
  - *Publicize and highlight the positive behaviors of colleagues, including managers and supervisors*. When managers display positive behaviors such as performing interactive safety walks and investing in safety, ensure that people are aware of it. When a work team reports positive safety suggestions, share them. Research has shown that group norms and management attitudes predict noncompliant behaviors.
- *Reward desired behaviors and discourage undesired behaviors*:
  - *Consider reinforcement of the desired behavior and extinction of the undesired behaviors*. This can be done by the application of discouragement for intentional noncompliance while providing recognition for compliant behavior. Caution is required to ensure that the attitudes required to perform the behavior correctly are reinforced rather than just forcing people to comply with the behavior without a deeper understanding or desire to do so. For example, rewarding the quality of performance is more appropriate than rewarding performance alone. This approach is useful when there is a widespread compliance issue, when there is a high number of transient workers, or when other attempts to engineer out intentional noncompliance have proved unsuccessful.
  - *Ensure supervisors reinforce compliant or exemplary performance*. Ensure supervisors tackle noncompliant behaviors. Recruit and develop supervisors so they are able to perform these behaviors with the right level of competence. This is more appropriate when there is an acute compliance issue or when there are high numbers of short-term or transient contract staff.
- *Highlight the key messages*:
  - *Set clear standards and demonstrate strong management commitment to these standards*. Setting the expectations for safety behaviors and rule compliance is part of a strong safety culture, and if the necessary standards are not set and their importance demonstrated by management, changing the behaviors of others will be less effective, or indeed may be unlikely to succeed.

- *Ensure that important information and key messages are highlighted and stand out above other messages.* For example, put key messages on the front page of newsletters, use simple language, highlight information in bold, and avoid multiple messages.
- *Use personal language and messages.* Mass mailed messages do not make as big an impact as personalized messages that seem salient to the individual. For example, toolbox talks that are generic (and not always relevant) do not have the same impact as tailored talks on relevant topics.

- *Prompt safety compliance at the key moments*:
  - *Most people are honest and want to behave in a way that is consistent with their values.* Providing a prompt at a crucial time (such as just before starting work) enhances the likelihood that people will comply with the rules as it makes this link stronger. If there is no reminder and people are unsure, it can become more tentative.

- *Highlight the risk and impact of noncompliance*:
  - *Explain the purpose of rules and procedures.* The likelihood of compliance is improved if people understand why they are required to comply. For example, the basis of the regulatory requirement or the results of an internal investigation leading to the introduction of a procedure (Box 7.9).

- *Involve staff in improving work conditions and behaviors*:
  - *Involve staff in rule and procedure development.* Involvement is an important cornerstone for a strong safety culture and is also beneficial for buy-in to new or amended procedures. Experience of doing the task can be applied to development, which may be missed if procedures are written by management or technical authors. Personnel are also more likely to comply with procedures they have been involved in developing compared to a procedure that has been imposed on them without consultation.
  - *Use a behavioral based safety program* to identify and encourage positive behaviors and help staff avoid unsafe behaviors. In order to be successful, such programs need to be well managed to encourage the quality of performance rather than encouraging the number of observations gathered. This can be adopted as a proactive approach for improving safety culture as well as a reaction to a specific incident. It is essential to ensure that engineering and procedural safety systems are in place and effective before applying this type of program.

BOX 7.9 CASE STUDY: INTRODUCTION OF NEW PROCEDURES AT A STEEL MILL

Workers at a steel mill arrived at work one day to find that a new procedure had been introduced to prevent crossing a storage yard. A barrier had also been put in place. There was good reason for this because some of the steel could be very hot, and tripping in that area could result in serious injury. This reason had not been communicated. As a result, many workers were discovered climbing over the new barrier to cross the yard until the reasons for the new procedure were explained.

SUMMARY: INTENTIONAL NONCOMPLIANCE

Intentional noncompliance was described in this chapter as humans consciously deciding to perform a task in a way they know is incorrect. Different techniques were discussed, including an outline of the key underpinning theories in this area of human factors.

The next chapter discusses human factors in incident investigation and uses similar techniques to those outlined in this chapter.

KEY POINTS

- Accuracy is important to distinguish intentional behaviors from other forms of behavior. It requires care to avoid confusing the intentionality of the behavior with the intentionality of the outcome.
- Analysis of the intentional behavior requires a technique that is capable of providing a simple yet effective means of understanding the reasons for the behavior and how it can be addressed in the future.
- There is evidence to suggest that intentional behaviors are linked to the safety culture of the organization and peoples' perception of risk. Care should be taken to consider both of these aspects when trying to influence behaviors.
- The hierarchy of control should be applied when trying to influence intentional unsafe behaviors. Consider ways of designing the behavior out before considering more administrative approaches such as education or procedural changes.

REFERENCES

Ajzen, I., 1991. The theory of planned behavior. Organ. Behav. Hum. Decis. Process. 50, 179–211.

Bagozzi, R.P., Kyu, H.L., 2000. Intentional Social Action and the Reasons Why We Do Things with Others. Unpublished Manuscript.

Cohen, H., Cleveland, R., 1983. Safety program practices in record-holding plants Professional Safety, March, 26–33.

Conner, M., Armitage, C.J., 1998. Extending the theory of planned behavior: a review and avenues for future research. J. Appl. Soc. Psychol. 28, 1429–1464.

De Winter, J.C.F., Wieringa, P.A., Kuipers, J., Mulder, J.A., Mulder, M., 2007. Violations and errors during simulation-based driver training. Ergonomics 50 (1), 138–158.

Fogarty, G.J., Shaw, A., 2010. Safety climate and the theory of planned behavior: towards the prediction of unsafe behavior. Accid. Anal. Prev. 42 (5), 1455–1459.

Genaidy, A., Salem, S., Karwowski, W., Paez, O., Tuncel, S., 2007. The work compatibility improvement framework: an integrated perspective of the human-at-work system. Ergonomics 50 (1), 3–25.

Huang, Y.-H., Ho, M., Smith, G.S., Chen, P.Y., 2006. Safety climate and self-reported injury: assessing the mediating role of employee self-control. Accid. Anal. Prev. 38, 425–433.

Johnson, S.E., Hall, A., 2005. The prediction of safe-lifting behaviour: an application of the theory of planned behaviour. J. Saf. Res. 36 (1), 63–73.

Maddux, J.E., 1993. Social cognitive models of health and exercise behaviour: an introduction and review of conceptual issues. J. Appl. Sport Psychol. 5, 116–140.

Mearns, K., Whitaker, S., Flin, R., 2003. Safety climate, safety management practices and safety performance in offshore environments. Saf. Sci. 41, 631–640.

Mitchell, J. 2008, The necessity of trust and 'creative mistrust' for developing a safe culture. Symposium Series No. 154, IChemE, Hazards XX: Process Safety and Environmental Protection, Harnessing Knowledge, Challenging Complacency. 15–17 April, 2008.

Perugini, M., Conner, M., 2000. Predicting and understanding behavioural violations: the interplay between goals and behaviours. Eur. J. Soc. Psychol. 30, 705–731.

Skinner, B.F., 1953. Science and Human Behaviour. The MacMillan Company, New York, NY.

Wilde, G.J.S., 2014. Target Risk 3—Risk Homeostasis in Everyday Life. PDE Publications—Digital Edition, Toronto.

Zohar, D., 2000. A group-level model of safety climate: testing the effect of group climate on microaccidents in manufacturing jobs. J. Appl. Psychol. 85 (4), 587–596.

Zohar, D., 2002. Modifying Supervisory Practices to Improve Sub-unit Safety: A Leadership-based Intervention Model. J. Appl. Psychol. 87, 156–163.

REFERENCES

Kirwan, B., 1994. A Guide to Practical Human Reliability Assessment.. Taylor & Francis.

Rasmussen, J. 1983. Skills, rules, and knowledge; signals, signs, and symbols, and other distinctions in human performance models. In: IEEE Transactions on Systems, Man and Cybernetics, vol. smc-13 No. 3 May 1983.

Reason, J.T., 1990. Human Error. Cambridge University Press, Cambridge.

Shorrock, S.T., Kirwan, B., 1999. The development of tracer: a technique for the retrospective analysis of cognitive errors in ATC In: Harris, D. (Ed.), Engineering Psychology and Cognitive Ergonomics, vol. 3 Ashgate, Aldershot, UK.

Wickens, C., 1992. Engineering Psychology and Human Performance. HarperCollins, New York, NY.

Williams, J.C., 1986. HEART—a proposed method for assessing and reducing human error 9th Advances in Reliability Technology Symposium. University of Bradford.

Human factors in incident investigation

E. Novatsis and J. Wilkinson

LIST OF ABBREVIATIONS

ABC	Antecedents, Behavior, Consequences
HSE	Health and Safety Executive
PSFs	Performance Shaping Factors
SMS	Safety Management System

Despite the wide availability of many incident investigation reports, incidents continue to recur. While incidents may appear to be different, the underlying causes are often the same. There is clearly a failure to learn effectively from investigated incidents. It is argued that this is largely because human factors is often not properly considered and integrated within the investigation process.

This chapter provides an overview of integrating human factors within investigation processes and highlights common failures. Investigation methods are discussed, and in particular, the strengths and limitations of mainstream methods. It is argued that several methods, some more and some less constraining, may be required to accommodate different purposes and experience levels. The chapter then outlines how to adequately include human factors analysis within such methods.

Another focus in this chapter is the investigators themselves and how organizations manage their capability. The pressures and biases investigators are subject to are considered, as well as the critical qualities and skills they require, including interviewing skills. The different ways that organizations can manage investigation capability are discussed, along with practices that can help organizations to do this effectively. Finally, the potential of investigation for broader organizational learning is explored.

THE INVESTIGATION PROCESS

It is important to ensure quality investigations to optimize opportunity for learning. Over time, systems have become more complex and basic engineering, technical, and mechanical failures have become rarer (Strauch, 2004). However, complex systems are more vulnerable to human and organizational influences and these vulnerabilities often remain misunderstood. Human factors contribute to 80% of accidents across a wide range of industries and sectors (Strauch, 2004, p.xi). It is therefore even more important that investigations identify contributing human factors in a structured and reliable way, and with sufficient competence to recognize and understand what to do about them when they are found.

Investigations are carried out for a number of reasons. These include the following (see also Health and Safety Executive (HSE), 2015):

- in response to incidents, to understand causal factors and prevent reoccurrence;
- regulatory, insurance, and organizational requirements;
- as part of monitoring the Safety Management System (SMS) to uncover risk management failings;
- to identify where standards are drifting from what is specified in the SMS;
- as part of showing commitment to a safe culture and continuous improvement;
- to learn how work is actually being done as opposed to what procedures and other formal SMS arrangements require;
- to enable wider organizational learning, for example, for other site departments and areas, between different company sites and groups, and for sharing with other organizations.

Where failings of an individual are identified, they often reflect deeper organizational failings. These are important to identify and address, otherwise they will likely contribute to future incidents. Often, investigations are not undertaken effectively or consistently and do not uncover organizational failings. This situation may reflect a lack of investigator training and experience, the choice of investigation methods, investigation process failures, or other reasons.

KEY STAGES OF AN INVESTIGATION

The key stages of any investigation are the detection and reporting of events, data gathering, analysis of findings, writing recommendations, and sharing lessons learned. These stages can be broken into the detailed steps illustrated in Fig. 8.1.

There are common failings at each stage that can compromise investigation quality (detailed in Table 8.1). Notably, all of these investigation activities take place within the organizational culture concerned. Ideally, this is a "fair" or "just" culture, where the need to understand and learn from events is accepted and any allocation of individual or shared responsibility is done subsequently, transparently, consistently, and fairly. Without a fair or just culture, the workforce will quickly learn to keep quiet and not report events.

INVESTIGATION METHODS

Understanding incidents involves two main steps: (1) developing a timeline of events to reconstruct *what* happened and (2) conducting a thorough analysis of the events in

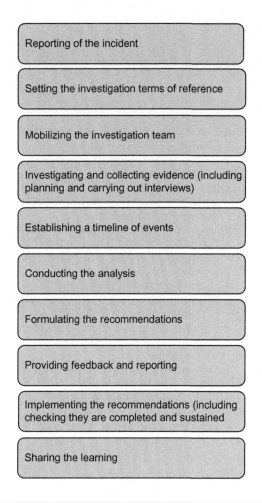

FIGURE 8.1

Detailed steps of an investigation process.

a systematic manner to understand *why* they happened, that is the root causes. This results in recommendations to address the root causes and prevent similar incidents. Investigators do not need to use a specific investigation method to undertake these steps, and many do not. Experienced investigators are often able to determine root causes without using any method. This approach may be suitable when investigation capability in an organization is very mature. However, organizations do not always have the luxury of a plentiful supply of experienced investigators and often advocate a preferred investigation method to support people in this role.

A variety of investigation methods are used by organizations operating in hazardous industries to help understand what happened and why. Many of them are proprietary commercially available methods. This chapter does not provide an evaluation of the methods or the extent to which they address human factors. A review

Table 8.1 Common Failings at Different Investigation Stages Related to Human and Organizational Factors

Investigation Stage	Common Failings
Reporting	• Not reported quickly enough, compromising accurate information being collected; for example, due to people's memory fading
Terms of reference	• Exclude looking at human and organizational issues • Findings are specified in advance
Mobilization	• Human factors expertise is not included in or available to the investigation team
Investigation and evidence collection	• Major assumptions are made on why people behaved as they did and these are not challenged within investigation team • Interviews are not planned • Key people not interviewed or badly interviewed • Evidence is not corroborated with other data
Timeline and analysis	• Intentional/unintentional distinction not made • Search for blame conducted • Human factors that influenced performance are not identified
Recommendations	• Poor symmetry between behavioral causes and recommendations • Unrealistic or impractical recommendations made • Reasons for the recommendations are not made sufficiently clear
Feedback/reporting	• Findings challenged without due cause, especially when they highlight management failures
Sharing learning	• Lessons not shared with all relevant parties • Little thought given to the method used to share the lessons • Underlying organizational factors are not identified or shared

of 28 mainstream methods is presented within guidance from the Energy Institute (2008) with a summary of their key features. This chapter does provide discussion on general advantages and disadvantages of mainstream methods and the key considerations in selecting a method to take account of investigation experience.

ADVANTAGES OF MAINSTREAM METHODS

Some mainstream methods provide a process for how to assemble a timeline, such as TapRoot® (Paradies and Unger, 2000) or Kelvin Topset® (Topset, 2015). This helps investigators present evidence about the sequence of events leading up to and after the event itself. In its simplest form, this task can be done by the investigation team with a set of stick-on notes arranged on a meeting room wall. Methods that include this guidance can offer tips on what type of issues to include in the main timeline, and what factors are associated with each of those items to help reconstruct the incident.

A structure for analyzing information is also given in some mainstream methods. This is particularly important for less experienced investigators or for those experiencing the pressures of investigation (discussed later in the chapter). For example, the Incident Cause Analysis Method (Safety Wise Solutions, 2015) leads the investigator to consider absent or failed defenses, individual or team actions, task or environmental conditions, and organizational factors in a structured way. Within each of these elements a set of examples is provided to assist investigators with their analysis. Other methods such as TapRoot® contain a list of potential root causes presented as a flowchart. Such lists can help ensure that the investigation team considers a comprehensive range of potential human and organizational contributions.

At an organizational level, a mainstream method with prescribed root cause categories can help organizations to track trends in root causes including human and organizational factors issues. Some organizations link root cause categories in their selected tool to the incident database, enabling reporting of root cause trends. The same can be done at a higher level, for instance when multiple operators use the same tool the regulator can collate this information.

A mainstream method can also provide a reference point for investigation teams to check where they are within the process and what is still left do to. This means of progress checking throughout investigations, particularly ones that occur over an extended timeframe, can be helpful.

DISADVANTAGES OF MAINSTREAM METHODS

The Energy Institute (2008) review of investigation methods identified that only 15 of the 28 methods reviewed can be considered a "complete" method for incident analysis. However, their definition of "complete" did not encompass the need for the method to include *how* to analyze the human factors. Some methods do provide a human failure taxonomy and some include checklists of human factors issues (as explained above), such as "procedure not available" or "supervision not present." Key human factors topics are often represented in this manner but guidance is not provided on how to examine why the issue occurred. For example, "why was the procedure not available?" or "why was supervision not present?" In other words, behavior analysis guidance is not provided. This means that the effectiveness of the root cause analysis performed when using most of the mainstream methods relies on an investigator being sufficiently experienced in human factors analysis to uncover the underlying reasons of the behaviors in question. A notable exception in the Energy Institute's review is the Human Factors Analysis Tools (HFAT®) (Lardner and Scaife, 2006). This method is designed specifically to analyze different behavior types, but does not include the timeline reconstruction guidance and critical factor identification that other methods provide. HFAT® is often used in conjunction with another method.

Another disadvantage of mainstream methods is that the provision of a defined structure can "box in" thinking. Investigators can become more concerned about which category in the method the evidence fits into rather than applying the open and inquisitive mind-set needed to examine why the event occurred. The investigation method can then end up creating bias and leading the investigation, rather than acting as a supporting tool; in other words, "the tail wagging the dog." Even if structured root cause checklists (that include a range of human factors) are available, they need to be well supported by training and guidance to be used reliably and consistently by the investigators. They also need good support to help determine appropriate recommendations to fix them.

Moreover, for highly complex incidents, even the more comprehensive investigation methods will not be sufficient to explore the detail and complexity of the issues at play. Such investigations require investigators to draw on a range of tools, experience, and concepts to understand the issues involved. Often less prescriptive methods are more suitable for such incidents.

CHOICE OF METHOD

Some methods are less constraining and simpler than others. For instance, in a method like "5 Whys" (see, e.g., Bulsok, 2015 for a good explanation of the method), the investigator asks why the incident occurred and then for each factor identified asks "why" several times until the root cause has been determined. If a simpler tool like this is used to understand human and organizational factors, there is a higher requirement for experience, training, and coaching to ensure the investigator can perform effective analysis. For example, the investigator would need to understand human factors concepts so that they are properly considered and explored in the process of asking "why." By contrast, a more constraining method can involve the user following predefined categories to assign root causes and therefore match evidence to where it best fits into the tool. If the tool is not too complex, this structure can provide direction to less experienced investigators, including consideration of human factors root causes.

One size does not fit all when it comes to choosing investigation methods. Investigation competency in organizations and within a given team can vary. Organizations might therefore be best placed to provide and teach several methods; some more simple and open, and some that are more constraining. Investigation teams then have the flexibility to select the method(s) from a "toolbox" that best suits their purpose and experience levels. One example of such a different purpose may be the investigation of a minor versus a major event.

Importantly, the methods selected should include a simple way of establishing the timeline and key events and a simple way of analyzing this with appropriate human factors input or additions as required. How to integrate human and organizational factors analysis into mainstream methods is covered in the following section.

INCLUDING HUMAN FACTORS WITHIN EXISTING INVESTIGATION METHODS

Regardless of the investigation method used, organizations can benefit from incorporating human factors analysis, or appropriate links to standalone methods, for analyzing key behaviors. Any human factors analysis as part of an investigation requires two things: (1) a taxonomy (a structured way of breaking down and classifying human failures) and (2) methods for subsequent analysis of the identified human failure(s). In this way, reasons for the behavior(s) can be understood and appropriate solutions identified.

There are many taxonomies for understanding human error, explained in Chapter 7.1, Human Error. What matters, is having an agreed taxonomy and understanding its strengths and limitations rather than fixating on choosing the perfect one. In this chapter, the tools provided as part of the HFAT® method are discussed. For human error analysis this is based on Wicken's model of information processing (Lardner and Scaife, 2006, see Chapter 7.1: Human Error), and Antecedents, Behavior, Consequences (ABC) analysis (described in Chapter 7.2: Intentional Noncompliance) for managing violations.

A simplified traditional model of incident investigation is presented in Fig. 8.2 and shows how and when human failure analysis is incorporated. First, evidence is gathered, and the timeline of events established to identify critical factors and causes. Where these factors involve a human behavior, and if there is sufficient information to specify it adequately, the behavior can be analyzed with a validated tool. In the figure, human error analysis is applied for unintentional behaviors. Chapter 7.1, Human Error, describes human error analysis in detail, including the identification of error types, the underlying causes (mechanisms), and Performance Shaping Factors (PSFs). ABC analysis is applied for an intentional behavior. The initial categorization as intentional or unintentional is not necessarily critical, because it becomes obvious during the analysis if the incorrect tool is being applied (Box 8.1).

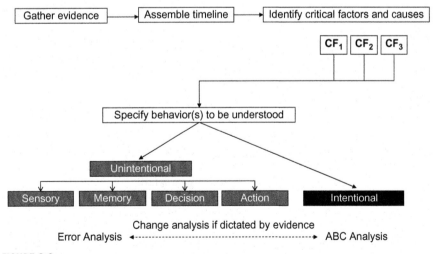

FIGURE 8.2

Diagram from HFAT® showing a traditional investigation process and the human factors analysis component.

> ### BOX 8.1 HUMAN FACTORS ANALYSIS OF BP TEXAS CITY, 2005
>
> The investigation of the 2005 BP Texas City explosion identified a range of human errors and violations. Understanding these behaviors led to invaluable lessons about human causal and contributory factors and also the underlying organizational ones. For example, one critical factor was the failure to control the unit start-up. An example of intentional behavior identified for this critical factor was the duty supervisor leaving site on the morning of the accident without a replacement. There was a clear BP requirement for a supervisor to be present at all times during start-ups and therefore either for the supervisor to find a replacement or for the start-up to be temporarily suspended until one could be found. Neither of these was done so in principle this was intentional behavior (a violation). When this behavior was analyzed using the ABC method, the underlying organizational factors emerged, including the safety culture, leadership expectations, and lack of adequate supervisor training. If the real hazards of start-up had been understood, effective arrangements put in place to manage them, and all of this reliably communicated through training and awareness, then either a replacement supervisor would have been found or the start-up delayed or paused. The investigation established that BP, as an organization, was not paying adequate attention to process safety issues, and so the supervisor's behavior during start-up had to be considered in that context. The last step of ABC analysis is to define the behavior that is actually wanted (that a supervisor is always present during start-ups) and design appropriate solutions to ensure this behavior happens.
>
> An example of a key human error was that the board operator did not notice the raffinate tower was overfilling. When this error was examined more closely, it became apparent that the combination of PSFs such as the poor alarm design, poor layout of the relevant control system display graphic, and poor information availability for start-up made the error much more likely. The organizational factors identified included a series of previous start-ups carried out under similar conditions reflecting the overall poor process safety culture and arrangements. Human error analysis allows each specific error to be analyzed and appropriate solutions identified for different types of error mechanism and PSFs.

HUMAN FACTORS CONSIDERATIONS FOR INVESTIGATORS

An often overlooked, yet significant, influence on the quality of internal investigations are the human factors that can influence investigators, investigation teams, and others involved in the investigation process. In many organizations investigators are tasked with completing an investigation while still performing their "day job." Accordingly, they themselves can experience a range of demands, such as time pressure, increased workload, performing an infrequent task, and stress. These factors can influence the reliability of their performance, especially given the type of analysis and decision making required for the investigation task.

Flin et al. (2008) describe four main types of analysis and decision making, which are shown in Fig. 8.3. These decision making types vary with the level of mental effort involved, speed, and the extent to which they are influenced by stress. The demands of the investigation task may encourage investigators to use less effortful processes, such as *recognition-primed* and *rule-based* decision making. Yet to investigate thoroughly, especially more serious, complex, or unusual incidents, it requires

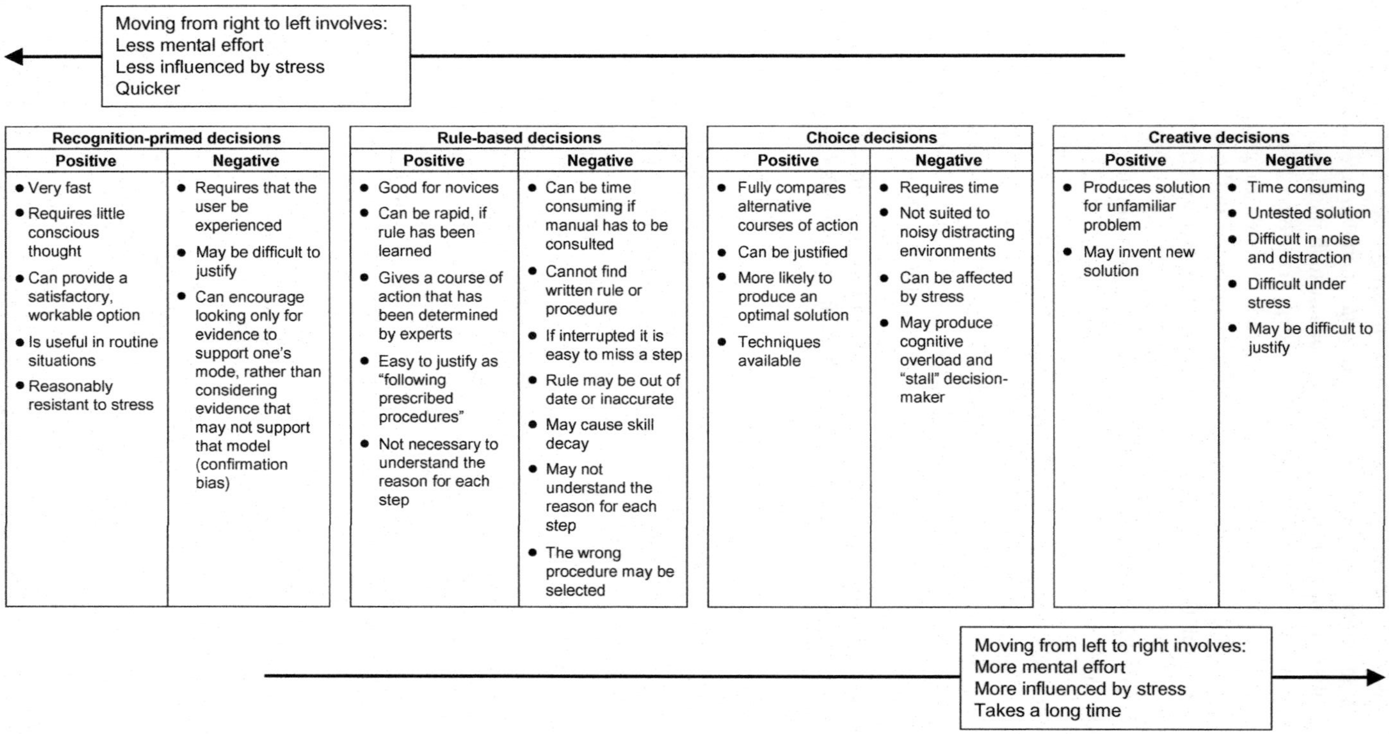

Recognition-primed decisions		Rule-based decisions		Choice decisions		Creative decisions	
Positive	Negative	Positive	Negative	Positive	Negative	Positive	Negative
• Very fast • Requires little conscious thought • Can provide a satisfactory, workable option • Is useful in routine situations • Reasonably resistant to stress	• Requires that the user be experienced • May be difficult to justify • Can encourage looking only for evidence to support one's mode, rather than considering evidence that may not support that model (confirmation bias)	• Good for novices • Can be rapid, if rule has been learned • Gives a course of action that has been determined by experts • Easy to justify as "following prescribed procedures" • Not necessary to understand the reason for each step	• Can be time consuming if manual has to be consulted • Cannot find written rule or procedure • If interrupted it is easy to miss a step • Rule may be out of date or inaccurate • May cause skill decay • May not understand the reason for each step • The wrong procedure may be selected	• Fully compares alternative courses of action • Can be justified • More likely to produce an optimal solution • Techniques available	• Requires time • Not suited to noisy distracting environments • Can be affected by stress • May produce cognitive overload and "stall" decision-maker	• Produces solution for unfamiliar problem • May invent new solution	• Time consuming • Untested solution • Difficult in noise and distraction • Difficult under stress • May be difficult to justify

Moving from right to left involves:
Less mental effort
Less influenced by stress
Quicker

Moving from left to right involves:
More mental effort
More influenced by stress
Takes a long time

FIGURE 8.3

Main types of decision making and associated cognitive effort.

more conscious, effortful analysis and decision making (see Chapter 7.1, section: Human Limits and Susceptibility to Bias). The *choice* and *creative* decision making types are essential to uncover the underlying issues in such incidents.

Unfortunately, this conflict is common in organizations. It is also one reason why organizations should provide structured methods to guide a thorough investigation and help set expectations for others involved in the process, such as the incident sponsor (the person accountable for the investigation). Better still, organizations should identify what is needed to set investigators up for successful performance and ensure the appropriate working environment is provided. For example, how will:

- investigation teams be organized;
- investigator competency be managed;
- workload in the investigator's "day job" be managed;
- the organization educate investigation sponsors (typically line managers) about the conditions they need to create to support investigation teams?

Such actions help show staff that the organization values the investigation process and is creating the structures and practices where investigations can be performed effectively.

KEY QUALITIES OF INVESTIGATORS

Being an effective investigator requires a specific set of qualities and lots of practice. The key qualities listed in Box 8.2 include a mix of skills, knowledge, and motivational style. This list does not include all of the skills and attributes investigators need, but rather the ones critical to their ability to understand the human factors and organizational issues within the event. All investigation team members should possess these qualities. In some cases this may not be possible, especially if the organization's maturity for incident investigation is low or investigators are still developing their skills. However, as an investigation is usually a team-based activity, at the least all qualities should be represented between team members. Investigators should be selected and developed against these key qualities.

BOX 8.2 KEY QUALITIES OF INVESTIGATORS

- Knowledge of human factors issues
- Knowledge of cognitive biases and how they influence decision making
- Interviewing skills
- Data analysis skills
- Able to use systematic investigation methods, checklists, and prompts

- Seeks corroboration of evidence
- Avoids making assumptions
- Willing to suspend judgment until analysis is complete
- Keen interest in understanding human behavior and organizational issues

INTERVIEWING SKILLS

Interviews occur at the evidence collection stage of an investigation. Highly developed interviewing skills are essential to uncover the underlying human and organizational contributions to incidents. Interviewing takes a great deal of practice to do well. Interviewees typically enter an interview with high expectations of how they want to be treated, and rightly so. For example, interviewees expect the interviewers to show respect and empathy, remain unbiased and impartial, listen carefully, check their understanding is correct, capture an accurate summary of what they have heard, and understand the investigation process. Meanwhile, interviewers need to juggle their own demands. For instance, they need to manage potentially anxious or defensive interviewees, establish rapport, lead the questioning process, think on their feet to ask the necessary probing questions, and avoid asking leading questions. Given this challenging task, a structured process for planning and executing the interview is essential. This goal can be achieved through the use of cognitive interviewing, a validated set of techniques that help to achieve accurate and thorough memory recall from an interviewee (Memon and Bull, 1991). Some of the key techniques involved in conducting a cognitive interview are explained in this section. Interviewers should spend time planning the interview using the following structure and tips.

Preparation
- *Review evidence*—ensure that any witness statements, initial accounts of what happened, and other relevant data available are reviewed prior to the interview.
- *Check for similar previous incidents*—it can be helpful to have an understanding of whether similar incidents have occurred in the organization and what human factors contributed to them.
- *Visit the scene of the incident*—if the interviewers are not familiar with the site, area, and equipment involved with the incident, visiting the scene can assist them to visualize what interviewees are talking about at interview. This will also allow the interviewer to develop better questions for the interview itself and to understand better what was being done at the time.
- *Understand how jobs should be performed*—examining the procedures for relevant jobs and the key responsibilities of relevant roles can assist interviewers to compare what should have been done to what actually occurred.
- *Establish timing and order of interviews*—interviewers should plan to interview people as soon as possible after an event as memory can deteriorate; they should also decide the best order in which to interview people (e.g., it may be appropriate to interview people further removed from the event before interviewing the person directly involved, or to interview the person directly involved before interviewing their supervisor).
- *Identify topics*—the topics that need to be explored at interview must be identified. The topics should be ordered appropriately; for example, more sensitive areas can be left to later in the interview. Some specific questions

should also be developed for each topic. Additional questions might still be asked during the interview; however, preparing questions ensures that key issues are not missed and that they are asked in effective manner.

- *Decide who will conduct the interview*—plan to conduct the interview with two people; a lead and a second interviewer. The lead interviewer directs the interview, guiding the process and leading the questioning. The second interviewer takes notes of the interview, looks for gaps and inconsistencies in the evidence being collected, and prepares additional questions. They only participate in the interview when asked to do so by the lead interviewer, typically at the end of each topic being discussed. This format creates a predictable flow for the three parties involved and enables both interviewers to perform their discrete tasks effectively. It can also be helpful to have one "technical" person and one "human factors" trained person as they would likely look out for and pick up on different things.

Setup and Rapport Building

- *Organize the venue*—wherever possible identify a quiet, neutral, and uncluttered venue, free from interruptions. The venue should reflect the value placed on the interviewing process.
- *Take time to put the interviewee at ease*—ask some opening background questions to put the interviewee at ease, explain the purpose and format of the interview, and show empathy for what they are experiencing. It is also important to address any immediate presenting issues. For example, if the interviewee is thinking "Will I lose my job?" or "Is my mate OK?" this needs to be addressed first as otherwise it will be difficult for them to focus on the interview. It is also important to "transfer control" to the interviewee by deflating your importance and enhancing theirs. For instance, by explaining that you are not an expert in their field so they should not assume any knowledge on your part and should explain things in detail to you.
- *Uninterrupted recall*—one of the most important techniques of cognitive interviews is in first allowing an interviewee to recount their experience uninterrupted (Waddington and Bull, 2007). Allowing interviewees to explain first in their own words what happened on the day without interruption is essential. This is because people do not typically recall their experience of events chronologically but rather in the order most salient to them. Therefore interjecting while they are explaining their story impedes their memory recall and the quality of information being revealed. Once the initial account is conveyed the lead interviewer can then proceed with the planned topic questions. The principle of remaining silent while the interviewee is recalling their experience or thinking about their response should also be observed throughout the interview.

Questioning Techniques

- A range of question types should be employed to elicit information during an interview. Open questions (e.g., "tell me about the shift handover that day") are critical to interviewing and encourage a detailed response from the interviewee in their own words. Closed questions are those that can be answered with "yes" or "no," or those pertaining to quantity (e.g., "How many operators were on that shift?"), identification (e.g., "What is your position?"), or time (e.g., "What time did the shift handover start?"). Closed questions can help to determine specific details. Probing questions (e.g., what, where, when, why, how, and who) are useful to look further into information that the interviewee has provided.

● Finally, the interview is not a court room and the interviewer should not try to allocate blame, comment unnecessarily, or conclude on the information that emerges (except to clarify or to acknowledge and encourage the interviewee). The purpose of the interview is to gather all the relevant evidence. Any judgmental comments will likely cause the interviewee to clam up.

Summary and Next Steps

● Offer thanks to the interviewee and explain the next steps in the investigation process and when they can expect feedback.

Conduct a "Walk Through"

● If possible and practical, conduct a "walk through" of the incident with the interviewee at the worksite where the incident occurred. This is useful, as being back where the event happened can help the interviewee recall additional information.

Ord and Shaw (1999) explain that when interviews are not well prepared or planned, key evidence can be missed and inconsistencies not identified. Additional data collection may therefore be required and moreover, interviewers may lose control of the interview. Planning interviews carefully can also help avoid unintended leading questions (e.g., "it appears that shift was under pressure to get back up to full production, tell me about your responsibilities on that busy day"). When in the moment and potentially under pressure in an interview, for example, from a defensive interviewee, interviewers can easily revert to poor questioning techniques. Ineffective and effective questions within an interview on the topic of shift handover are illustrated in Box 8.3. The importance of well-developed interviewing skills for understanding the underlying human factors cannot be overstated.

COGNITIVE BIAS IN THE INVESTIGATION PROCESS

A number of cognitive biases are known to contribute to ineffective investigations with the main problem being that it is easy to slip into a person-centered rather than systems approach. That is, to focus on ascertaining responsibility and blame first without considering the (usually more important) job and organizational factors. Examples of biases have multiplied in recent years, but there are a few "big hitters" for investigation. The main one is the fundamental attribution error (Reason, 1997; Strauch, 2004). This is our tendency to underestimate the contribution of external

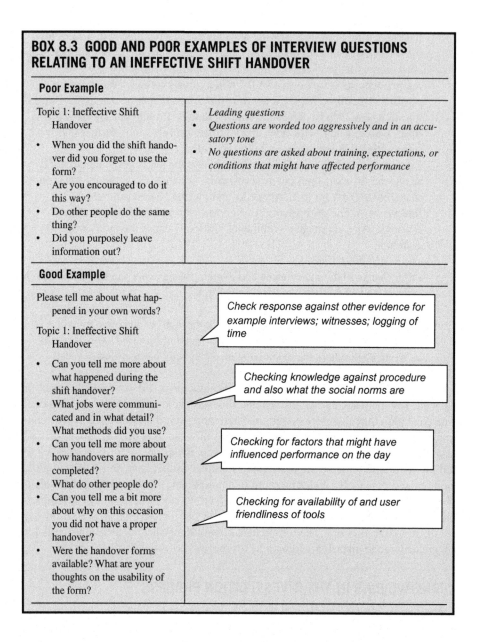

BOX 8.3　GOOD AND POOR EXAMPLES OF INTERVIEW QUESTIONS RELATING TO AN INEFFECTIVE SHIFT HANDOVER

Poor Example

Topic 1: Ineffective Shift Handover

- When you did the shift handover did you forget to use the form?
- Are you encouraged to do it this way?
- Do other people do the same thing?
- Did you purposely leave information out?

- *Leading questions*
- *Questions are worded too aggressively and in an accusatory tone*
- *No questions are asked about training, expectations, or conditions that might have affected performance*

Good Example

Please tell me about what happened in your own words?

Topic 1: Ineffective Shift Handover

- Can you tell me more about what happened during the shift handover?
- What jobs were communicated and in what detail? What methods did you use?
- Can you tell me more about how handovers are normally completed?
- What do other people do?
- Can you tell me a bit more about why on this occasion you did not have a proper handover?
- Were the handover forms available? What are your thoughts on the usability of the form?

Check response against other evidence for example interviews; witnesses; logging of time

Checking knowledge against procedure and also what the social norms are

Checking for factors that might have influenced performance on the day

Checking for availability of and user friendliness of tools

factors to someone else's behavior and overestimate their contribution to our own behavior. In other words, I am more likely to attribute *your* behavior in an accident to a personal/character failure, but when explaining my *own* behavior I will appeal first to external and environmental factors. So investigators need sufficient time, and a structure to stand back and allow a proper balance. The truth, as Reason and Strauch say, generally lies somewhere in between these two positions.

A good investigation process will allow the investigators that time and equip them to ask better questions and collect better evidence. It will also integrate human and organizational considerations to make sure the investigators understand what behavior(s) are involved and apply appropriate analyses to these. So for example, applying the ABC model to what appears to be an obvious violation requires the investigators to suspend judgment and to try and put themselves "in the shoes" of the person concerned. This in turn requires empathy and understanding, but not approval of course. Finding ways of changing such behaviors requires the behaviors to be understood from the point of view of the person concerned in the incident. What they did is likely to have made sense to them at the time and in those circumstances and given what they knew or did not know. Direct sabotage or deliberate acts in the full knowledge of the consequences are rare. Usually people are well-intentioned and want to get the job done. So considering the full range of evidence and looking at the personal, job, and organizational factors equally are the first steps in obtaining this understanding and correcting our own biases.

Hindsight bias is common (see Dekker, 2014). It is a tendency to find causal explanations in events by arguing backward from what happened (the outcome) but not recognizing that these insights were not available to those present at the time. Those present display "local rationality." For instance, what they did or did not do, made sense to them at the time, as did what they did or did not pay attention to. What investigators may pay attention to after the event can be equally selective. A typical after-the-event hindsight explanation might be "The operator(s) lacked situational awareness" or "The operators were complacent." These are not explanations and they are not causes. Investigators should therefore focus on understanding how events made sense to those involved (at the personal, job, and organizational levels), and not selectively work backward from the known outcomes.

Investigators should also be aware of confirmation bias. It is the tendency to look for evidence that supports what they expect to find; that is, their initial view of what went wrong. Evidence that confirms what the investigator expects is prioritized over other evidence that is neutral or contradicts this. Alternatively, investigators may simply not look for such evidence, or having found evidence to support what they expect, may stop looking. Just "keeping an open mind" is not enough to overcome this tendency. An investigator needs a structure that encourages and supports such openness to discover any alternative explanations that may still be available. Investigators can then ask better questions and consider more of the evidence before forming their views on what went wrong.

Investigator training needs to include sufficient awareness of these main biases so that they can be vigilant for them. Organizations need to allow investigators sufficient time and resources to enable them to properly adjust for these.

MANAGING INVESTIGATION CAPABILITY

There are several ways that organizations manage investigator capability for human factors in investigations. Most commonly, organizations have a pool of trained investigators who can be called on when there is an incident. In addition to being provided with knowledge of the investigation process and methods, investigators should have been provided with additional skills in human factors analysis. They should understand their competency boundaries, and when to ask for specialist help. This approach affords the organization some flexibility where it can draw on a range of people. For this to work well, organizations should keep a database detailing the training and experience of its investigators, allowing line managers a practical "go to" point to nominate investigators for incidents in their area.

Managing investigation capability requires investigators to put aside their "day job" to perform the investigation. Organizations should be prepared to redistribute or postpone other work to allow investigators to focus on the important investigation task. It may be a long time between investigations for any one person, which can make it hard to develop skills and experience. Accounting for staff turnover and the reality that it takes time to develop deep experience, maintaining a pool of trained investigators can take considerable effort. In organizations with established human factors teams, a variation of this approach is to deploy a human factors lead to the investigation team. So while all investigators may not have in-depth knowledge of human factors analysis, the nominated human factors lead can guide the team through the analysis.

Alternatively, or in addition, organizations can establish dedicated investigation teams. The investigators are trained to a high level and are experienced in the organizations' preferred investigation methods and in human factors analysis. This approach allows investigators to hone their skills and build a bank of experience which also provides efficiency in not having to "relearn" the process or methods each time an investigation occurs. It also allows them to gain a picture of common failings across the organization. A criticism of this approach is that the investigators will not have enough to do since the organization (fortunately) does not have enough incidents for them to investigate. However, this is short sighted. There are a great many productive tasks a dedicated investigation team could undertake in the absence of investigations. For example, the team can:

- examine trends from past events;
- create engaging methods to help the organization learn from its serious events;
- match delivery of key lessons to operations or maintenance teams in advance of work where those lessons are needed (e.g., lessons from past isolations' incidents in advance of planned commissioning work);
- educate others on how human factors can affect performance;
- apply their skills proactively to examine potential issues in the organization; a topic covered in the next section.

This type of arrangement could work well as a secondment post for 2–3 years, to develop investigation skills more widely within the business and to maintain "fresh eyes." This experience will also enrich what the employee can bring to other roles.

Regardless of the approach(es) used to manage investigation capability, the following practices can help organizations ensure that human factors are effectively examined within incident investigations:

- Ensure the organization's investigation procedure and supporting documents define expectations for when to include human factors analysis within the investigation process. Logically, where humans are working in organizations, such factors are likely to be involved. However, where there are resource constraints, lower maturity in investigation capability, or when organizations are testing new human factors analysis tools, human factors analysis is sometimes only made a selective requirement (e.g., for serious or repeat events).

- Develop selection criteria for investigators which include the key qualities for investigations discussed earlier. Select and develop investigators against these criteria. Investigation should not just be added to the job description for everyone in certain roles; people should be formally selected for this important role.
- Manage and maintain competency of investigators, for example, by conducting refresher courses, peer checking, promoting good examples, and establishing a focal point to manage competency. Of course human factors analysis skills should be refreshed within these practices.
- Establish a process to review investigation quality to check that human factors are being sufficiently investigated. On a practical level, reviews may be restricted to serious, high potential, and repeat incidents otherwise this task may be too time-consuming.
- Include key behaviors within culture models that support effective investigation. For example, specified manager behaviors might include "I will ensure that the human factors that affect health and safety are being managed effectively" and "I will not shoot the messenger that raises organizational issues in my area." Appropriate behaviors could be defined at different levels, for example, managers/executives, supervisors, and other personnel.

Organizations wishing to introduce human factors analysis to their existing investigation method should also consider the following points to aid successful implementation:

- Educate senior leaders before introducing human factors analysis methods. Senior leaders should be aware of the new terminology and be prepared for a deeper analysis that will point to organizational failings and opportunities for improvement. Leaders also need to be ready to endorse and act on improvement actions aimed at an organizational level.
- Educate health and safety, and human resource personnel before introducing human factors analysis methods. These professionals play a key support role for people and people processes. Therefore they should be made aware of the introduction of analysis methods that will help uncover organizational issues that influence human reliability. Moreover, if any misunderstanding about the purpose of human factors analysis emerges, these professionals can clarify the purpose.
- Link the human factors analysis taxonomy used in the new analysis method to the organization's "Just or Fair Culture" model if this exists. Using the same terminology for different error types and intentional behaviors across complementary tools can avoid the potential confusion.
- Add key elements of the human factors analysis taxonomy to the incident reporting database fields. This can enable the corresponding causal trends to be identified.

PROACTIVE APPLICATION OF HUMAN FACTORS ANALYSIS

As part of developing, maintaining, and improving an investigator's capabilities it makes sense to identify as many opportunities to practice and consolidate the learning as possible. For example, investigating or reinvestigating "cold cases," near-misses, quality issues, plant and equipment failures, and difficult behaviors. It also makes sense to "investigate" what is going well because much can also be learned from understanding why this is so. For example identifying and sharing good practices. Sometimes closer examination shows that all is not as well as was thought or that while things are working well, there are work-arounds being used to make this happen. Such work-arounds are not necessarily bad things. They show the human capacity for innovation, creativity, and improvement, but they need to be understood and managed.

It therefore also makes sense if these investigation skills and knowledge can be applied routinely in advance of a failure and even better if this is integrated into the normal SMS assurance processes, such as monitoring, auditing, and review. An example would be to actively look for PSFs as part of routine audits. An experienced investigator has a lot to offer as an auditor and may be able gain a deeper understanding quicker than their fellow-auditors. Such an approach is more likely to move on to consideration of system issues including organizational factors so that these can be addressed before something actually goes wrong. In addition, there is often in fact a business efficiency payback in getting these issues right.

Organizational factors can be referred to as the wallpaper, "the way things *are* round here" (Wilkinson and Rycraft, 2014), as opposed to the traditional safety culture definition of "the way we *do* things round here." Becoming aware of the wallpaper, challenging it, and doing something about it are difficult to do inside an organization and so these issues are often simply tolerated and left unaddressed.

SUMMARY: HUMAN FACTORS IN INCIDENT INVESTIGATION

Integrating human factors in investigations can help elicit underlying organizational factors more effectively and help prevent incidents recurring. This approach avoids blaming individuals and can help eliminate similar behaviors by others in the organization. It can also identify key improvements to move past plateaus in safety performance. Equally, a human factors focus for investigators, their training, and capabilities can have significant benefits including for wider organizational learning.

Investigations often find issues related to insufficient design as a causal factor that influenced human performance. The next section discusses human factors in design, which can enable robust solutions in response to investigation findings. More importantly, good human factors design can reduce the likelihood of human failure or reduce the potential for human failure to lead to an incident.

KEY POINTS

- Investigating incidents is an important contributor to organizational learning; investigation skills and approaches can also be applied proactively as part of the larger SMS.
- Increasing system complexity and reliability have increased human and organizational contribution to incidents and the importance of identifying and addressing them.
- Organizational factors often recur as incident causes because they are harder to identify and address.
- Mainstream investigation methods can support investigators in developing a timeline and identifying root causes although they have potential to "box in" thinking.
- Mainstream investigation methods do not generally provide tools to analyze human behaviors of interest so these need to be incorporated or standalone methods used.
- Organizations should provide several investigation methods; some more simple and flexible and some more constraining to accommodate varying experience levels and incident complexity.
- Investigators who are effective at identifying human and organizational contributions possess key qualities and organizations should select and develop investigators against these qualities.
- Strong interviewing skills are essential to uncover the human factors issues.
- Organizations should proactively manage their human factors in investigation capability and establish supporting processes and the right conditions to set investigators up to perform effectively.

REFERENCES

Bulsok K., 2015. An Introduction to 5-Why. Available from: http://www.bulsuk.com/2009/03/5-why-finding-root-causes.html (accessed 19.11.15.).

CSB, 2007. Refinery Explosion and Fire (15 Killed, 180 Injured), BP Texas City, Texas, March 23, 2005. Report No. 2005-04-I-TX. Available from: http://www.csb.gov/assets/1/19/csb-finalreportbp.pdf (accessed 19.11.15.).

Dekker, S., 2014. The Field Guide to Understanding 'Human Error', third ed. Ashgate, Farnham, ISBN 978-4724-3905-5.

Energy Institute, 2008. Guidance on Investigating and Analysing Human and Organisational Factors Aspects of Incidents and Accidents. Energy Institute, London.

Flin, R., O'Connor, P., Crichton, M., 2008. Safety at the Sharp End: a Guide to Non-Technical Skills. Ashgate Publishing Limited, Hampshire, England.

HSE, 2015. Investigating Accidents and Incidents. Available from: http://www.hse.gov.uk/managing/delivering/check/investigating-accidents-incidents.htm (accessed 19.11.15.).

Kelvin Topset, 2015. Kelvin Topset© Methodology. Available from: http://www.kelvintopset.com/about/methodology (accessed 19.7.15.).

Lardner, R., Scaife, R., 2006. Helping engineers understand the human factors at work. Process Saf. Environ. Prot. 84 (B3), 179–183.

Memon, A., Bull, R., 1991. The cognitive interview: it's origins, empirical support, evaluation and practical implications. J. Community Appl. Soc. Psychol. 1, 291–307.

Ord, B., Shaw, G., 1999. Investigative Interviewing Explained. The New Police Bookshop, Surrey.

Paradies, M., Unger, L., 2000. TapRooT®: The System for Root Cause Analysis, Problem Investigation, and Proactive Improvement. System Improvements, Inc., Knoxville.

Reason, J., 1997. Managing the Risks of Organisational Accidents. Ashgate, Farnham ISBN 978-1-84014-105-4.

Safety Wise Solutions, 2015. ICAM. Available from: http://www.safetywise.info/SWS_Incident_Investigation.php (accessed 19.11.15.).

Strauch, B., 2004. Investigating Human Error. Ashgate Publishing Ltd., Aldershot.

Waddington, P.A.J., & Bull, R., 2007. Cognitive Interviewing as a Research Technique. Social Research Update, Issue 50. Department of Sociology, University of Surrey, Guildford, UK.

Wilkinson, J., & Rycraft, H., 2014. Improving organisational learning: why don't we learn effectively from incidents and other sources? Proceedings of the Institute of Chemical Engineers' Hazards 24 Conference. Edinburgh.

Human factors within design and engineering III

Overview of human factors engineering

J. Edmonds

LIST OF ABBREVIATIONS

ABS	American Bureau of Shipping
HCD	Human-Centered Design
HFE	Human Factors Engineering
HFI	Human Factors Integration
HTA	Hierarchical Task Analysis
HRA	Human Reliability Analysis
ISO	International Organization for Standardization
NORSOK	Norsk Sokkels Konkuranseposisjon
OGP	International Oil and Gas Producers
TAD	Target Audience Description
TNA	Training Needs Analysis
TTA	Tabular Task Analysis
UCD	User-Centered Design

Poor design of the human interface is a primary cause of human failure, whether it is the way information is presented, the layout of controls, the physical workspace, or aspects of the environmental design. Several examples were cited in Chapter 1, What is Human Factors?, demonstrating how poor design has contributed to major accidents. Specific design flaws for four major accidents are explained in Box 9.1.

Human performance is directly influenced by design. Something that is well designed achieves its purpose, is motivating and enjoyable to use and there is a lower risk of injury. Poor design on the other hand can lead to human failure, poor performance, frustration, and greater risk of injury. The more effort that is given to achieving an effective design, the less likely the need for "work-around" fixes during operations, the more reliable the human performance and the cheaper the overall cost of ownership. This case is argued in Chapter 10, Human Factors Integration Within Design/Engineering Programs.

A well-designed system requires an understanding of the end users, what they need to achieve, and effective design of the human interface to meet the end user needs. This includes their tasks, tools and equipment, workspaces, work environment, and

---

BOX 9.1 SPECIFIC DESIGN FLAWS CONTRIBUTING TO MAJOR ACCIDENTS

Three Mile Island nuclear radiation disaster, 1979—Plant operators failed to diagnose the loss of coolant for several hours due to an ambiguous control room indicator on the user interface. It indicated the stuck valve as closed (the light only indicated the power status of the solenoid providing a false indication, on this occasion, of a closed valve).

BP Texas City refinery explosion, 2005—The board operator did not have a clear indication of fluid flows in and out of the isomerization tower or the actual level in the tower and consequently did not know about the dangerously high level.

Formosa chemical plant explosion, 2005—The plant operator erroneously tried to open the bottom valve on a reactor in process and then overrode the safety interlock releasing the reactor contents. The plant design contributed to both of these human failures.

Kegworth air crash, 1989—The pilots shut down a healthy engine due to no clear indication of which engine had failed. There had been changes in the design of the aircraft from the previous version of the Boeing 737, including the operation of the auto-throttle but also the design of a vibration meter which would have indicated that the wrong engine had been shut down.

---

organizational structures. All of these aspects of the system design need to enable the end users to execute the functions they need to perform.

Human Factors Engineering (HFE) is the application of human factors to the *design* of systems to enable the end users' needs to be met. There are four iterative phases:

1. gain an understanding of what needs to be taken into account during the design using a toolkit of analysis techniques;
2. use human factors scientific knowledge embodied within standards and regulations in conjunction with the first phase to derive the design criteria;
3. design of the human interfaces within the system;
4. evaluate design to ensure it meets the needs and requirements of the end users.

Within this chapter, relevant terms and boundaries are defined prior to describing the complexity of human interactions and the approach for applying HFE to the design of new work systems. This is continued within Chapter 10, Human Factors Integration Within Design/Engineering Programs, to define how HFE needs to be integrated within engineering programs. The remaining chapters in this section are dedicated to specific HFE topics and areas of application. The section includes the following chapters:

- Chapter 9—Overview of Human Factors Engineering
- Chapter 10—Human Factors Integration Within Design/Engineering Programs
- Chapter 11—Building and Control Room Design
- Chapter 12—Workstation, work area, and Console Design
- Chapter 13—Control System Interface Design
- Chapter 14—Plant and Equipment Design

DEFINITIONS AND BOUNDARIES

Within Chapter 1, What is Human Factors?, a working definition of human factors was provided, effectively describing it as the application of a body of knowledge from the human sciences to the aspects of the work system where there are human interactions. A distinction was made between the phases of a system where human factors is applied, simply described as:

- *Design*: where it is conceptualized, designed, developed, and built;
- *Operation*: where it performs the function it was designed to perform;
- *Decommissioning*: where it is taken back to its starting point (i.e., where it no longer exists as a system).

Human factors relates to the human interactions within all of these phases, whereas HFE specifically focuses on *human factors within the design phase*. In reality design can occur during the other phases, such as would be required for mid-life system or equipment updates. HFE is specifically relevant to capital projects.

The purpose of HFE is to optimize the human contribution to system performance and minimize the potential for design-induced risks to health and personal or process safety. Within this book, the term HFE is used synonymously with the term "ergonomics."

There is also a need to make a distinction between HFE and Human Factors Integration (HFI). HFI is a design management activity specifically related to the *process* of applying HFE within the system design and engineering process. This is discussed in Chapter 10, Human Factors Integration Within Design/Engineering Programs.

UNDERSTANDING HUMAN INTERACTIONS

The "humans within the work system" represent a web of interactions with other aspects of the system. This may relate to interactions with the hardware or software elements, the work area or environment, or other human beings within the organizational and social context. The scope of the application for HFE is illustrated in Fig. 9.1.

The areas of application illustrated in Fig. 9.1 can be used as a useful "mental" checklist to ensure adequate coverage of the human factors which need to be taken into account within design. It should be recognized that these areas are inter-related and can affect each other. For example, an excellent control panel design becomes

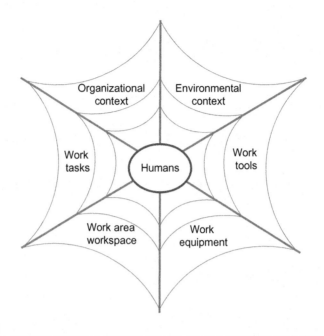

FIGURE 9.1

Multiple human interactions within the work system.

less effective if the end user cannot see it from their operating position or the lighting creates reflections on the indicators. Human Factors Engineers take a systems perspective on design because the human interactions within the system are broad as well as deep. It is not just about designing the control panel, the control room, the procedures, or the even organizational aspects such as the training. It is about seeing the "big" systems 'picture', as well as being able to deal with the minute detail such as whether a control indicator needs to be yellow, 8 mm in diameter, and positioned above or below a specific control.

The level of complexity of the human interactions may vary. There can be simple one-to-one relationships, for example, one person interacting with one piece of equipment. An example of a physical task is manual loading of a hopper with powder chemicals. In this scenario there is the person retrieving a sack of raw material to place on the machine conveyor. The conveyor moves the sack to the "operator station" where the top edge is cut open ready for presentation to the hopper.

- *Work Tasks*: The operator needs to undertake monitoring of the hopper level, perform manual handling of the sacks, operate the conveyor controls, and cut open the sack.
- *Work Equipment*: The specific physical interfaces are the sacks, the machine controls and displays (including the level indicator), and the sack slitter.

- *Work Area/Workspace*: The work area is a specific area of the plant area (such as the mud mixing room, the reactor floor, or other similar area). The workstation is comprised of the conveyor and the operator station where the sack slitting is undertaken and the space in front of the machine control panel.
- *Environmental Context*: The environment is subject to some noise and vibration, lighting may be natural or artificial or a combination of both. The thermal environment includes temperature, humidity, and ventilation control and there may be inherent hazards related to the chemicals in use.
- *Organizational Context*: This includes the team work and communication which needs to occur with other team members and the training and competency development required for the job. There may be other organizational influences on performance which may need to be controlled.

There are more complex one-to-one work scenarios such as a person driving a forklift truck. There is one person performing a number of tasks including operating the vehicle controls, monitoring the vehicle information systems, reading the route, and navigating objects outside the vehicle. He or she may also add tasks like talking to a colleague, visualizing, and planning the work activity. This is an example of a more cognitive task with a lot of information being processed by the driver, even though the skilled person may find the task easy. The workspace is the truck cockpit, the work area is the production floor, the environment may be internal or external or both, and on the same plant, there may be similarities in the organizational context with the previous task.

Whatever the task, the person receives information from the senses (auditory, visual, touch, and possibly smell). This is perceived by the brain to add meaning to the information. This may require retrieval of information from memory and the use of working memory as the information is mentally manipulated. As the information is manipulated, decisions are made and the person performs psychomotor actions in response. This is known as the model of information processing. It relates to the person's interaction with the world around them and it occurs on a continual basis. This is illustrated in Fig. 9.2.

So far, the examples described have only considered the work scenarios as one-to-one relationships, and there is a need to apply HFE to the design of each aspect of the interaction. However, within a system, the level of complexity is greater as one role may have several different interfaces within the system. For example, the field operator who performs valve line ups, performs manual start-ups of equipment, undertakes line walks, and takes samples as well as other activities. The different parts of the system may have many different roles interacting with it for different purposes (such as the field operator, the control room operator, the mechanical technician, the electrical technician, or the vendor). So the interactions within the system can also be one to many, many to one, and many to many. There will also be human-to-human interactions. This is illustrated in Fig. 9.3.

Each relationship needs to consider the elements shown in Fig. 9.1, albeit some of these may be shared, such as the work environment and the organizational context.

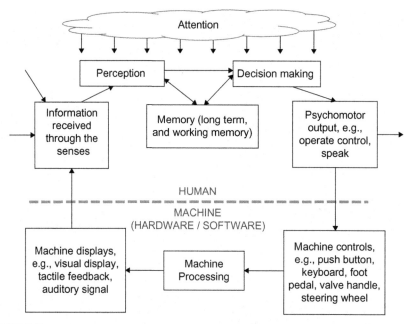

FIGURE 9.2

Information processing in relation to machine operation.

HUMAN-CENTERED DESIGN APPROACH

Understanding and defining the human interactions within the system is normally recognized as one of the initial steps in the application of HFE. It is effectively the starting point for what is known as User-Centered Design (UCD) or Human-Centered Design (HCD). There are a handful of models for HCD such as the International Organisation for Standardisation (ISO) 6385 (2004) which uses a six-phase approach as follows:

1. formulation of goals;
2. analysis and allocation of functions;
3. design concept;
4. detailed design;
5. realization, implementation, and validation;
6. evaluation.

This standard can be applied to any system. There are also standards using a similar approach for specific purposes, such as ISO 11064 for control centers (introduced in Chapter 11: Building and Control Room Design) and ISO 9241-210 (2010)

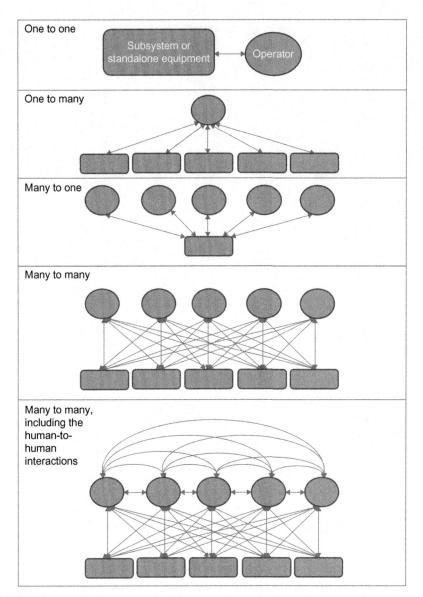

FIGURE 9.3

Different levels of complexity of human interactions.

Key: ● = operator; ▬ = subsystem or equipment

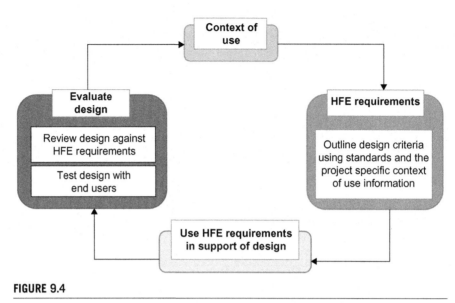

FIGURE 9.4

Human-centered design process.

(formerly ISO 13407, 1999) which uses four interdependent stages aimed at computer-based interactive systems:

1. understand and specify the context of use;
2. specify the user and organizational requirements;
3. produce design solutions;
4. evaluate designs against requirements.

A generic HFE process is broadly outlined in Fig. 9.4 which captures the essence of these various models. It is described in the following subsections.

CONTEXT OF USE

There is a need to analyze and define the human interactions within the system. Firstly, there is a need to know who the end users of the system and equipment will be, and their characteristics, limitations, and capabilities. This is sometimes documented in a standalone document known as a Target Audience Description (TAD) or as part of a larger user specification or context of use document.

In the early stages of design, there may be an opportunity to review the system functions and perform a functional allocation of tasks to humans, hardware, software, or both. Performance variability can generally be reduced with well-designed and engineered automated systems. However, the automated systems (process control or otherwise) are not infallible, and indeed, optimal performance is not necessarily achieved from a "system" perspective for several reasons, for example:

- Automation is rarely without a human interface at some point, whether it is during design, installation, maintenance, monitoring, and so on.
- Automation may simplify one task, but shift the complexity elsewhere in the system. For example, the operation may be "simplified" but the maintenance complexity greatly increased.
- Humans sometimes have to override automated safeguards (typically for good reason such as during maintenance). People may behave in ways that were never anticipated. This may create different (unknown) opportunities for error.
- Automation is often applied to tasks that are easy to automate, leaving complex tasks for the human. This can increase human error.
- More automation may cause reduced performance and error through monotony. "Machine minding" is not a human strength and so the person is unlikely to be at their optimal performance.
- Automation may not make the best use of the skills within the workforce. Natural human performance variability makes them potentially better than "machines" for several reasons, mostly related to dealing with anomalies and making deductions from little data or unusual/unexpected circumstances. Designers cannot foresee all such situations and the emergent properties of complex systems may make this more challenging.
- Automation or semi-automation of tasks may not make the best use of the skills because of an increased dependency on specialists (who may not be available) or specialist tools.
- More automation may trigger over-reliance and complacency.
- More automation may cause a reduction in situational awareness. Using the example of flying a plane, the task can be fully automated. What this does is take the "human out of the loop" so the person's cognitive awareness of the situation may be reduced at a given time and is thus less capable of responding effectively when rapid action is required, such as when the engineered system fails.
- Automation can lead to an inability to cope in unforeseen circumstances.

Modern technology provides more opportunities to achieve a good allocation of function, and automation no longer needs to be viewed as a fixed allocation of tasks to humans or to machines. A more appropriate approach is to use the philosophy of HCD and recognize that the hardware and software elements are there to support human goals. There is opportunity to consider dynamic allocation, which changes on the basis of evolving scenarios. For example, a pilot may fly in auto pilot at cruising altitudes but switch to manual for takeoff and landing. Likewise, a chemical process may be automated until there is a need or desire to switch to "manual." Functions can also be allocated to different roles and this can also be fixed or dynamic. Indeed functions can be allocated to different aspects of the hardware and/or software which again can be fixed or dynamic.

The traditional approach was mainly concerned with allocating functions on the basis of strengths and weaknesses, such as the machine being quicker or better than humans at repetitive tasks or complex calculations and the human being better at inductive reasoning, judgement, and sensing unusual or unexpected patterns. This method is useful but too simplistic in the context of complex and evolving systems and it risks the human being left with the "bits" that are not easy to automate and undermining the human (and overall system) performance.

A better approach to allocation of function is as follows (Railway Safety and Standards Board, 2008):

- Where there is a distinct requirement to allocate the function to the human or the hardware/software (such as health and safety reasons or either is clearly more capable of performing the function) then those functions could be allocated.
- The remaining functions need to be reviewed to identify options on the basis of factors including: whether or not the person needs to be present, designing a cohesive and satisfying job role, cost, and retaining situational awareness.
- The scale of fixed to dynamic functional allocation can also be included in the judgment, whether it is human controlled (such as switching between auto and manual) or machine controlled (such as machine operation instigated by the detection of an error).

In addition to allocating functions to human or machine elements, there is also a need to allocate tasks to job roles to enable a good job design.

The context of use document typically includes task analysis to define the tasks that will be undertaken by the end user. There will be different tasks for different end users so the task analysis for an operator will be different from the maintainer and other system end users. A task analysis may also include data about the information, controls, communications, tools, workspaces, environment, and other factors related to performing the required tasks. This enables a clear picture of the human interaction to emerge. There are various task analysis techniques which are useful for different purposes. Hierarchical Task Analysis (HTA) defines the structure of a task, Tabular Task Analysis (TTA) defines the detail for a task, link analysis is used to define the human-to-human or human-to-machine interactions, and there are multiple other techniques.

An HTA or a TTA is often used as the baseline for other techniques, such as:

- Human Reliability Analysis (HRA) to identify how someone might fail and identify risk controls (discussed in Chapter 6: Human Factors in Risk Management).
- Timeline and workload analysis for answering the question of whether someone can cope with the workload they are being asked to perform and how they might manage tasks over time (discussed in Chapter 20: Staffing the Operation).
- Staffing assessment to identify the optimal number of people for different operational scenarios (discussed in Chapter 20: Staffing the Operation).
- Training Needs Analysis (TNA) for identifying skill gaps (discussed in Chapter 20: Staffing the Operation).

HFE analyses are broadly undertaken for two reasons:

1. to identify specific *end user needs* which need to be captured within the design criteria, such as the link analysis to determine the optimal co-location of personnel;
2. to identify potential *human factors issues* which could occur and need to be resolved within the design stage, such as the potential for human failure, excessive workload, or high training burden.

The context of use stage is fundamental for the application of HFE because it presents a design baseline for supporting human performance. The level of detail of the analyses can vary and this may be determined by safety criticality or other reasons such as the novelty of the design. One important aspect of this stage is to maintain a high level of end user involvement to avoid erroneous assumptions about how the system might be used. Effectively the analyses provide the opportunity for the designers and end users to agree on and define how the system will be used. This saves considerable time and effort at later stages in the design.

It is rarely the case that a new system is totally designed from scratch. It is more common that it at least incorporates some element of "off-the-shelf" equipment or components. Despite this limiting the opportunity to create the ideal ergonomic solution, there is still a need to gain a good understanding of the context of use and use this to assess the strengths and weaknesses of the off-the-shelf elements.

HFE REQUIREMENTS

The next stage in the process is to develop HFE requirements. These become the template against which the design can be progressed and tested against. There are two main categories of HFE requirements (as defined in International Oil and Gas Producers (OGP) 454, 2011 and Norsk Sokkels Konkuranseposisjon (NORSOK), S-002, 2004):

1. *Goal Orientated*—these requirements are determined by undertaking analysis of proposed future operations, for example, using task analysis, link analysis, and the other activities undertaken in support of the context of use. Goal-oriented requirements are related to understanding the end users' needs for the system particular to the scope of the project.
2. *Prescriptive*—these are requirements derived from specific criteria that need to be met as defined within HFE standards and regulations. Prescriptive requirements are translated into specific design criteria relevant to the scope of the project.

Together the goal-oriented and prescriptive requirements are effectively the "technical HFE design requirements." Examples of these different types of requirements are provided in Box 9.2.

It can be seen how the context of use is needed to actually define how many of the prescriptive requirements need to be met, even if the requirement has initially been

---

BOX 9.2 EXAMPLES OF HFE REQUIREMENTS

Goal Orientated

On the basis that the maintenance task related to equipment 'x' will typically be '16' hours, the procedure for this task shall include a checklist to support the handover between shifts.

"On the basis that the instrument panel 'x' must be used in conjunction with the 'xxx' valve, these shall be co-located so that both are within sight and reach simultaneously." *Note that the relevant HFE standard will be used to define acceptable control/instrument locations and may need this statement to be supplemented with prescriptive requirement statements.*

Prescriptive

Floor-space allocation in the control room shall allow for 9 m² to 15 m² per working position.
Based on ISO, 11064.

The visual display for a standing operator shall be mounted between 1040 mm and 1780 mm from the floor unless it requires precise and frequent or emergency use, in which case it shall be mounted between 1270 mm and 1650 mm above the floor.
Based on American Bureau of Shipping (ABS), 2014.

---

derived from a specific standard. This is an argument for not bypassing the context of use stage and believing that HFE is just about meeting HFE standards.

HFE IN SUPPORT OF DESIGN

The HFE requirements are intended to support and drive design and provide a baseline for enabling that to happen. Within this stage there is more detailed HFE activity such as the design of the control panel interface, defining the graphical user interface hierarchy, defining the actual dimensions of a console and the control room layout.

It is generally within this stage that design trade-offs occur and where the HFE effort needs to resort to first principles. An example of a design trade-off is presented in Box 9.3.

EVALUATE DESIGN

The HFE evaluation of design involves two aspects:

1. *Verification of design*—The HFE requirements are used to measure how well the design meets the criteria set within the HFE requirements.
2. *Validation of design*—This is a measurement of how well the design meets the operational needs of the end users. This is typically undertaken through gaining feedback on the design. In the early stages of design, this may be as simple as discussing a proposed floor layout with operators. During the detailed design this may progress to using simulations of the human interface

> ### BOX 9.3 EXAMPLE OF A DESIGN TRADE-OFF
>
> The HFE requirement derived from a standard might state that a manual valve handle needs to be 838–1143 mm above the deck. However, it might be the case that this is not possible because there is insufficient gradient on the pipeline for adequate flow which means that the valve needs to be at a lower height.
>
> This needs to be assessed and even calculated to understand what the implications are of not achieving the HFE requirement. Who is affected? How are they affected? What options exist to optimize human performance in this situation?
>
> It may be acceptable for the valve handle to be lower, or it may not. It may mean that at the lower height only 30% of the crew are able to actually operate it because they cannot apply sufficient force in that position, or that an injury-prone posture is dictated and/or that there is insufficient clearance for undertaking maintenance removal of the valve.
>
> It would not be acceptable to just automate the valve because it does not necessarily resolve the maintenance issue. If the deck was lowered in that position, it could create a hazard and implications for the system elsewhere. A simple solution might be to increase the valve stem, if that is mechanically feasible. So just because there is a trade-off, it does not mean that the solution is necessarily costly or complex. It may mean that the design does not progress with an obvious flaw which has implications for the end users.

or three-dimensional models. At an appropriate stage in design, performance measurements may be taken under controlled user testing conditions for specific scenarios.

Design modifications should be made in response to the results of the HFE evaluation. It may be possible to design the system (or elements of it) in a different way, or there may be justifiable reasons for not making a complete modification, rather a trade-off solution may prove adequate. The solutions need to be assessed on the basis of the results.

PRACTICAL APPLICATION

In reality the context of use, requirements derivation, support to design, and evaluations occur iteratively throughout the project stages. HFE analyses may be performed to understand an issue in more detail after the requirements have been derived and the design has already commenced. Likewise, an evaluation may occur early on in the process and be used to drive the design in a certain direction or test certain situations. The HCD process is not a single iteration or a linear approach.

There are agreed principles across the models advocated in the different standards:

- HFE should be integrated within engineering practice and the project management process from the start of the project and continue throughout.
- End user participation should be incorporated in a structured manner throughout the design process.

- The process should be implemented using an interdisciplinary team which includes disciplines such as ergonomics, engineering, architecture, and industrial design. The combined skill and knowledge will enable a more optimal design to be achieved.

SUMMARY: OVERVIEW OF HUMAN FACTORS ENGINEERING

There are several well-known disasters where poor human factors design has made a significant contribution to the accident. This chapter has focused on the application of human factors to the *design* of systems to enable a better design. The HFE approach to design advocates that the system should be designed to meet end users' needs rather than expect the user to adapt.

In the next chapter, the management activity for integrating human factors within engineering programs is discussed.

KEY POINTS

- HFE is the application of human factors to the *design* of systems and the term is used synonymously with the term "ergonomics."
- The purpose of HFE is to optimize the human contribution to system performance and minimize the potential for design-induced risks to health, personal, or process safety.
- The "humans within the work system" represent a complex web of interactions. The HFE scope includes the tasks that people perform; the tools and equipment that they use to perform the task; the workspace/work area where the work is undertaken; and the influence of the environmental and organizational context.
- The complexity of human interactions varies, from simple one-to-one interactions through to many-to-many interactions.
- A HCD approach is advocated, whereupon the context of use is analyzed and defined, HFE requirements are derived, and the design solutions are developed and evaluated. The process is not a single iteration, nor a linear approach.
- HFE should include end user participation, should be integrated within engineering practice, and should be undertaken by an interdisciplinary team.

REFERENCES

American Bureau of Shipping (ABS), 2014. Guidance Notes on the Application of Ergonomics to Marine Systems.

International Oil and Gas Producers (OGP), 2011. Report 454—Human Factors Engineering in Projects. OGP454.

International Organisation for Standardisation (ISO) 6385, 2004. Ergonomic Principles in the Design of Work Systems.

ISO 9241, Part 210, 2010. Ergonomics of Human System Interaction—Human Centred Design Processes for Interactive Systems (formerly ISO 13407, 1999).

ISO 11064. Ergonomic Design of Control Centres. Part 1, Principles for the Design of Control Centres, 2001; Part 2, Principles for the Arrangement of Control Suites, 2000; Part 3, Control Room Layout, 1999; Part 4, Layout and Dimensions of Workstations, 2013; Part 5, Displays and Controls, 2008; Part 6, Environmental Requirements for Control Rooms, 2005; Part 7, Principles for the Evaluation of Control Centres, 2006.

Norsk Sokkels Konkuranseposisjon (NORSOK), S-002, 2004. Working Environment.

Railway Safety and Standards Board, 2008. Understanding Human Factors—a Guide for the Railway Industry.

Human factors integration within design/ engineering programs

10

J. Edmonds

LIST OF ABBREVIATIONS

BoD	Basis of Design
CAPEX	Capital Expenditure
CIEHF	Chartered Institute of Ergonomics and Human Factors
DTC	Defence Technology Centre
EHFA	Early Human Factors Analysis
EHS	Environmental Health and Safety
FEED	Front End Engineering Design
HAZID	Hazard Identification
HAZOP	Hazard and Operability
HFE	Human Factors Engineering
HFEMP	HFE Management Plan
HFI	Human Factors Integration
HFIP	Human Factors Integration Plan
HFIR	Human Factors Issues Register
HRA	Human Reliability Assessment
OGP	International Oil and Gas Producers
OPEX	Operating Expenditure
RAIDO	Risks, Assumptions, Issues, Dependencies, Opportunities
SCTA	Safety Critical Task Analysis
SEMP	Systems Engineering Management Plan
SMP	Safety Management Plan
TNA	Training Needs Analysis
WEHRA	Work Environment Health Risk Analysis

The scope and approach to Human Factors Engineering (HFE) was discussed in Chapter 9, Overview of Human Factors Engineering, where the concept of human centered design was introduced. This chapter focuses on Human Factors Integration (HFI), a design management activity specifically related to the *process* of applying HFE within the engineering project life cycle.

HFI is a systematic process which has been developed over several decades to closely align with systems engineering and project management activities. It started its development within the military in the 1980s and has evolved and spread across different industries ever since. The process varies by industry and by individual organizations because of the different approaches to systems engineering and project management.

Regardless of the variability in process, the principles remain similar and the aim is to ensure that there is systematic and comprehensive consideration of the human elements in the design of the system. Failure to adequately consider the human element can, at best, lead to a failure to optimize human performance. At worst it could undermine safety and lead to a major accident. Unfortunately, there is an abundance of examples of poor human factors design which have cost people, organizations, and societies dearly.

Barriers to using an HFI approach are evident; for example, the perception that HFI represents additional cost to the project with limited benefits. The cost–benefits are discussed within this chapter and positive arguments for implementing an HFI approach are discussed. There is another barrier which is the relative lack of maturity of HFI within the chemical and process industries (other than nuclear). The International Oil and Gas Producers (OGP) has sought to redress this with the publication of guidance (OGP454, 2011) within the oil and gas domain. Ideally, this should challenge erroneous assumptions that organizations already "do it" or at least already "do it comprehensively" or that they already have sufficient expertise to tackle all HFE issues adequately. It is hoped that other industries within this sector will follow this lead and develop HFI processes relevant to their own domains.

As well as discussing the cost–benefits of HFI, this chapter provides a discussion about the organization of HFI and HFI within the project life cycle covering management, technical, and assurance activities.

COST–BENEFITS OF HFI

The most obvious barrier to implementing HFI is the perceived cost of the required effort for what some might consider as relatively intangible benefits. How easy is it to measure and directly relate a lack of accidents and injury to the HFI effort, or indeed, the improved productivity of the system? Most projects do not take the extra time and effort to measure and assess the direct cost–benefits of implementing HFI. However, there has been some research in this area, mostly conducted in the defense industry and some examples are cited in Boxes 10.1 and 10.2, taken from the HFI Defence Technology Centre guidance (HFI DTC, 2006).

An activity that the defense industry does meticulously is the analysis of through-life cost, including both Capital Expenditure (CAPEX) and Operating Expenditure (OPEX). This is probably because ultimately the funding comes from the same

---

BOX 10.1 EXAMPLES OF "COST–BENEFIT" SUCCESSES DUE TO THE IMPLEMENTATION OF HFI

British Nuclear Fuels integrated HFI within a project to design a new thermal oxide reprocessing plant. The HFI program included consideration of the user interface design, training, maintenance, staffing levels and emergency response capabilities. The cost of the program was substantial, at around 15 person years of effort. However, the safety issues identified by the human factors team would have amounted to costs leading to the economic ruin of the design organization had they remained (Kirwan, 2003).

The DD(X) is a family of US naval surface combat ships. HFI was implemented to analyze and assess changes to the crew roles, responsibilities and complement. The through-life cost saving estimated just prior to construction was $600 million per ship, representing a total saving for 32 ships of $18 billion (United States General Accounting Office, 2003).

BOX 10.2 EXAMPLES OF "COST–BENEFIT" FAILURES DUE TO A LACK OF HFI

An inquiry was held into the failure of a new computer-aided dispatch system for the London Ambulance Service. The failure of the system was attributed to key areas of human factors; namely inadequacies in training, processes and procedures, working practices and system usability (South West Thames Regional Health Authority, 1993).

Modifications to the Minehunter submarine costing £1.9 million were required due to operability problems related to remote controlled recovery in high sea states. It was identified that human factors analyses would most likely have avoided the problems using standard HFI processes and activities to analyze operating conditions and evaluate the user interface (National Audit Office, 2000).

---

source. An early quantitative cost review of three specific projects was undertaken by the US Army Research Laboratory taking account of costs avoided and investment savings (Booher, 1997). This demonstrated the cost saving to investment ratio for incorporating human factors was between 22:1 and 43:1 for these three projects showing a major return on investment.

Within the chemical and process industries the budget owners for CAPEX and OPEX can be multiple and varied, so there can be a motivation to minimize CAPEX, regardless of the effect on OPEX.

The CAPEX budget can also be structured quite differently from major government projects. Within some chemical and process industries, value engineering activities are undertaken at different design stages to assess the cost viability of proceeding with a project. There are financial merits for a project in keeping the cost of engineering low for as long as possible. This can drive budget pressure on HFI, particularly during the early stages of a project which is exactly the time of greatest benefit of an HFI approach.

A study was conducted in the air traffic management domain to review several projects to assess the cost–benefits associated with the timing of the HFI implementation. The conclusion was that early insights and identification of HFE risks can help to direct expenditure on the right developments and make significant savings before making major investment decisions. The study showed the increased cost of detecting and resolving human performance issues as a factor of the progression of design stages at which issues were identified. In other words, projects with little or no HFE typically resulted in substantially more design flaws being identified during operation, by which time the cost of redesign or mitigation was either unfeasible or significant (Eurocontrol, 1999). Late changes may also require changes to training and procedures as well as the design. This work provides a strong argument for considering HFI at an early stage. However, even with a late realization that HFE is necessary, it is still possible for a project to benefit from integrating HFE within the design, or at least its evaluation.

The primary cost–benefits of proper integration of HFE in projects are:

- reduction in CAPEX by contributing to:
  - more efficient and inherently safer design;
  - avoiding the need for expensive changes and/or rework late in design or during or after construction;
- reduction in life cycle costs of operating and maintaining facilities (OPEX);
- reduction in accident costs (and Environmental Health and Safety (EHS) risk);
- enhanced user commitment ("buy in").

ORGANIZATION OF HFI

HFI cannot succeed in isolation because the human element pervades through the whole system (i.e., the engineering aspects, the health and safety aspects, and the operational aspects). HFI requires the involvement and engagement of the engineering management and design disciplines, operational experts, and health and safety disciplines. HFI is not a single role responsibility even if there is substantial HFE expertise on the project.

The roles and responsibilities may be relatively simple for a small project, but more complex for a larger project, dependent on how the project is organized.

There are roles which are *not* project-specific but take a *corporate* cross-project focus. These typically include the "head of projects" or "head of project safety" who would hold responsibility for ensuring that HFE is considered within all capital projects. This role requires a competence level 1, as defined in Table 10.1. An HFE design authority would be responsible for providing an assurance role for HFE

Table 10.1 Recommended Competencies for HFI

Competence Level	Description
1	Knowledge and awareness of HFE, its scope, relevance, applicable standards, and requirement within the project
2	In addition to level 1, greater knowledge of relevant standards, HFE terminology, and the organization of HFI and able to implement simple HFE design analysis techniques
3	In addition to level 2, ability to provide advice on many HFE technical matters, implement many HFE design activities, able to translate HFE standards, and resolve most HFE design issues. Should hold a relevant basic qualification in HFE and be assessed as competent for the HFE activities undertaken by an HFE professional of level 4 or 5 competency
4	In addition to level 3, ability to resolve significant non-standard and complex HFE issues, ability to develop HFE processes and practices and have sufficient credentials to assess the competency of levels 1–3. Should hold a relevant professional HFE qualification (such as a degree in ergonomics) and have achieved accreditation from a recognized HFE professional body (such as the Chartered Institute of Ergonomics and Human Factors (CIEHF)) with a minimum of 10 years HFE practitioner experience
5	In addition to level 4, ability to own and approve HFE standards and assess the competence of levels 1–4. Should have at least 20 years HFE practitioner experience

across projects and have a competence level of 4–5. The recommended competencies shown in Table 10.1 are based on OGP454 (2011).

Within a project there may be a number of layers of organization, including:

- The client organization (the "to be owner" of the new system. The client typically assesses and specifies the requirement for the system);
- The engineering contractor (the organization(s) with the responsibility for designing, developing, and implementing the system); and
- The subcontractor organizations (those responsible for specific sub-systems/components or aspects of the design, development, or implementation of the system).

The *project-related* HFI roles are described in general terms in Table 10.2.

Table 10.2 Project-Specific HFI Roles

Role	Client	Engineering Contractor	Sub-contractor	Responsibilities	Competence Level
Project manager	X	X	X	Responsible for project implementation of HFE within their scope for the project and integrating HFI within the project program	1
EHS or engineering manager	X	X	X	Organizes the resources for HFI, is the recipient of human factors issues/risks, and supports the HFI technical work within their scope for the project	1
HFI coordinator	X	X		Responsible for the management and organization of the HFI program. The role may be assumed by or shared with the EHS manager or the HFE expert within their scope for the project	2–3
HFE expert	X	X	(X)	Responsible for the technical aspects of the program. More critical/complex projects will require greater competency	3–5
Operations experts	X			Responsible for providing operations expertise (e.g., process operations, maintenance, and emergency response) as required during the HFE technical activities	1
Project and discipline engineers	X	X	X	Responsible for supporting the technical program, implementing HFE requirements, and reporting HFE issues during design	1

HFI WITHIN THE PROJECT LIFE CYCLE

There are differences in project life cycle models from one industry to another and even from one organization to another within the same industry. There are the traditional waterfall models whereby each life cycle stage is undertaken in succession. These are typically delineated by design stage gates where design is reviewed and either accepted or issues addressed before passing to the next stage. Concurrent engineering is where some of these stages occur in parallel. V-models of systems engineering are requirements driven which include verification and validation activities undertaken to "accept" design. HFI has to fit within whatever project life cycle model is used. The human centered design process (as described in Chapter 9, Overview of Human Factors Engineering, and illustrated in Fig. 9.4) is better described as a spiral life cycle model. This is due to its iterative nature and the expanding level of detail at each stage within the life cycle.

Broadly speaking (using a generic illustration from the process industry), a project will go through the stages as described in Fig. 10.1. Alternative terminology for the stages is presented with a general statement about the objectives of HFI at each stage.

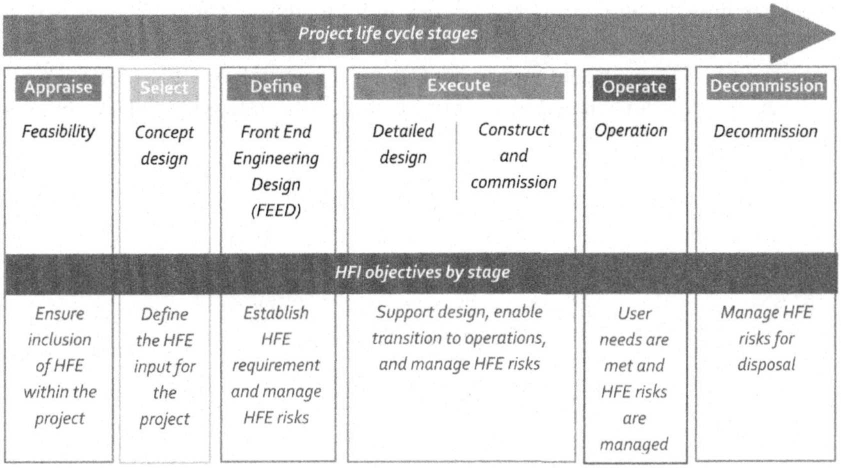

Project life cycle stages						
Appraise	**Select**	**Define**	**Execute**		**Operate**	**Decommission**
Feasibility	Concept design	Front End Engineering Design (FEED)	Detailed design	Construct and commission	Operation	Decommission
HFI objectives by stage						
Ensure inclusion of HFE within the project	Define the HFE input for the project	Establish HFE requirement and manage HFE risks	Support design, enable transition to operations, and manage HFE risks		User needs are met and HFE risks are managed	Manage HFE risks for disposal

FIGURE 10.1

Broad description of project life cycle stages.

There are three main aspects to an HFI program (as with any other discipline);

- The management of the HFI program which includes the development of a systematic and structured plan to manage the key human factors issues within each life cycle stage;
- The technical program which includes the HFE analyses and studies, requirements derivation, and design support for the system being designed; and
- The assurance of design through verification and validation and support to safety demonstration.

These three aspects are interrelated but are divided out in the next sections to help clarify the focus for the HFI program.

MANAGEMENT OF HFI

TYPICAL HFI MANAGEMENT ACTIVITIES BY LIFE CYCLE STAGE

In the early stages of a project the key requirement is to secure a budget for HFE. As a general heuristic in the absence of the detailed requirement, this may be 0.5–2% of the total development effort on an engineering project (excluding capital items). It could be higher once the HFE screening has been undertaken. There is also a need to ensure that HFE is included in key project documentation such as the Basis of Design (BoD). The low level of HFE involvement shown in the first stage is only commensurate with the typical lack of other project activity during that stage. Should the project be structured differently the HFE involvement should be brought forward from the other stages.

Still fairly early on, the scope of HFE will need to be determined and a plan derived for implementation. As the project unfolds, the HFI management activity will be focused on the implementation and management of the plan, recording and closing out of HFE issues and ensuring adequate integration of HFE within the project. These aspects are discussed in more detail in the following subsections and summarized in Fig. 10.2.

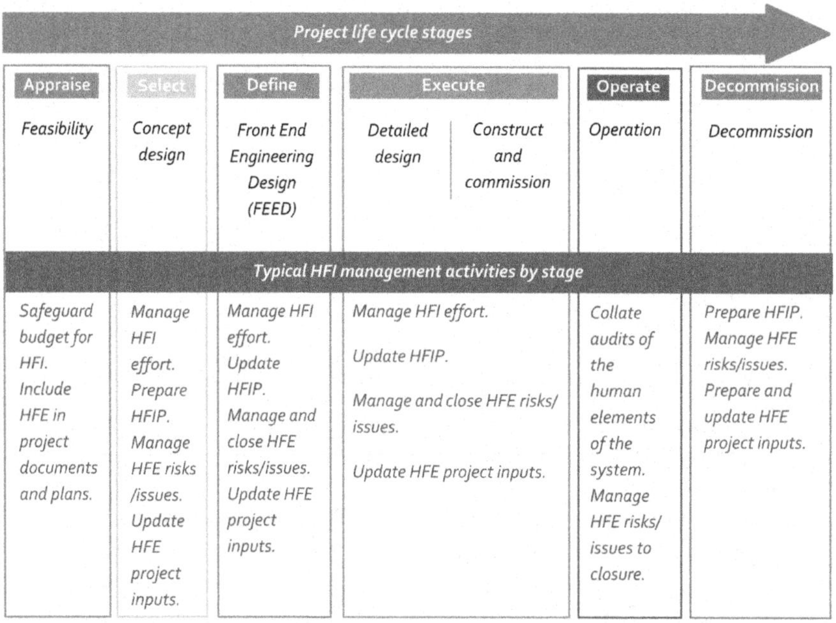

Project life cycle stages						
Appraise	**Select**	**Define**	**Execute**		**Operate**	**Decommission**
Feasibility	Concept design	Front End Engineering Design (FEED)	Detailed design	Construct and commission	Operation	Decommission
Typical HFI management activities by stage						
Safeguard budget for HFI. Include HFE in project documents and plans.	Manage HFI effort. Prepare HFIP. Manage HFE risks /issues. Update HFE project inputs.	Manage HFI effort. Update HFIP. Manage and close HFE risks/issues. Update HFE project inputs.	Manage HFI effort. Update HFIP. Manage and close HFE risks/ issues. Update HFE project inputs.		Collate audits of the human elements of the system. Manage HFE risks/ issues to closure.	Prepare HFIP. Manage HFE risks/issues. Prepare and update HFE project inputs.

FIGURE 10.2

HFI management activities by project life cycle stage.

Although not included in the diagram and dependent on the existing level of HFE knowledge and capability within the project, there may be a need for HFE awareness training. In the initial stages, this is most likely to focus on the client project team and later the engineering contractor and subcontractor project teams as relevant to the stage of the project.

SCOPING AND STRUCTURING OF THE HFI PROGRAM

There are different approaches to determining the level of commitment to HFE for a project. One method used in the past was to set criteria based on the value of the project. For example, a project over a certain value of say £1 million would warrant a specified level of HFI program. This approach has its limitations as the value of a project does not indicate the complexity or criticality of HFE issues. Even a small value project will have its HFE risks.

A risk-based approach can be used such as using safety critical task screening as a starting point, as described in Chapter 6, Human Factors in Risk Management. This is likely to need to be done anyway but does not necessarily capture the breadth of the issues that need to be addressed.

An alternative approach is to screen for HFE issues. Within the defense sector this is undertaken using Early Human Factors Analysis (EHFA) which asks questions relating to the Risks, Assumptions, Issues, Dependencies, Opportunities (RAIDO), and constraints associated with the human component of the system. In its simplest form, the EHFA is intended to clarify any concerns of these aspects in relation to the domain areas of HFI (as described in the defense industry) (Table 10.3).

An HFE screening tool specific to the oil and gas industry is documented in OGP454 (2011). The HFE screening needs to be undertaken in a workshop setting with representatives from across the project, including project management, engineering, safety, and operations to discuss the following factors for the project:

- criticality and the potential for major accident hazards;
- complexity of the HFE issues;
- level of manual interaction from operations personnel;
- novelty of the design for the asset or organization;
- status of design and scope of HFE influence; and
- known issues with predecessor equipment.

The justification for the extent to which HFE will need to be included within the project is captured by this process and from there the detailed plan and cost of the HFI program can be determined.

HUMAN FACTORS INTEGRATION PLAN

The Human Factors Integration Plan (HFIP), also known as the HFE Management Plan (HFEMP) and various other terms, is used to plan and coordinate the HFI effort. It is typically viewed as a live document updated at the start of each life cycle stage.

Table 10.3 Defense Domain Areas for HFI

Domain	Description
Manpower	Determination of staffing requirements (numbers of people required to operate, maintain, and support the system). It also covers activity related to identifying whether personnel can cope with the workload.
Personnel	Identification of the skills required for the operation, maintenance, and support. It includes the identification of skills within the current "resource pool" and matches this against requirements.
Training	Analysis of the training requirements for all end users of the system.
Human engineering	The identification of user tasks, functional allocation between human and machine, specification of human machine interfaces, work area, and environmental requirements. It is concerned with the specification, design, and assessment of the human interface within the system. It can include the accommodation of the user within their working environment, ensuring that it is sufficiently "habitable."
Health hazards	The identification, analysis, and reduction of health and safety hazards created and imposed by the operation of the system. This includes environmental factors such as noise, climate, vibration, lighting, chemical/substance hazards, radiation, and so on.
System safety	The identification, analysis, and reduction of safety hazards created and imposed by the system, particularly those related to human failure. It includes survivability, that is, the mitigation against threats which may actually occur (e.g., ensuring that personnel survive an adverse event).
Social and organizational	The identification of social influences on health, safety, and performance and the need to consider these aspects within the design.

It is intended to sit at the same level as the Systems Engineering Management Plan (SEMP) and the Safety Management Plan (SMP) but also to be integrated with and cross-referred to by these other management plans. An HFIP typically includes the following elements:

- objectives and scope of HFE for the project;
- background to the project;
- key findings of the HFE screening and conclusions arising;
- standards and legislation to be used in support of the HFE activity;

- management and organization of the HFE effort;
- project milestones and HFE integration activities across the project;
- HFE activities (management, technical, and assurance), including the detail on the:
    - definition and structure of each work package,
    - how each work package will be undertaken, including its inputs and outputs,
    - duration of each work package,
    - effort required for each work package,
    - required interfaces with the rest of the project;
- HFE program schedule (which will then be incorporated into the project schedule).

The starting point for the first HFIP will be the HFE screening results. The subsequent HFIPs will be determined by the work being undertaken by the project during that next project stage, taking account of the HFE work already undertaken during the previous stage.

PROGRESS UPDATES AND COORDINATION WITHIN PROJECTS

As with any management activity, there is a need to actively manage the plan, allocate resources, and monitor progress. One key success factor is the degree to which the HFE effort is integrated with the rest of the project. This requires careful planning and engagement of all relevant parties within the program and must be regarded as a two-way activity. HFE will support engineering, operations, and safety activities and likewise engineering, operations, and safety need to support the HFE program.

HUMAN FACTORS ISSUES REGISTER

The Human Factors Issues Register (HFIR) is a core recording mechanism of the HFI program. Any and all HFE issues are recorded within the register and are regularly monitored until they are satisfactorily resolved and closed out. The HFIR can be stand-alone but it is necessary to ensure that it is properly integrated with the project, engineering action, risk, and/or hazard registers used across the project. Ideally, the HFIR will be incorporated within the "project register" so that it gains the visibility and appropriate focus at the same level as other project issues.

TECHNICAL WORK PROGRAMS

Typical technical activities are outlined by project stage in Fig. 10.3. This is not an exhaustive list or presented in detail but it provides a reasonable overview of the types of activities that would be undertaken.

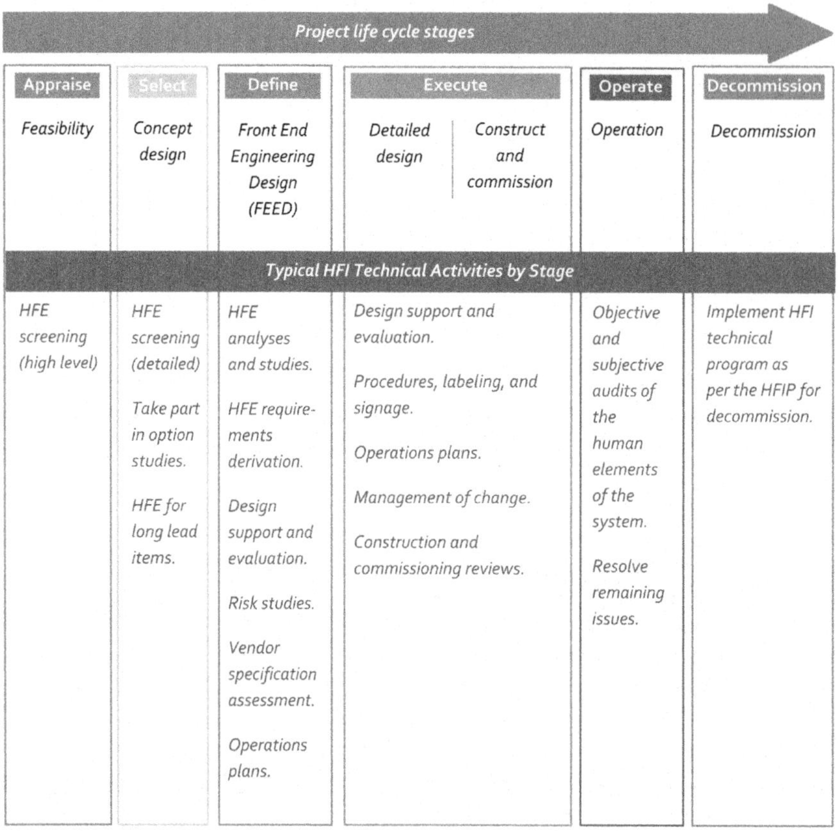

Project life cycle stages						
Appraise	**Select**	**Define**	**Execute**		**Operate**	**Decommission**
Feasibility	Concept design	Front End Engineering Design (FEED)	Detailed design	Construct and commission	Operation	Decommission
Typical HFI Technical Activities by Stage						
HFE screening (high level)	HFE screening (detailed) Take part in option studies. HFE for long lead items.	HFE analyses and studies. HFE require-ments derivation. Design support and evaluation. Risk studies. Vendor specification assessment. Operations plans.	Design support and evaluation. Procedures, labeling, and signage. Operations plans. Management of change. Construction and commissioning reviews.		Objective and subjective audits of the human elements of the system. Resolve remaining issues.	Implement HFI technical program as per the HFIP for decommission.

FIGURE 10.3

HFI technical activities by project life cycle stage.

Initially the effort is focused on technically scoping the program on the basis of the issues and concerns relevant to the project, as identified during the HFE screening work. The interpretation of the HFE screening results and scoping of the HFE activity is a skilled task and needs to ensure that the coverage is com-prehensive and relevant given the multitude of different HFE activities that could be undertaken. The task is best undertaken by an HFE expert of level 4 or 5 com-petency. The areas of involvement fall into three categories: engineering design, health and safety, and operations, as illustrated in Fig. 10.4. It is preferable for many of the stated activities to be undertaken during earlier stages. However, given the need for minimizing project costs to inform "go/no go" decisions a compromise is presented.

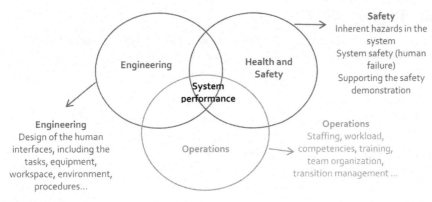

FIGURE 10.4

Crossover between HFE and other disciplines within the project.

ENGINEERING DESIGN

There will be discussions in the concept stages relating to different options for the system solution. This might include, for example, decisions regarding the level of automation or remote control. These are fundamental decisions that benefit greatly from HFE involvement and analysis/studies.

It is also possible that long lead items may be procured during an early stage in design. The identification and specification of HFE requirements for these items should not be overlooked as there is potential to miss critical design issues.

Entering Front End Engineering Design (FEED), the HFE effort should be focused on implementing the human centered design process. The process is described in detail in Chapter 9, Overview of Human Factors Engineering, and illustrated in Fig. 9.4. It involves four key stages (which are iterated throughout the design stages):

- specify the context of use—what is the human involvement? What form does it take? What are the risks? What are the human-related needs?;
- HFE requirements—outline what the design needs to meet and how this should be achieved;
- HFE in design—develop the human interfaces to meet the HFE requirements and resolve HFE trade-off situations;
- HFE evaluation of design—ensure that the design meets the HFE requirements and end user needs.

During the detailed design stage and beyond, the HFE involvement continues to cycle through this process. As the design matures, design input and evaluation happen at a greater level of detail until the HFE requirements are satisfied.

HEALTH AND SAFETY

There is overlap with the health and safety disciplines within the project. Some of the focus may be similar but there are aspects which are unique. The focus should be on ensuring comprehensive coverage and using the HFE expertise where it adds most value.

The HFE role should contribute by undertaking studies that assess risks related to the human–system interface. Examples include Safety Critical Task Analysis (SCTA) and Human Reliability Assessment (HRA) as described in Chapter 6, Human Factors in Risk Management. These do have a different focus from the safety engineering analyses and assessments and should therefore not be regarded as "already being done." The HFE role should also support safety studies such as Hazard Identification (HAZID) and Hazard and Operability (HAZOP) studies. This is because there is a need to consider human failings that could contribute to risk and challenge assumptions about human-related risk controls.

The HFE focus with regard to inherent system hazards is described in Box 10.3. These are normally assessed as part of a "Work Environment Health Risk Analysis" (WEHRA) to eliminate or reduce the risks. These issues are described in more detail in Chapter 16, Environmental Ergonomics

BOX 10.3 INHERENT SYSTEM HAZARDS TYPICALLY COVERED BY HFE

Lighting	Chemical hazards
Indoor climate	Biological hazards
Outdoor climate	Radiation hazards
Altitude/atmospheric pressure	Mechanical hazards
Psychosocial and fatigue	Noise
Space arrangement and layout	Vibration/motion
Demanding work	

OPERATIONS

The focus of involvement of HFE with operations is related to the organization of staff to operate, support, and maintain the system. There is a large crossover with the design of the system, as many operational issues can be resolved through the design. The operational planning is often seen as being the recipient of design information, for example, "this is the system, now let us resource it," or assumptions are made that the way systems have been resourced in the past is suitable for the future system. However, this is typically the opportunity to reduce the operating costs and by assessing the issues earlier within design, adverse impacts on operating costs can be mitigated. The operational aspects therefore need to feed into the design and work in tandem with the design so that the optimal solution can be realized. Key considerations include:

- *Staffing and workload*: The topic of staffing and workload is discussed in more detail in Chapter 20.1, Staffing and Workload. There are various techniques available to identify the staffing requirement and to assess its suitability, that is, the numbers of the right personnel to operate, maintain, and support the system. A key issue is to ensure that the workload of staff is well balanced as both underload and overload can lead to poor performance and human error. Workload assessment is important for all operational scenarios, not just normal steady state;

- *Skills and training*: Alongside the size of the workforce, the technical and non-technical competencies of personnel need to be assessed to determine the extent of the new or different requirements for the system. This is an opportunity to realize unrealistic competency requirements and challenge design decisions. A Training Needs Analysis (TNA) should be undertaken to understand the requirements for development and the mechanisms by which training can be delivered. This is discussed in more detail in Chapter 20.2, Training and Competence.
- *Team organization*: Performance is not just restricted to individual roles, although getting this aspect right is fundamental. The team complements and team skills will affect performance as people have to communicate, cooperate, and coordinate to make critical decisions and achieve the system goals (Flin et al., 2008). Given that many teams are not static entities and individuals have to work with different people at different times, there is a need to consider the team aspects and include focus on developing "team" skills in the individual;
- *Transition planning*: Transition planning is critical to the success of the system at the start of operations and clearly needs to be well planned. The initial start-up or cut-over from one system to another is often focused on the technical aspects of the system, with insufficient attention to the human aspects. Sometimes the strategy can be to flood the site with people but with insufficient consideration of the required roles, how they will be coordinated and how risks will be managed in different operational scenarios. The management of organizational change is discussed in Chapter 19, Managing Organizational Change.

ASSURANCE ACTIVITIES

The assurance of HFE in design is borne out of the work undertaken within the management and technical HFI program. The focus is to ensure that the system design is fit for purpose and safe from the HFE perspective. The position is normally summarized at the end of each life cycle stage and used in support of the stage gate design reviews that affect the project-wide decision of whether the project can proceed to the next stage of design or not. This is illustrated in Fig. 10.5.

All of the activities under the remit of HFE in engineering, health and safety, and operations provide the confidence that the design is suitable or not. The HFE work is typically summarized in a project stage summary report. This includes reference to the HFIP and progress made against the plan, the outcomes of the work undertaken within the technical program and also the close out status of the HFIR. Any residual issues are added to the HFIP for the next stage of design, assuming that they do not

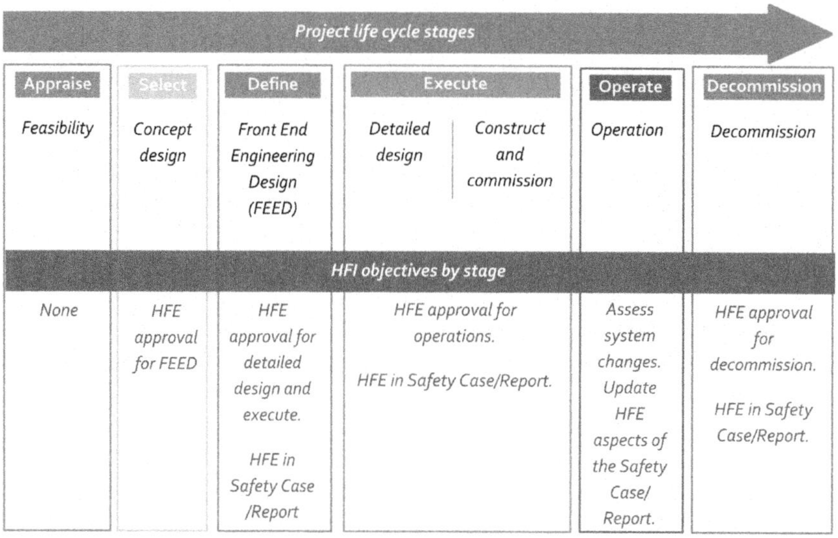

Project life cycle stages						
Appraise	**Select**	**Define**	**Execute**		**Operate**	**Decommission**
Feasibility	Concept design	Front End Engineering Design (FEED)	Detailed design	Construct and commission	Operation	Decommission
HFI objectives by stage						
None	HFE approval for FEED	HFE approval for detailed design and execute. HFE in Safety Case /Report	HFE approval for operations. HFE in Safety Case/Report.		Assess system changes. Update HFE aspects of the Safety Case/ Report.	HFE approval for decommission. HFE in Safety Case/Report.

FIGURE 10.5

HFI assurance activities by project life cycle stage.

adversely affect the project stage gate decision, in which case they would need to be resolved prior to the next stage.

The HFE expert should be called upon to develop safety demonstration arguments in support of the safety case or safety report. This is a specific vehicle for providing HFE assurance and is integrated with the project assurance activity.

SUMMARY: HFI WITHIN DESIGN ENGINEERING PROGRAMS

This chapter presented cost–benefits arguments for HFI and guidance for the organization of the HFI effort. Three elements of the HFI process were discussed: management, technical, and assurance to help explain the focus during different project life cycle stages, taking account of how HFI integrates with different project disciplines. In the next chapters the technical areas of HFE application are discussed in more detail.

KEY POINTS

- This chapter specifically focuses on HFI, a design management activity related to the *process* of applying HFE within the engineering project life cycle.

- The aim is to secure optimal human performance and safety from the outset of design.
- The cost–benefits of HFI are often challenged but research from other industries has demonstrated its importance not just in cost reduction but the overall safety and success of a project.
- HFI requires early project involvement to ensure spending on the right developments.
- It needs to be applied and resourced as a complete process to be effective and efficient. This is about having HFE expertise and also about ensuring it is adequately integrated with the rest of the project activities.
- HFE crosses over with engineering, operations, and safety disciplines and needs to be applied to varying degrees of detail at different project life cycle stages.
- The HFI effort can be described as including management, technical, and assurance activities.

REFERENCES

Booher, H., 1997. Human Factors Integration: Cost and Performance Benefits on Army Systems. Army Research Laboratory., ARL-CR-341.

Eurocontrol, 1999. Human Factors Module: A Business Case for Human Factors Investment. Report Number: HUM.ET1.ST13.4000-REP02.

Flin, R., O'Connor, P., Crichton, M., 2008. Safety at the Sharp End. A Guide to Non-Technical Skills. Ashgate.

HFI DTC, 2006. Cost Arguments and Evidence for Human Factors Integration. Issue 1.

International Oil and Gas Producers (OGP), 2011. Report 454—Human factors engineering in projects. OGP454.

Kirwan, B., 2003. An overview of a nuclear reprocessing plant human factors programme. Appl. Ergon. 34 (5), 441–452.

National Audit Office, (2000). Accepting equipment off contract and into service. Report by the Comptroller and Auditor General, Ministry of Defence.

South West Thames Regional Health Authority, Communications Directorate, 1993. Report of the Inquiry into the London Ambulance Service, February 1993. ISBN: 0905133706. <http://www0.cs.ucl.ac.uk/staff/A.Finkelstein/las/lascase0.9.pdf>.

United States General Accounting Office, 2003. Navy actions needed to optimize ship crew size and reduce total ownership costs. GA0-03-520. <http://www.gao.gov/new.items/d03520.pdf>.

Building and control room design

E.J. Skilling, C. Munro and K. Smith

LIST OF ABBREVIATIONS

CRO Control Room Operator
FEED Front End Engineering Design
FO Field Operator
HFE Human Factors Engineering
ISO International Organization for Standardization
PO Plant Operator

The chemical and process industries operate a variety of large-scale processes in which materials undergo chemical reactions and/or physical changes to produce products. The processes typically involve a series of unit operations and these are often controlled and monitored from a central control room. A control room may be local to the plant, remote from an onsite location, at an offsite location, or within an operations center that controls and monitors several sites.

The design is dependent on the nature of the processes being controlled and monitored but needs to accommodate the control operators and other roles along with the control equipment. This may include the process control screens, communications equipment, and the hardwired safety system control panels such as the emergency shutdown functions and emergency detection system panels. Some control rooms use off-console displays such as large process or activity overview screens which are wall-mounted. A typical example of this, say a power control room, would need to provide an overview of the whole distribution network as well as real-time displays, grid frequencies, and voltages. This type of control room, using wall mounted displays is illustrated in Fig. 11.1.

The control room may be surrounded by offices and administration facilities, as shown in Fig. 11.2. There may also be server rooms, telecommunications rooms, mechanical rooms, and electrical rooms which accommodate the marshalling cabinets for the various systems.

Within this chapter, the control building means the structure housing the control room for a process or facility. The control room refers to the facility used to house the personnel, equipment, and systems to control, monitor, and respond to the process. The control center refers collectively to both the control building and control room.

Human Factors in the Chemical and Process Industries.

FIGURE 11.1

Wall-mounted overview screens.

Created By BAW Architecture for the purpose of the Keil Centre book.

FIGURE 11.2

Control room support functions.

Created By BAW Architecture for the purpose of the Keil Centre book.

The increased awareness of workplace health, along with the arrival of new technologies, and the latest ergonomic standards has stimulated a striking shift in control center design. The functions performed within the control room are typically safety critical which means that a key focus of design is to facilitate human performance.

As the scale of automated solutions has increased, the operator has retained the critical role in monitoring and supervising complex automated/semi-automated systems. The consequences of operator error can be catastrophic and so the integration of human factors within the design is essential. The goal of human factors is to eliminate or minimize the potential for human error and enhance the effectiveness and efficiency with which the work is conducted.

This chapter discusses the approach to control center design and relevant Human Factors Engineering (HFE) tools and techniques for building and control room layout. Console design and control system interface design are discussed separately in Chapter 12, Workstation, Work Area, and Console Design, and Chapter 13, Control System Interface Design, respectively. This chapter does refer to the environmental design and this is discussed further in Chapter 16, Environmental Ergonomics.

HUMAN FACTORS APPROACH TO CONTROL CENTER DESIGN

The key reference for ergonomics in control center design is the International Organization for Standardization (ISO) 11064; parts 1–7; 2000–2013. It advocates a human-centered design process in the manner described in Chapter 9, Overview of Human Factors Engineering. It provides guidance for specific aspects of the control building design: control room layout; workstation design; control system interfaces; environmental design; and the evaluation of the design. The standard (and the human-centered design process in general) is applicable to the design of new control centers as well as to the expansion, refurbishment, and upgrade of existing control centers.

Part 1 of ISO 11064 sets a framework for designing and evaluating control centers, modified for illustration in Fig. 11.3.

STEP 1—DEVELOPMENT OF THE INITIAL DESIGN

It is recommended that the following should be included within step 1:

Clarification of goals

The operational goals, relevant requirements, and constraints for the control building and/or control room design need to be clarified. This should include the purpose, functional relationships, and the known constraints associated with the design. The design typically needs to be suitable for 24 hour, 7 days a week operation.

Consideration needs to be given to the prospect that the control building may not only be used for monitoring and control of plant process, but also have to function as the following:

- communications and visitor center;
- training facility for operators;
- incident command center;
- security station.

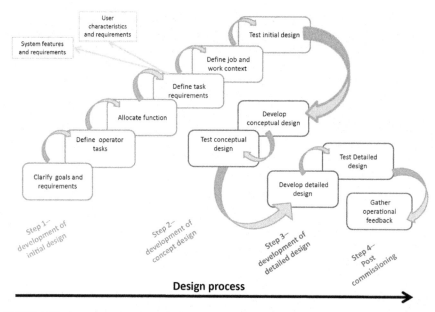

FIGURE 11.3

Framework for design and development.

There are always design constraints and trade-offs. However, human performance and the well-being of operators should not be adversely affected. Both new builds and remodeling of existing facilities can be impacted by constraints but arguably there are fewer constraints during new building projects. Examples are:

- the location of a control center may already be predetermined and this may impinge on the space available for the building;
- existing control centers may be constrained by existing equipment, architectural features, or structural support members. For example, it may be undesirable to move an existing fire and gas panel due to the need for uninterrupted functioning. Windows can restrict the position of equipment, whilst pillars can cause obstructions for the field of vision. Other constraints may include ceiling heights, existing doors, and emergency exits that are potentially fixed and may not be movable.

The information gathered during this step should form the basis of a working document that contains:

- a statement of the operational goals;
- applicable requirements and constraints;
- relevant information regarding the users, tasks, and environment.

Define tasks and allocate functions

A functional allocation review should be undertaken to determine which functions and tasks will be the responsibility of operators and which should be partially or fully automated (Sheridan, 2002). Inappropriate allocations can lead to problems such as increases in cognitive demands, reduced situational awareness, and loss of knowledge and expertise. It can also lead to a reduced ability to cope when called upon in unforeseen circumstances. Technological solutions can not match humans with regard to the consideration of the multiple and often unpredictable conflicting goals which arise frequently within a work setting. Poor automation design can lead to operators having to "work around" the technology and compensate for its constraints (Cook and Woods, 1996).

Functional allocation should form an explicit stage early in the design and involve a multidisciplinary team incorporating different perspectives, expertise, and knowledge.

Define the task requirements

Once the allocation is known, a high level task analysis for the control center functions should be derived which includes all relevant operational scenarios. Hierarchical and tabular task analyses are useful techniques to define tasks for different roles. They can also be used to support other analyses, such as Human Reliability Analysis.

Define the job and work organization

Each role should be examined to determine specific requirements. A simple generic role breakdown is presented in Tables 11.1–11.3. Within these tables a distinction

Table 11.1 Typical Roles and Responsibilities of Residents of a Control Room

Typical Roles	Typical Team Allocation	Typical Responsibilities
Control room operator (CRO)	Operations	Monitoring and control of the process and the individual functions included within the process, including process optimization, support functions, such as utilities and other functions such as loading and unloading of raw materials and products. The functions may be divided between more than one CRO. The control room often becomes the communications hub and so additional functions may be undertaken because of this. During plant upsets and emergencies, the CRO typically manages the process control aspects of restoring control.
Shift supervisor	Operations	Supervises plant operations during steady-state operation. During plant upsets, the shift supervisor typically performs a key role in supporting the restoration of control. During emergencies, he/she will be a member of the incident command team, such as acting as the event recorder, muster controller, or emergency response team lead. Supervisory functions can be partly or fully distributed.

Table 11.2 Typical Roles and Responsibilities of Regular Nonresident Visitors to a Control Room

Typical Roles	Typical Team Allocation	Typical Responsibilities
Field operator (FO) or Plant operator (PO)	Operations	The FO/PO is often effectively the eyes and ears on the plant. The typical tasks include plant monitoring and line walks to ensure steady-state functioning, sampling, manual valve line ups, checking instrument readings, manual additions of chemicals, and preparations for maintenance. It may include processing of samples in a laboratory. The FO/PO can often act as a back up to the CRO within the control room.
Maintenance technicians	Maintenance	Perform the maintenance on plant equipment including: • mechanical maintenance, such as the repair or maintenance of pumps, motors, valves, pipework, and vessels. This can often include mechanical handling of heavy items of equipment being removed and replaced; • electrical maintenance includes the maintenance of switchboards, generators, switches, function testing of motors and other equipment, integrity checks, power sockets and supplies, appliance testing, and lighting; • control and instrumentation maintenance includes the inspection, test, repair and calibration of field instruments, controllers, control loops, and control system administration.

Table 11.3 Typical Roles and Responsibilities of Residents of a Control Building

Typical Roles	Typical Team Allocation	Typical Responsibilities
Operations management	Operations	Oversees plant operations and take a lead role in emergencies as part of an incident command team
Maintenance supervisor	Maintenance	Supervises and organizes plant maintenance. Maintenance includes planned preventative maintenance, inspections, tests, and repairs. During emergencies he/she may be a member of the incident command team
Permit coordinator	Maintenance	Coordinates and manages permits to work for maintenance activities. This role may be subsumed within another role

(Continued)

Table 11.3 Typical Roles and Responsibilities of Residents of a Control Building (Continued)

Typical Roles	Typical Team Allocation	Typical Responsibilities
Technical support	Support	Provides detailed engineering technical advice and system administrator functions. They may be resident at the site or remote, or a combination
Administration support	Support	Provides administrative functions which may include the administrative management of materials, coordination with other sites and agencies. There is likely to be a role related to health and safety and medical services in the case of offshore platforms. Other support functions may include helicopter administration and training

is made between roles which commonly reside in the control room, those which are frequent users of the control room but are not generally resident, and roles which commonly reside in a control building. In reality, there are large variations from site to site.

This can be a useful preliminary breakdown to identify staff numbers, functions, and equipment and enable allocation of roles and equipment to different spaces within the control building. The organizational structure, role interactions, and training requirements should be documented. The interactions within the control building can be derived using techniques described later in the chapter.

STEP 2—DEVELOPMENT OF CONCEPT DESIGN

An early consideration in design is the location of the control building. It can be a requirement for the control building to be located at a distance from the process plant to provide the opportunity to maintain control of the process in the event of an emergency scenario and remove critical operators from the blast zone. Other factors include:

- whether the plant needs to be directly visible to CROs;
- team communication and coordination;
- control building accessibility, walking routes, and emergency exits;
- environmental aspects such as temperature, noise, and vibration.

HFE analyses are used to develop the conceptual design. Typically, this will include:

- layout of the control building;
- layout of the control room;
- initial control interface and console design.

Other design decisions may include communication methods, security and access arrangements, and environmental aspects.

It is vital that there is early and regular testing of the control center design concepts against operational goals, relevant requirements, and design constraints. Verification methods should indicate whether the design continues to meet the requirements and specification. Validation methods should determine the design's effectiveness, that is, whether the end user's needs are being met.

The choice of methods for evaluation is dependent on the stage of the design and the level of detail required. Methods include the following:

- verifying design features against design criteria (derived from ISO 11064 and HFE analyses) to determine compliance;
- validating functions and usability;
- conducting link analysis;
- walk through/talk through;
- use of drawings, photographs, and mock ups for user testing.

Verification and validation methods should be a continuous part of the design process and enable early identification of human factors issues.

STEP 3—DEVELOPMENT OF DETAILED DESIGN

Design specifications continue to evolve and, as they do, they should continue to be tested using HFE methods such as compliance audits, walkthroughs/talk-throughs, simulations, mock ups, and user trials.

STEP 4—POST-COMMISSIONING

Gathering operational feedback is useful for future learning and can be obtained by observation and measurement, interviews and questionnaires, and data review.

CONTROL BUILDING LAYOUT

The purpose and functions of the control building and its relationship with relevant subsystems needs to be established. An adjacency matrix is a simple but effective method for identifying the degree of proximity required between the building functional areas, as illustrated in Fig. 11.4. The development of an adjacency matrix requires input from operational end users, project management, architects, HFE specialists, and engineers. This is typically undertaken within a workshop setting.

The functional areas to be accommodated within the building are established on the basis of design requirements documents and through discussion with relevant stakeholders. This is typically at least partly established prior to the workshop. For each functional area, the degree of adjacency required to all the other

	Reception/security	Permit office	Break room	Main toilet	Production sup/office	Senior office	Field op area	Meeting room	Training room	Services area	Control room	CR toilet	Alertness recovery
Reception/security													
Permit office	o												
Break room	o	o											
Main toilet	1	o	1										
Production sup/office	1	o	o	o									
Senior office	o	1	o	o	o								
Field op area	o	1	o	o	o	1							
Meeting room	1	o	o	1	o	o	o						
Training room	o	o	o	o	o	o	o	o					
Services area	o	o	o	o	o	o	o	o	o				
Control room	o	2	2	o	2	2	2	2	1	o			
CR toilet	o	o	o	o	o	o	o	o	o	o	2		
Alertness recovery	o	o	o	o	o	o	o	o	o	o	2	o	

Adjacency required	
-1	Avoid adjacency
o	None
1	Preferred
2	Essential

FIGURE 11.4

Example of an adjacency matrix.

Created By BAW Architecture for the purpose of the Keil Centre book.

functional areas is rated using a simple scale such as the one shown in Fig. 11.4. The adjacency matrix can then be translated into a functional adjacency plan as shown in Fig. 11.5.

A key is used to illustrate the level of relationship; in the example shown a bold line reflects a "required" adjacency and a thin dotted line reflects a "preferred" adjacency. "No adjacency requirement" uses no line at all and requirements for undesirable adjacencies are represented by separating the "bubble" diagrams.

The functional adjacency plan and analyses are used as a template to derive concept options for the control building layout. These can subsequently be used

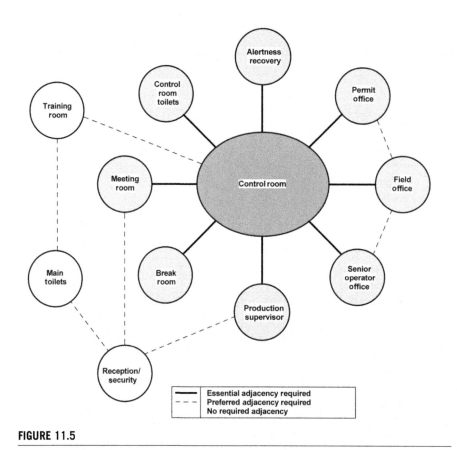

FIGURE 11.5

Example of a functional adjacency plan.

Created By BAW Architecture for the purpose of the Keil Centre book.

for reviewing different concept designs to ensure that the functional needs are met. Feedback from all stakeholders is essential to reach an agreement on the optimal solution. This stage is typically iterative and several concept options may be considered but the functional plan remains a core reference point during the process. An example of a control building block plan layout based on the information attained from the adjacency matrix is presented in Fig. 11.6. The schematic design is then derived from the block plan to define actual space arrangements and relevant scaling.

A similar technique, often referred to as a functional-role analysis maps the functional areas to the different job roles. This can be used to identify who the users will be for the different functional areas and how this may affect the traffic flows within the building.

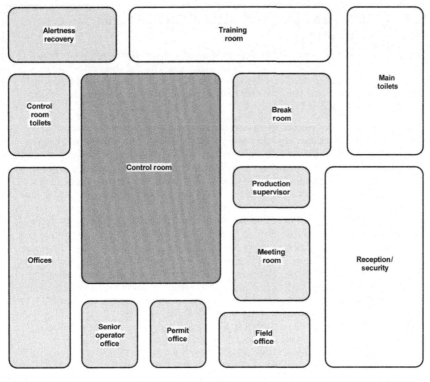

FIGURE 11.6

Example of a building block plan layout.

Created By BAW Architecture for the purpose of the Keil Centre book.

CONTROL ROOM LAYOUT

SPACE DESIGN

The guidance in ISO 11064 recommends a space allocation of 9–15 m² usable area per working position within the control room and a minimum of 3 m floor to ceiling height (ISO 11064 Part 3, Control room layout, 1999). A comment on these two specific aspects of the control room is provided in Box 11.1.

FUNCTIONAL LAYOUT

The focus for the functional layout of a control room is to optimize team working opportunities and social interaction. It should avoid noise-distracting task activities at adjacent workstations, but allow direct verbal communication between personnel (ISO 11064 Part 3, Control room layout, 1999).

BOX 11.1 COMMENT ON SPACE ALLOCATION AND FLOOR TO CEILING HEIGHT

The Keil Centre has worked on several control building projects with BAW Architecture who specialize in control building design (see the acknowledgment at the end of the chapter for more detail). Based on actual real world experience, BAW has found that, in practical application, the space recommendations from ISO 11064 can be too restrictive and BAW recommends more space for the following reasons:

1. *Scalability*: ISO 11064 requires the space to accommodate a 25% growth trajectory. The workflow processes change, as do the number of people utilizing the control room. The plant is a dynamic environment and additional operators, engineers, trainees, and management often end up occupying portions of the space originally devoted to operators. Greenfield projects take 5–10 years minimum to complete. Larger projects can take decades to come to fruition. The danger of starting with a fixed minimum square meter is that it tends to get etched in stone, without anticipating future needs.

2. *Acoustics*: The trend toward large consolidation of many consoles in one room poses significant acoustic challenges. ISO 11064 requires that the decibel level be not more than 45 dBA and not less than 30 dBA. Acoustic engineering is needed to mitigate sound, and a flat, low ceiling does not support those mitigations. Articulation is needed to capture and isolate noise, adding to the square meter footprint. The ISO 11064 suggestion of a minimum of 3 m of finished floor-to-ceiling height in reality would limit the design of a centralized control room and infers a low, flat ceiling. A large space with multiple operators, alarms, and other typical activities will not be optimized for acoustic best practices with this minimum ceiling height.

3. *Lighting*: The minimum of 3 m ceiling height may lead to lighting concerns. The optimum lighting in a control room demands quality, ambient, uniform illumination which is a combination of fixtures, indirect, task, and suspended lighting. If all lighting comes from the ceiling, intense glare may make screens much more difficult to read; therefore suspended lighting is also needed. These fixtures need space or they will likely become obstructions.

4. *Lines of sight*: In some cases, management needs an overview of all plant operations, necessitating secondary overview screens. This requires adequate vertical space within the control room.

The Keil Centre concur that, for some types of control room, the space allocation can be restrictive (albeit that this is intended to be usable space, not gross space). However, there is a balance to be found between ensuring enough space whilst supporting necessary communications. Face to face communication between people who are too far apart can lead to an increase in ambient noise levels as operators have to speak in raised voices to be heard.

The following real-life scenarios illustrate the ramifications of space per operator in control room design.

Design Scenario One

A petrochemical company in North America had an outdated control building featuring small control rooms in poor condition. These were in various field centers throughout the plant in proximity to the refinery, putting the operators at levels of significant safety risk. A large, centralized, 12-console control room was proposed. A cross-functional team was formed to ensure sufficient input was obtained for the design to support both steady state and abnormal operation scenarios.

An iterative approach utilizing HFE best practices led to the design which featured a floor plan that incorporated 41 m^2 per operator, in contrast to the ISO 11064 recommendation of 9–15 m^2. Additionally, due to the multiple operators under one roof, the designed ceiling height was 6 m, in contrast to the ISO 11064 recommended 3 m minimum.

The new ergonomically designed, centralized control building housing all 12 operators in one control room allowed collaboration, improved safety, and was deemed successful. The space, completed in 2013, also has the capacity to accommodate growth and embrace future development.

Design Scenario Two

A plant expansion was required to accommodate growth for a petrochemical company in Africa and an existing control building was to be upgraded in conjunction with the new larger footprint. The engineers developed plans for the refurbished control room but little thought was given to HFE best practice.

During Front End Engineering Design (FEED), operators recognized that the proposed design did not comply with many human factors requirements for control buildings/rooms. They sought control building subject-matter experts to evaluate the existing layout and provide a control building design that was based on operator input and HFE best practice. An alternative layout was developed and proposed. Unfortunately, the alternative design cost approximately 10% more than the original proposal and was rejected. The original layout proposed by the engineers was constructed and operators moved in. Within 2 years of completion and occupation, the operators refused to work in the new control building. The response by the operations team was so strong that the building has since been deemed uninhabitable. Millions of dollars in construction costs and productivity were needlessly lost.

(BAW Architecture)

Link analysis is an effective technique to define the interactions among components (human or non-human) of a work system (Chapanis, 1996). The frequencies and/or the importance of the component interactions are identified either through observation for an existing system or through expert judgment for new buildings. For the purpose of control room layout, role to role interactions are established by systematically weighting the strength of the relationship. A simple rating scale might be as follows:

- high—almost continuous interaction throughout a shift;
- medium—intermittent interaction throughout a shift;
- low—interaction may or may not occur within a shift;
- none—the roles do not interact with each other.

This is recorded in a matrix similar to the adjacency matrix but using role descriptions. This should be undertaken for all relevant operational scenarios: normal steady state; plant upset; emergencies; and planned start-up. The interaction requirements will change from one scenario to another.

A role-to-equipment link analysis is undertaken in a similar manner and is used to determine the layout of equipment in relation to the roles.

This information can then be used to guide design layout of the control room (and console design when used at that level).

CONSOLE ORIENTATION

As well as the functional placement of consoles, console orientation is also a factor. The most appropriate orientation can be a trade-off, for example, face to face orientations can increase noise interference between operators.

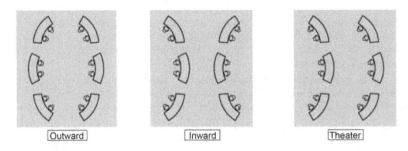

Outward   Inward   Theater

FIGURE 11.7

Illustration of console orientations.

Created By BAW Architecture for the purpose of the Keil Centre book.

The console orientation needs to be determined between operators of the same group, between groups of operators, and between supervisors and groups. Some basic arrangements of intergroup console orientations are shown in Fig. 11.7.

Several factors are relevant to console orientation:

- the functions and tasks for operators, hardware, and software components that are necessary for process monitoring and controlling;
- team communication interactions and team links with other personnel;
- the number of people required to operate the system in all plant states;
- viewing requirements;
- sharing of equipment;
- maintenance access;
- supervisory input, including viewing and communication requirements;
- support storage.

In general an inward facing design is beneficial for verbal communication and eye contact: an outward design reduce noises interference. If viewing of shared wall displays is required by all operators, a theater style may prove the best design.

ENVIRONMENTAL DESIGN

Comprehensive guidance is presented within ISO 11064 Part 6, Environmental requirements for control rooms (2005). Environmental factors can contribute to effective operator performance and reduce discomfort and ill-health. The key environmental factors associated with control centers are:

- thermal comfort;
- lighting;
- noise;
- vibration;
- aesthetics.

THERMAL COMFORT

There is a combination of factors that can affect thermal comfort including: room temperature, humidity, air flow, equipment heat, the tasks that are conducted, and the clothing that is worn. Lower temperatures can affect operators' ability to process information and may reduce fine dexterity which, in turn, can impact on the use of control devices. Higher temperatures can affect cognitive functions such as concentration and this may contribute to fatigue.

Thermal comfort surveys and physical measurements can be taken in an existing control room to ascertain if there are any issues for operators.

LIGHTING

The suitability of the lighting environment is dependent on the tasks to be conducted. Typically control room tasks involve the use of visual display screens and paper work which benefit from different lighting levels. A reasonable compromise is 200–500 Lux (unit of illumination). Positioning of light sources (including natural light) is important to avoid glare and reflection issues. Lighting is assessed through lighting surveys and computer modeling dependent on the stage of design.

NOISE

The issue of noise in a control room focuses on noise distraction, interference of communications, and noise as a stressor, rather than noise-induced hearing loss. Noise levels can be predictively modeled during design and measured once the control room is in operation.

VIBRATION

Depending on the type of industry and relative location of the control center, there may be vibration sources. Vibration is unlikely to lead to direct ill health such as vibration white finger or whole body vibration issues associated with high or long-term exposure (dependent on the building type, structure and/or location). It may however cause distraction and/or lead to general discomfort.

AESTHETICS

Good aesthetics within the control center can assist with providing a calming and pleasant working environment. The guidance in ISO 11064 discusses surface reflectance levels, colors and contrast, materials, and other aspects to support operator well-being.

SUMMARY: BUILDING AND CONTROL ROOM DESIGN

There are many factors that need to be taken into account when designing control rooms. Primarily, the design needs to support the functionality required of the building, control room, control workstations, and control interfaces. These aspects are integral to each other and the integration of them is important to consider. ISO 11064 provides guidance on the human-centered design approach to ensure that human factors issues are adequately integrated within the design activity. Simple techniques can be used to support the integration process.

The next chapter discusses workstation and work area design, both of which are integral to good control room and building design. The focus is broader than the specific design of control consoles, but it provides an outline of the key considerations for any type of workspace and work area.

KEY POINTS

- The work undertaken within a control building and a control room in particular is typically safety critical.
- Poor design can have consequences for operators' health and well-being as well as the safe operating and monitoring of the facility.
- Integrating human factors into the design of control buildings and/or control rooms reduces the potential for human error and enhances the effectiveness and efficiency of human performance.
- A human-centered design process is recommended by ISO 11064.
- Both prescriptive and goal-orientated HFE requirements need to be established from the start of the project.
- Adjacency matrices and link analysis techniques, amongst other techniques, can be used to support building layout and control room design.

ACKNOWLEDGMENT

Acknowledgment and gratitude are extended to BAW Architecture for the provision of figures and support in writing this chapter. BAW Architecture is a world leading specialist in control building design with over 100 built projects on five continents including within the chemical and processing control industries. The company is exemplary in how it embraces the human-centered design approach and uses a multidisciplinary team to achieve the highest standards and compliance in design. The Keil Centre has worked with BAW Architecture on several control (and other) building projects where the benefits of human factors in design are clearly understood. Further information can be obtained at www.bawarchitecture.com.

REFERENCES

Chapanis, A., 1996. Human Factors in Systems Engineering. John Wiley & Sons, Inc., New York, NY.

Cook, R., Woods, D.D., 1996. Adapting to new technology in the operating room. Human Factors 38, 593–613.

ISO 11064, Ergonomic design of control centres; Part 1, Principles for the design of control centres, 2001; Part 2, Principles for the arrangement of control suites, 2000; Part 3, Control room layout, 2000; Part 4, Layout and dimensions of workstations, 2013; Part 5, Displays and controls, 2008; Part 6, Environmental requirements for control rooms, 2005; Part 7, Principles for the evaluation of control centres, 2006.

Sheridan, T.B., 2002. Humans and Automation: System Design and Research Issues. John Wiley and the HFES, Santa Monica, CA.

Workstation, work area, and console design

12

E.J. Skilling

LIST OF ABBREVIATIONS

ABS   American Bureau of Shipping
ANSI   American National Standards Institute
ASTM American Society for Testing and Materials
DSE   Display Screen Equipment
FEA   Federal Aviation Administration
HFES Human Factors and Ergonomics Society
MSD   Musculoskeletal Disorder
LAN   Local Area Network

There are two key factors relevant to designing a work area or workspace: the dimensions of the space and the layout/arrangement of the equipment within the workspace. This should take account of tools, equipment, and furniture as well as the users and their tasks. Anthropometry is the study of body size and can be used to determine the dimensions of the space.

A poorly designed work area or workspace can lead to discomfort, musculoskeletal disorders (MSDs), errors, and reduced efficiency.

This chapter introduces the general principles of work area and workspace design and includes the following:

- workstation dimensions;
- work area space requirements;
- workspace arrangement;
- working posture;
- seated workstations;
- seating design;
- standing workstations;
- visual requirements;
- vehicle workstations;
- plant side work areas.

The principles apply to any type of workstation or work area related to the chemical and process industries, including work benches, control consoles, laboratory areas, crane cabs, equipment and machinery workspaces, storage areas and warehouses, drillers cabs, forklift truck cabs, and assembly workstations, to name but a few.

WORKSTATION DIMENSIONS

The physical dimensions of work areas, workspaces, and equipment need to take account of the end user and the tasks he/she performs. Body dimensions vary by age, gender, and ethnic origin. Most anthropometric variables conform to a normal distribution with few people at either end of the extreme (small or large). The aim is to accommodate as many people as possible within the end user population; the parameters used to do this are typically from 5th to 95th percentile or 1st to 99th percentile. Designing for the "average" or the 50th percentile is typically the least accommodating as it is too restrictive. However, designing for the extreme or providing adjustability is not always practical. So in this instance, dimensions should be calculated to determine the largest number of people to be accommodated.

There are two types of body measurement: static and dynamic. Static dimensions are measurements taken in static postures whereas dynamic dimensions are taken with the body engaged in physical movement.

More static anthropometry data exists (e.g., Peebles and Norris, 1998) than dynamic data, even though functional dynamic data are more representative of the actual requirements for the task. Sanders and McCormick (1992) suggest that the following translation from static to functional dimensions may be helpful if not exact:

- heights—reduce by 3%;
- elbow height—no change;
- knee height—no change except with high-heeled footwear;
- forward or lateral reaches—decrease by 30% for convenience, increase by 20% if able to extensively move the shoulder and trunk.

Clothing adjustments also need to be made when applying anthropometric data. Table 12.1 provides some examples of adjustments related to clothing.

Workspace design requires the application of principles relating to clearance, posture, and reach.

Table 12.1 Recommended Clothing Adjustments

Description	Adjustment (mm)
Safety helmet	35
Hat	25
Heavy outdoor clothing	10–50
Light indoor clothing	10
Everyday shoes	25 ± 5

CLEARANCE

Clearance is required to avoid unwanted contact with the work environment by providing adequate space for head room, knee room, and access space. Dimensions for clearance are determined by using the largest population dimension. If the largest person can fit in or through a space, then everyone smaller is also accommodated. In general the aim is to accommodate at least the 95th percentile (typically male dimension, other than hip breadth). However, there may be a need to accommodate more than 95% of the population. A typical doorway of 1.95 m high exceeds the worldwide 95th percentile male stature.

REACH

Reach is concerned with the positioning of controls or items used by the hands and feet. It includes considerations such as the height of a seat where it is necessary for the user's feet to touch the floor. Reach should be determined by the dimension of the smallest user. If the smallest person can reach the item, everyone larger than that person is also accommodated. Reach is generally determined by using the 5th percentile population dimension (typically female).

POSTURE

It is necessary to enable the user to adopt a good working posture where the least amount of strain is placed on the musculoskeletal system and where the strongest position is enabled. Designing for posture is a careful balance. For example, a bench surface that is too low will cause the user to stoop and may lead to discomfort in the lower back. One that is too high may cause the user to work with their arms abducted (raised) placing more strain on the shoulders and mid-back. In this case there are three options: provide work surface height adjustment; provide work surfaces of different heights; or calculate a "compromise height" which maximizes the number of users accommodated.

Best practice is to provide adjustment for the range from the 5th percentile female to the 95th percentile male of the relevant population dimension. For example, seat height adjustment is determined using popliteal height and standing work surface height is determined using standing elbow height.

Posture is integral to reach and clearance as the aim is to enable the necessary tasks to be performed in suitable postures. A general guideline for reach distances for both seated and standing work is to locate the most important/frequently used items in the primary zone. Lower frequency/less important items should be located in the secondary and tertiary zones as shown in Fig. 12.1 (Illustration by BAW Architecture, Denver, USA; adapted from Dul and Weerdmeester, 2008).

Anthropometry and good design principles are important to derive a design in the most efficient way but there are often emergent features, opportunities, and trade-offs that can be optimized through effective testing. Therefore, user trials are necessary to test the adequacy of design for the range of users who may use the space.

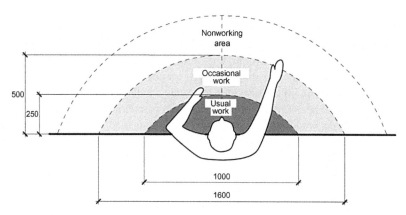

FIGURE 12.1

General guidelines for work reaches.

Created By BAW Architecture for the purpose of the Keil Centre book.

BOX 12.1 CONTROL OPERATOR WORKSPACES

Control operator work spaces need to take account of the following:

- *additional apparatus*—console functionality requires many additional features, which may at a minimum include radios, phones, multiviewer video conferencing screens, a second computer and screen for Local Area Network (LAN) connections, printers, log books, specialized keyboards, task lighting, personal storage, and a meeting area;
- *screen size*—standard screen size has increased to 580 mm and the trend is for larger. In some control rooms the console furniture may include overview screens in addition to either single or dual-tier screens. These screens must have the ability to raise and lower in accordance with the sit/stand furniture. Therefore additional space and depth is required;
- *sit/stand*—ISO 11064 recommends that operators have sit/stand workstations. These are large, robust, pneumatic lift, height-adjustable stations. They take up space which results in additional width and depth allocations related to the increased size of the console footprint;
- *hard wire shut down*—despite technical and computer advances, operations have not, and likely will not, migrate away from hard wire shut down switches. This is an additional equipment space requirement that is often overlooked.

(Summarized by BAW Architecture—acknowledgment at the end of the chapter)

WORK AREA SPACE REQUIREMENTS

Work area space is dependent on the number of people who may work in any particular area, their tasks, and the space taken up by furniture, fittings, and equipment. The work area height should also be taken into account. Specific guidance is often area/task specific. An example is presented in Box 12.1 in relation to control operator workspaces within a control room.

WORKSPACE ARRANGEMENT—WORKSTATIONS AND CONSOLES

McCormick (1970) initially formulated the key principles for workspace arrangement. These are as follows:

- *importance principle*—the components that are most essential to safe and efficient operations should be in the most accessible positions;
- *frequency of use principle*—the components that are used most frequently should be in the most accessible positions;
- *function principle*—components with closely related functions should be located close to one another;
- *sequence of use principle*—components used in sequence should be located close to each other and their layout should relate logically to the sequence of the operation.

Link analysis is an effective technique to define interactions between components (human or nonhuman) of a work system (Chapanis, 1996). For the purpose of workstation design, interactions are established by systematically weighting the strength of the relationships to each item of equipment. The frequencies and/or the importance of the component interactions are identified either through observation for an existing workstation or through expert judgment for new workstation. An example of a link analysis for a control operator use of equipment within the control room during normal operations is shown in Fig. 12.2. The solid and dashed line shows the frequency of use of different equipment. Similar analyses should also be undertaken for different operational scenarios, where relevant, for example, abnormal and emergency scenarios. The analysis can be used in support of the assessment of an existing console layout or in the development of a new console. Where equipment is shared or used in sequence this should also be identified to assist with layout and positioning.

The key considerations are that the user should be able to:

- maintain an upright and forward facing posture;
- avoid unbalanced postures;
- adopt several different, safe postures;
- be seated or standing, dependent on the nature of the task;
- undertake tasks with body joints at their midpoint (in a neutral posture) for the majority of the time;
- use the largest appropriate muscle group where force has to be exerted and if repetition is necessary, the ability to exert force with either arm or either leg;
- avoid work tasks that are performed consistently above heart level;
- view the task with the head and the trunk upright or with a slight forward incline (adapted from Corlett and Wilson, 1995).

These principles are explained further in the following sections.

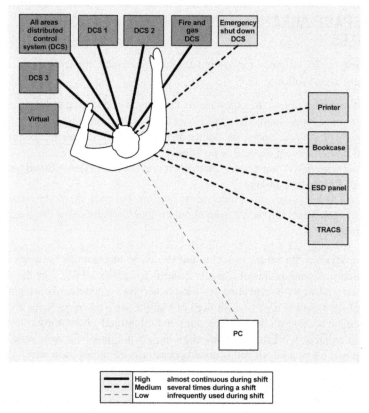

FIGURE 12.2

Diagrammatic of role to equipment link analysis.

Created By BAW Architecture for the purpose of the Keil Centre book.

WORKING POSTURE

The height of a work surface is dependent on the task being conducted. The aim is to enable an upright and forward facing posture to be adopted that will not lead to musculoskeletal discomfort and/or disorders.

Tasks typically involve the use of the hands and eyes. As a general rule, where the hands are a more dominant aspect of the task, the "task position" may need to be toward elbow height; if the visual aspects are dominant the "task position" may need to be closer to eye height. A compromise between these two positions is required if the use of the hands and eyes are fairly equal. Consequently, it is important to distinguish between working height and work surface height. Additional devices can be used such as jigs or monitor risers to improve the adjustability and flexibility of the working height, regardless of the work surface height. General guidance for seated and standing work height is provided in Table 12.2 below (adapted from Pheasant and Haslegrave, 2006).

Table 12.2 General Guidelines for Work Height in Sitting and Standing

Types of Task	Work Height
Manipulative tasks involving moderate force and precision	50–100 mm below elbow height
Delicate manipulative tasks (e.g., writing with wrist support)	50–100 mm above elbow height
Heavy manipulative tasks	100–250 mm below elbow height
Lifting and handling tasks	Between mid-thigh and mid-chest levels
For hand operated controls	Between elbow height and shoulder height

An early decision to be made is whether the workspace will be designed for sitting or standing or both. This should be determined by the nature of the tasks to be performed.

SEATED WORKSTATIONS

A seated workstation is preferred when the task requires:

- fine manipulation;
- high visual attention;
- high degree of stability;
- precise foot control;
- fixed postures for extended periods.

Sitting is preferable to standing as it is less fatiguing. The body is better supported by the seat, the backrest, and the armrests whilst the floor supports the feet. In addition the body has some protection if there is vibration within the workspace.

RECOMMENDED DIMENSIONS FOR SEATED WORKSTATIONS

Guidance for preferred dimensions for seated workstations is shown in Fig. 12.3 (summarized from Federal Aviation Administration (FEA), 2012). Work surfaces should be in the region of 720 mm above the floor, for example, for Display Screen Equipment (DSE) tasks. Exact measurements are dependent on the work task to be undertaken and the work population undertaking it.

When seated, if it is not possible to adjust the work surface height and instead the seat has to be raised, the user's feet still need to be supported to prevent strain on the legs and back. A footrest should be made available. Sufficient legroom must also be provided under the work surface, as follows:

- lateral legroom (width clearance) of 790 mm is recommended;
- depth clearance at the knee level of 400–460 mm and 1000 mm at foot level (to allow the legs to be stretched out);
- to aid with leg room, the thickness of the work surface should not exceed 30 mm.

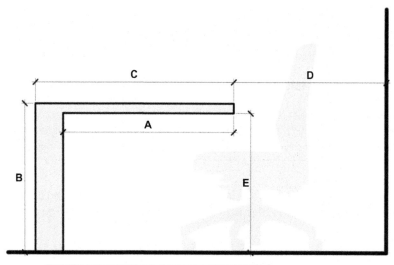

Key	Description	Guidance (mm)
A	Knee hole depth	460
B	Work surface height	720
C	Work surface depth	910
D	Work surface to wall	810
E	Knee hole height (Knee hole width not shown)	640

FIGURE 12.3

Guidance dimensions for seated workstations.

Created By BAW Architecture for the purpose of the Keil Centre book.

These dimensions allow the user to remain close to the task location and reduce the likelihood of trunk and neck bending.

SEATING DESIGN

The seat itself is an important component of any seated workstation, especially as many seated workers spend a high percentage of their working day seated. Seating to suit multiple users can present a challenge as different people find different attributes comfortable. When choosing new or replacement seating, a seating trial is always recommended to gain participation from the users and gather their opinions on certain features and design. Seating suppliers are usually willing to supply seating on approval to allow a variety of designs to undergo a seating trial. The following basic principles should be considered.

ADJUSTABILITY

Providing adjustability is a good human factors design principle and even when the workstation cannot be adjusted, the seating can. Multiple studies have shown that adjustable seating increases productivity. Dainnoff and Smith (1990) found a 17.5% productivity increase for people working in an ergonomically optimal setting compared to one which was ergonomically suboptimal. Harrison and Robinetter (2002) found a 17.7% productivity increase with a highly adjustable chair and ergonomics training. Shute and Starr (1984) demonstrated that seat adjustability reduces complaints of shoulder and back pain. However, adjustments need to be easy to use, easy to access, and obvious to the user. Users should be provided with written or pictorial information on how to adjust their seating and this information should also be incorporated into a relevant training program. A typical oversight is that users are not shown how to use seating properly which can undermine even the best designs on the market. They also need to be informed of why they need to make the adjustments.

SEAT HEIGHT

The seat height should be adjusted to reduce pressure on the underside of the thighs so allowing normal circulation to the legs. This means that the seat should be at or just lower than the distance between the floor and the underside of the thigh, referred to as the popliteal height. If the knee level is higher than the hip level, this can lead to a loss of the lumbar curve and cause back pain. Seating made available that is adjustable should accommodate the 5th percentile female to the 95th percentile male (this is not actually achievable in a single seat). The American National Standards Institute (ANSI, 2007) recommends a minimum seat height adjustability range of 405–520 mm. If the work surface height is not adjustable, a suitable footrest should be provided to ensure that the feet remain supported and the under surface of the thighs are not subjected to undue pressure. Seat height adjustment should be possible from the seated position.

BACKREST

The backrest should provide sufficient support especially to the lower back. The recommended adjustment range is between 150 mm and 300 mm above the seat height (Dul and Weerdmeester, 2008). Convex shaping to the lower section of the backrest will assist to maintain the lumbar curve when the user is seated. Many seat designs now also provide a lumbar pump which can increase the lumbar support as required. ANSI (2007) recommends a seat back angle with a minimum range of 90°–115° to the seat pan and a backrest width of at least 310 mm in the lumbar region.

SEAT PAN DEPTH AND WIDTH

For workstation seating, ANSI (2007) recommends a seat depth of 380–430 mm. A seat depth adjustment range is important to ensure that users have adequate thigh support whilst ensuring that the front edge of the seat does not meet the back of the lower legs. This can be assisted by the front edge of the seat having a "waterfall" contour, as shown in Fig. 12.4. Seat pan width should be a minimum of 450 mm with open sides between it and any armrests in situ (ANSI, 2007).

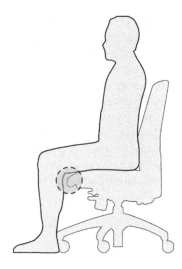

FIGURE 12.4

Waterfall edge contour.

Created By BAW Architecture for the purpose of the Keil Centre book.

ARMRESTS

The arms represent approximately 10% of our total body weight and can require muscular exertion of the upper back, shoulders, and neck. Supporting the weight of the arms reduces the stress on the spine. However, in order to work, the arm rests must fit the user and should be adequately padded. Armrests that do not adjust or produce contact stress in the vulnerable areas of the elbow and forearm can increase the risk of injury. Armrests need a considerable range of adjustability to be effective and avoid hindering the task. Armrests should be adjustable in height and have back to front adjustability as well. Non-adjustable armrests can prevent the user from attaining position close to the work surface. This may cause extended reaches or require the user to adopt a perched posture at the front edge of the seat. This is particularly common for individuals working in curved or corner desk/console configurations. Armrests may also be adjustable in width and pivot and this assists in accommodating the variation in hip width. It also allows smaller individuals to use the armrests without having to sit with their arms excessively abducted (away from the body).

MOBILITY

Swivel-action chairs provide flexibility when the user needs to conduct a variety of tasks and reduces torsional forces acting on the spine from twisting. Chairs with castors should not be able to slide away too easily when the user gets up or sits down and different types of castors are available for different floor types. Most new chairs are equipped with hard wheel castors and these should be used on carpet surfaces as they will slide too easily on hard floor surfaces. Hard floor surfaces (e.g., tile, wood, and cement) require soft wheel castors. Soft wheel castors are also quieter on hard floor surfaces. In certain situations glides are safer to use than castors. Glides enable the chair to slide but not roll. A foot replaces the castor to keep the chair stationary. Glides on seating are preferred for work requiring controlled hand manipulation or fine dexterity, for example, laboratory work or where the task requires the operation of a foot pedal where chairs with castors are impractical.

UPHOLSTERY

The seat and backrests should be padded to ensure that the body does not press on the frame of the chair. Density and thickness of padding affects the pressure distribution. Padding needs to be firm rather than soft and of good quality to ensure that the chair remains comfortable for a reasonable time. Lueder (1986) recommends a seat cushion thickness range from 40 to 50 mm.

Seating covers need to be nonslip, easy to clean, and of a fabric which "breathes" (permeable to moisture). Good quality and durable covers aid comfort and prolong the service life of the chair. Seating should meet BS 5852 (1990) or an equivalent standard to ensure that upholstery does not present an unacceptable fire risk.

STANDING WORKSTATIONS

A standing workstation is preferred when the task requires:

- significant amounts of lifting;
- force to be exerted;
- long horizontal reach distances (greater than 400 mm) required frequently;
- frequent walking or mobility;
- significant variety of tasks and postures.

It is not recommended that the user remains in a standing position for the duration of a shift. Standing for prolonged periods causes fatigue in the legs and back; if the head and trunk are bent it can lead to neck and back pain. Using the arms constantly or repeatedly in a raised position may also lead to discomfort in the shoulders. Tasks that require standing should be alternated with tasks that require sitting or walking and users should have the opportunity to sit down during breaks.

DIMENSIONS FOR STANDING WORKSTATIONS

Guidance for minimum and preferred dimensions for standing workstations including work benches are shown in Fig. 12.5 (summarized from FEA, 2012). Work surfaces should be in the region of 915 mm ± 15 mm from the floor (exact measurements are dependent on the work task to be undertaken and the work population) and at least 760 mm wide.

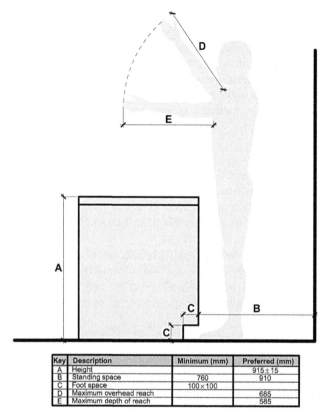

Key	Description	Minimum (mm)	Preferred (mm)
A	Height		915±15
B	Standing space	760	910
C	Foot space	100×100	
D	Maximum overhead reach		685
E	Maximum depth of reach		585

FIGURE 12.5

Guidance dimensions for standing workstations.

Created By BAW Architecture for the purpose of the Keil Centre book.

Sit/stand workstations can also be considered. For non-adjustable workstations, the work height should be suitable for standing work and a suitable chair provided. Sufficient leg room is required and a foot rest should be in situ for use when seated with a cut out to allow foot access when standing (see Fig. 12.6). The dimensions can be used for guidance but additional consideration should be given to the type of work being undertaken.

In recent years, adjustability has been incorporated into workstation design (see Fig. 12.7). Sit/stand height should be a range from 650–1300 mm. The control pad for motorized adjustment should be conveniently located at the front edge of the work surface and should not cause the operator to have to adopt awkward postures to access it. Manual adjustment systems, for example hand cranks, are also available but again should be easy to use and accessible.

Sufficient space should be available under the work surface for standing workstations to allow the user to get close to the task and also to ensure that they can change

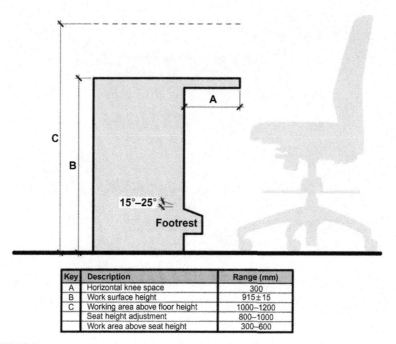

Key	Description	Range (mm)
A	Horizontal knee space	300
B	Work surface height	915±15
C	Working area above floor height	1000–1200
	Seat height adjustment	800–1000
	Work area above seat height	300–600

FIGURE 12.6

Nonadjustable sit/stand workstation dimensions.

Created By BAW Architecture for the purpose of the Keil Centre book.

position frequently. At work benches, for example, an open front design is preferable. Where this is not possible, a foot space cut out, sometimes known as a kick space, of at least 100 mm in height and 100 mm in depth should be provided at the base of the front surface.

VISUAL REQUIREMENTS

The primary visual zone is at or just below the horizontal line of sight and directly in front of the user to within 70° lateral position (35° each direction) (see Fig. 12.8).

The secondary zone is up to 40° above and 40° below the line of sight and ideally within the same lateral zone as primary displays. Displays should be in front of the user within 60° of the center line.

ADDITIONAL INFORMATION RELEVANT TO CONSOLE DESIGN

A console is defined as "*a group of controls and displays associated with one or more individual pieces of equipment or systems mounted together on a structure dedicated*

FIGURE 12.7

Sit/Stand workstation with adjustability (Honeywell's Experion Orion Console).

Created By BAW Architecture for the purpose of the Keil Centre book.

to the control and monitoring of those individual equipment or systems" (American Bureau of Shipping (ABS), 2013).

In some instances it may be necessary to design a console larger than the dimensions recommended earlier due to the need to accommodate more equipment.

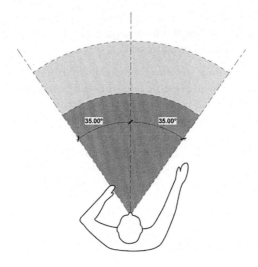

FIGURE 12.8

Primary viewing position.

Created By BAW Architecture for the purpose of the Keil Centre book.

A wraparound console may be considered to maintain consistent reading and visual distances. The right and left segments should be at an angle, measured from the frontal plane of the central segment, so that these segments can be reached by the 5th percentile operator. The total required left-to-right viewing angle should not exceed 190°. Frequently used controls should still be within a distance of 460 mm from the console's centerline and the less frequently used controls may be located within a distance of 800 mm from the console's centerline.

For standing and seated consoles where vision over the top is not required and lateral space is limited, the console height may be increased. In recent years, it has become common for the console to be made up of multiple, individually adjustable, display screens, often of varying sizes. The screen positions should meet the visual requirements and be user tested.

Users may need to see over the top of the console, for example, to see a wall-mounted visual display. In this case the vertical, lateral, and distance positioning of the wall display needs to be calculated in conjunction with the console dimension to enable the user to have an adequate posture and view. The equation for calculating this is defined in ISO 11064, Part 3 (2000).

VEHICLE WORKSTATIONS

Vehicular workstations include crane cabs as well as trucks, forklifts, and other forms of vehicle. These types of workstations share most of the same principles as any other workstation including the design need to enable reach, access, clearance, good posture, and effective workstation arrangement. However, there are some specific issues which may need to be addressed:

- the driving posture is typically more reclined and the body angles differ;
- the user uses foot and hand controls, some of which are in continuous use;
- the workspace is subject to movement and vibration;
- the user may be subject to additional noise, thermal comfort, and lighting issues;
- operating sight lines need to be accommodated;
- the vehicle may need to take account of providing a specific means of escape;
- maintenance interfaces and workstations may be less accessible (e.g., in the case of a crane cab).

PLANT SIDE WORK AREAS

The postures required of operators and maintainers in process plant areas may include seated and standing positions. However, they may also need to be in a variety of other postures such as kneeling or lying horizontal. The standards (such as ABS, 2013, and American Society for Testing and Materials (ASTM, 2013) include anthropometric data to derive workspace envelopes and access spaces for different types of posture.

Many of the principles relating to control room layout also apply to other types of wider workspaces such as supporting communication links, viewing arrangements, and access. There is also a need to maintain a logical layout (as far as possible) and consider traffic, work flow, and wayfinding. This enables the work area to be easy and quick to access and egress and reduces the potential for errors such as going to the wrong equipment (a common error). Overall building and control room design is discussed in Chapter 11, Building and Control Room Design. Application of these principles also allows for safer performance of tasks, such as materials handling and crane operations (for more detail refer to Chapter 15, Human Factors in Materials Handling). Detail regarding specific plant area design considerations can be found in Chapter 14, Plant and Equipment Design.

KEY POINTS

- Correctly selected anthropometric data assists in the design of work areas and workspaces to support neutral working postures and improve productivity. This helps to reduce error and injury.
- The principles of positioning should be applied to the functional arrangement of workspaces and work areas.
- The type of workspace will be dependent on the tasks to be undertaken including seated, standing, or sit/stand workspaces. These are equally relevant to plant side areas.

- For seated workstations, the type of seating provided should be carefully chosen relative to the tasks undertaken and with user input.
- The positioning of equipment needs to enable good viewing postures.
- Vehicles present some specific considerations.

ACKNOWLEDGMENT

Acknowledgment and gratitude are extended to BAW Architecture for the provision of figures and support in writing this chapter. The Keil Centre has worked with BAW Architecture on several control room projects where the console design has been within the project scope. Further information can be obtained at www.bawarchitecture.com.

REFERENCES

American Bureau of Shipping, 2013. The Application of Ergonomics to Marine Systems. ABS, Houston, TX.

American Society for Testing and Materials, 2013. Standard Practice for Human Engineering Design for Marine Systems, Equipment and Facilities (ASTM F 1166—2013). Author, West Conshohocken, PA.

American National Standards Institute, 2007. Human Factors Engineering of Computer Workstations. ANSI/HFES 100-2007, Washington DC.

BS 5852, 1990. Methods of Test for Assessment of the Ignitability of Upholstered Seating by Smouldering and Flaming Ignition Sources.

Chapanis, A., 1996. Human Factors in Systems Engineering. John Wiley and Sons, New York, NY.

Corlett, E.N., Wilson, J.R., 1995. Evaluation of Human Work, second ed. CRC Press, Boca Raton, FL.

Dainnoff, M.J., Smith, M.J., 1990. Ergonomic improvements in VDT workstations: health and performance effects Promoting Health and Productivity in the Computerized Office: Models of Successful Ergonomic Interventions. Taylor & Francis, London.

Dul, J., Weerdmeester, B., 2008. Ergonomics for Beginners. CRC Press, Boca Raton, FL.

Federal Aviation Administration (FEA), 2012. Human Factors Design Standards. FEA, Washington, DC.

Harrison, C.R., Robinetter, K.M., 2002. CAESAR: Summary Statistics for the Adult Population (Ages 18–65) of the United States of America. Air Force Research Laboratory, Wright-Patterson AFB, OH, (NTIS No. AFRL-HE-WP-TR-2002-0170).

ISO 11064. Ergonomic Design of Control Centres.Part 3, Control room layout, 2000.

Lueder, R., 1986. Workstation design. In: Lueder, R. (Ed.), The Ergonomics Pay-off: Designing the Electronic Office Holt, Rinehart and Winston, Toronto.

McCormick, E.J., 1970. Human Factors Engineering. McGraw-Hill, New York, NY.

Peebles, L., Norris, B.J., 1998. Adultdata—The Handbook of Adult Anthropometric and Strength Measurements. Department of Trade and Industry, London.

Pheasant, S., Haslegrave, C.M., 2006. Bodyspace, second ed. CRC Press, Boca Raton, FL.

Sanders, M.S., McCormick, E.J., 1992. Human Factors in Engineering and Design. McGraw-Hill, New York, NY.

Shute, S.J., Starr, S.J., 1984. Effects of adjustable furniture on VDT users. Human Factors 26 (2), 157–170.

Control system interface design

13

R. Scaife

LIST OF ABBREVIATIONS

ASM	Abnormal Situations Management
CRO	Control Room Operator
DCS	Distributed Control System
EEMUA	Engineering Equipment and Materials User Association
HCI	Human–Computer Interface
HMI	Human–Machine Interface
MoC	Management of Change
NUREG	Nuclear Regulatory Guide
P&ID	Piping and Instrumentation Diagram

The Human–Machine Interface (HMI) is a broad topic as it refers to any aspect of a system which the human operator or maintainer interacts with to perform their task. This includes physical interaction via hardwired control panels and input devices or via the Human–Computer Interface (HCI) which is the interface between a human and a programmable system. Computerized or hardwired interfaces are subsets of the HMI. Within a control room environment, the Control Room Operator (CRO) and other users interact with the equipment via information displays (such as the Distributed Control System (DCS)), controls (such as keyboards, mice, hardwired control panels, switches), and alarm systems displayed on separate alarm panels and the main DCS system.

Any HMI, regardless of its specific purpose, comprises both hardware and software components. When considering HMI design for a control room, it is necessary to consider both the physical characteristics of the hardware and the characteristics of the software interface. It is important that the interactions between the hardware and software aspects are designed to be intuitive and support the users' tasks. If the system is designed around the user, then the likelihood of errors is significantly lower. If the design is not intuitive the user has to adapt to the design, and this introduces opportunities for error as the user has to "unlearn" previously learned ways of working and adopt new ones. This issue is exacerbated for situations where two or more conceptually different systems are in use side by side, such as new and old systems.

This chapter focuses on the design and evaluation of the HMI, which is somewhat artificial. In a control room design project one would consider not just the HMI but also how the various interfaces between the human and the system interact with each other. For example, consideration would be given to the relative positioning of interfaces in the control room, equipment positioning in relation to the console or other control room furniture, and the interoperability of different HMIs. Other chapters in this book cover other issues that should be considered in conjunction with the HMI (including Chapter 11, Building and Control Room Design, Chapter 12, Workstation, Work Area, and Console Design, and Chapter 16, Environmental Ergonomics).

The key principles for the design and evaluation of HMIs used in process control and monitoring are discussed in this chapter. More detailed guidance on specific design considerations can be found in standards and guidelines on HMI design such as Nuclear Regulatory Guide (NUREG) 0700 (2002), Abnormal Situations Management (ASM) Consortium guidelines (2008), and Engineering Equipment and Materials User Association (EEMUA) 191 (2013).

AN INTERDISCIPLINARY EFFORT

Designing the HMI is an interdisciplinary design task. The designer must understand the needs of the user (in all modes of operation and maintenance) and design the displays and controls to meet these needs. Once the prototype system has been developed it is also necessary to test it to make sure that all user requirements have been met, and that the system allows for safe and effective operation of the process.

The disciplines involved in HMI design often work in isolation, but significant benefits can be derived from developing the HMI as an interdisciplinary team. Much research and practical effort has been applied to better integrate human factors in projects (see Chapter 10: Human Factors Integration within Design/Engineering Programs). The major benefits have been demonstrated in many case studies to have a significant positive impact on project cost and solution effectiveness.

COMPONENTS OF AN HMI

There are three main components of an HMI: information displays, controls, and management of the interaction.

The way that information is presented to the user primarily requires consideration of the visual, auditory, and tactile elements of the display to ensure that the design is matched to the capabilities of the user. This needs to include how the person detects information on the display, and how they interpret and use the information. For example, when considering visual displays, there is a need to assess size, layout, coding, and grouping of information. When considering auditory or tactile displays, it is also necessary to consider the coding and delivery of this information so that the user can detect and make sense of it.

The controls used to manipulate the system include hardware controls such as push buttons, switches and rotary controls, and software controls, such as menus, drop-down lists, and radio buttons that allow the operator to control the system. The focus also needs to include how the displays and controls integrate with each other to form the complete interface.

The management of interactions with the system relates to the switching between operational, training, and maintenance modes of operation, the detection and prevention of human errors, and the maintenance of system integrity and information security.

INFORMATION DISPLAY

The display of information to the user is essential to the design of any human-centered system. Several principles are fundamental to optimize user performance and reduce error. These include comprehensive derivation of user requirements, attention to the interface layout, and effective alarm management. These issues are discussed in the following subsections.

INFORMATION REQUIREMENTS

Prior to any design work it is essential to fully understand the users' needs for effective control and monitoring of a plant, that is, the user requirements. These are derived by analyzing the users' tasks using task analysis techniques. For information display design, the task analysis is used to identify the information requirements to conduct the task and how the interaction needs to be structured. The user requirements are also derived from human factors standards and guidelines.

End user participation in the analysis is essential. Even an experienced design engineer and human factors specialist cannot replicate the knowledge of an experienced user. For modifications to existing systems it may be possible to undertake a walk-through of the task to identify the information used in normal operations and other modes of operation, including upset and emergency conditions. Other methods should also be used, such as talk-through or simulation, particularly for variations to the existing system and new systems.

Information requirements are typically developed iteratively as the project progresses. In the early stages of a project high-level requirements can be defined with the users using walk-through and talk-through techniques. As the design develops, prototypes can be used to gather more detailed requirements. For example, the human factors team typically works with experienced operators to develop scenarios which can be used to inform the interface design. At key stages of the project, users test the prototype to conduct the tasks required of the scenarios and provide feedback on the usability of the prototype. This results in both familiarization with the system as it is developed for the users, but primarily clarifies the information that users need to complete the task and the form in which it needs to be presented. Prototypes can take many forms from simple two-dimensional screen sketches to high-fidelity simulations. The benefit in using this approach, even with the simplest prototypes is to identify critical issues before expensive design time has been spent and when changes become more difficult and costly. It is not appropriate to leave the HMI to the last stages of design. The author has experienced several projects where the HMI is seen as a finishing touch and not in need of attention until late in the project. Human factors input in the later stages of HMI design cannot be effective and can become an exercise similar to "flower arranging"—trying to make the best of the visual design but being unable to influence some of the underlying design assumptions. By the late stages of design, "human error" can already be "locked in to the design," and it is therefore much more effective for human factors expertise to be integrated from concept design and continue through the life cycle.

Information for process operations and plant is presented in the following ways:

- Schematic displays that represent the process configuration (typically, based on Piping and Instrumentation Diagrams (P&IDs));
- Textual displays with information for the user to read;
- Graphs showing information as a histogram or other form of graphical representation;
- Auditory information (such as sounds, speech);
- Plans of the plant (such as showing locations of fire and gas detection systems);
- Flowcharts.

Once the options for the display type have been identified, they need to be informed by and assessed against recognized human factors and ergonomics design standards and guidelines. Specialist knowledge and experience is needed at this stage as standards and guidelines represent best practice, and often there is a need to compromise different criteria to arrive at a pragmatic and optimal solution.

It is also desirable to involve the end users, especially when more than one design option meets best practice. A pragmatic approach often has to be applied to this too. Access to end users is often restricted due to operational demands. Every effort should be made to maximize end user involvement, not just during definition of the user requirements but throughout the design of the interface.

LAYOUT OF DISPLAYED INFORMATION

With the information requirements specified and the most appropriate means of presenting the information defined, the detailed design of the user interface can be undertaken.

Several key principles apply throughout the detailed design of the interface, but those of key importance are:

- *Consistency of information*—in terms of layout, coding, and methods of interaction with the system;
- *Grouping of information*—in a way that is logical to the users for example by process plant unit, priority of information, or function of the system;

- *Simplicity of information*—in displaying this to the user, keeping clutter to a minimum and ensuring that the required information is easy to see and understand.

Laying out visual information in a control system usually begins with deciding the overall structure of information represented on the system. This is defined as the display hierarchy which should be limited to four levels of detail to allow representation of the overall plant through to the detailed aspects of individual selected pieces of equipment such as pumps, valves, sensors. See Fig. 13.1 for an example.

Establishing a system hierarchy helps the user to navigate and understand the system, but to be effective the hierarchy must be applied consistently throughout the user interface.

Within this structure, the individual displays are constructed to represent the information layout for the overview screen, and the screens further down the hierarchy.

In order to ensure consistency between the screen designs it is advisable to produce a style guide informed by human factors and ergonomics standards, guidelines, and best practice. This should cover the template for the arrangement of information on the screen, and the detailed style guidance for font, colors, abbreviations, and other aspects of the fine detail.

Each window or screen within the overall system hierarchy can then be designed in line with best practice guidance and standards. The key principles continue to apply at this stage and any deviations from best practice due to operational or system constraints need to be assessed and documented.

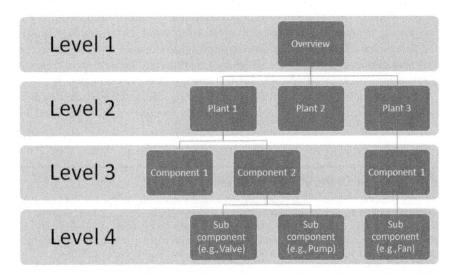

FIGURE 13.1

Generic HMI display hierarchy.

Consistency is aided by the use of a style guide, ensuring that a standard layout for each screen in the interface is achieved, including standard positions for any software controls such as command buttons, radio buttons, and checkboxes. Consistency for navigation controls is also important and needs to be replicated on each screen, including direct access to specific locations, such as the overview screen. Consistency within the interface enables the user to anticipate and know where to find the desired information with the minimal effort so that it is intuitive and obvious.

Grouping of information needs to be considered carefully, and where possible applied consistently. Guidance on the best way to group the information on a screen should be informed by the task analyses undertaken to derive the information requirements. This indicates which pieces of information need to be used together and which information is most important to the user. Groupings should be logical to the user population; it is always better to design around the user than to design something that requires the user to adapt. It is sometimes necessary to be pragmatic when applying this principle, for example, when users have been working with a badly designed interface for a long time and the design has been changed to be more logical. Errors may be introduced due to the new interface being different from the one that they have been used to. It is therefore necessary for human factors input to consider the long-term benefits of the change, including use by newly trained users and to consider the trade-offs against short-term adaptation challenges for existing users.

The grouping of information needs to be balanced against the principle of simplicity. Users will perform less efficiently if the amount of information displayed on a single screen is overwhelming or difficult to assimilate. So although it may be desirable to group all of the information relating to one plant component on a single screen, if this creates a highly complex set of information for the user to read, or is too cluttered, then it may be better to consider options for splitting information between screens. This requires careful consideration of information that needs to be referred to together so it is displayed on the same screen. For example, it may be possible to subdivide a gas turbine into compressor, combustor, and turbine screens rather than display everything on a single screen whilst still keeping an overview of the entire unit visible to the operator the next level up in the system hierarchy.

A good example of how the principle of *simplicity* has influenced the design of control system displays comes from the aviation industry and is explained in Box 13.1.

BOX 13.1 THE PRINCIPLE OF THE "DARK COCKPIT"

Displays in aircraft cockpits have for some years adhered to a philosophy that the pilot only needs to see their cockpit displays lighting up when there is something they need to deal with, otherwise too much information can act as a distraction. Therefore, in the "Dark Cockpit" philosophy, when everything is normal, the amount of information displayed to the pilot is minimal and in modern aircraft it is tailored to the phase of flight that they are in. So, for example, during a landing sequence the display would show the information required to land the aircraft only, unless another system develops a fault.

The "dark cockpit" principle has been applied at a simpler level within the process industries. For example, the ASM guidelines (2008) describe systems which minimize the use of color during normal operations. When there is a problem, color is then used to draw the operator's attention to pertinent information. The days of black background display screens with brightly colored lines showing lines carrying different materials seem to be largely a thing of the past.

Comprehensive and detailed HMI guidance on the design of information displays is covered in a wide range of design standards developed within industries ranging from nuclear power generation to defense. These include a lot of detail regarding how to ensure that information displayed on a screen will be readable by the operator. This takes account of specific issues, such as the position of the operator at a specific distance from the screen, the use of color coding, flash coding, standard symbology sets, and many other specific considerations. Some useful examples of design standards and guidelines are presented in the chapter references section. For the display of information via an HMI, the most relevant standards and guidelines are the ASM guidelines and NUREG 0700. There is also value in looking towards other industries where interface usability has been embedded as a central component of software design for a long time. The obvious example is consumer products, and some excellent examples of intuitive and effective design can be seen with the products of Apple Inc. There are of course great examples of poor user interfaces within the consumer product industry as well.

ALARM MANAGEMENT

Alarm management is included in this section as a special form of information display. An alarm is defined as a system indication that requires the user to take action. Some systems used in control rooms also have "alerts" which are operator assistance tools to tell them when a certain condition has been met. For example, an alert might be set by an operator to indicate when the level in a tank begins to increase during a product transfer. Such an alert does not mimic an alarm and should be displayed somewhere on the interface separate to the alarm list to avoid confusion.

It is difficult to design alarms that result in the operator taking timely and effective action during an abnormal situation. History clearly shows that getting alarm design and alarm management wrong can have disastrous consequences (e.g., Three Mile Island, 1979, and Texas City, 2005).

EEMUA presents a guide to design, management, and procurement of alarm systems (EEMUA 191, 2013). It states that alarms with the following characteristics are found in well-designed alarm systems:

- Relevant—not spurious or of low operational value;
- Unique—not duplicating another alarm;
- Timely—not too long before any response is required or too late to do anything;
- Prioritized—indicating the importance that the operator deals with the problem;
- Understandable—having a message which is clear and easy to understand;

- Diagnostic—identifying the problem that has occurred;
- Advisory—indicative of the action to be taken;
- Focusing—drawing attention to the most important issues.

In this section the general best practice principles of alarm design are discussed. For more detailed guidance the reader should refer to standards and guidelines for visual displays and alarm design such as EEMUA 191 and others listed in the references section of this chapter.

It is easy during the design of alarm systems to be driven by the need to deliver alarms to the operator based upon the design of the sensor systems. While sensor information is critical to the safe operation of the plant, good design should not simply ensure that all alarm conditions are delivered to the operator but should investigate *the best way* to present this information to the user and which user they should be presented to. Otherwise, there is a high risk of cognitive overload when plant function deteriorates.

Stanton (1994) developed a useful model of alarm initiated activities describing how operators manage information presented by an alarm system (see Fig. 13.2).

This provides a useful insight into human interaction when considering the design of an alarm system:

- *Observe*: This involves the detection of the abnormal condition. Designers of alarm systems need to consider how their design optimizes this process, including the use of color-coding and flash-coding to enhance search and monitoring of the system by the operator. Excessive use of color and flashing of information make new alarms more difficult to detect; auditory information, for example, aligned with the frequency and priority of alarms. If alarm sounds become annoying or frustrating the alarm management task can transform into one of trying to keep the alarm system quiet, which is not helpful.

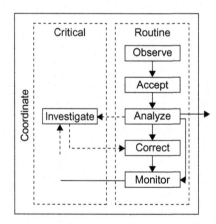

FIGURE 13.2

Model of alarm initiated activities (Stanton, 1994).

From Stanton, N. (1994). Alarm Initiated Activities. In N. Stanton (Ed.), Human Factors in Alarm Design. Taylor and Francis: London, pp. 93–118.

- *Accept*: This refers to the operator indicating to the system that he or she is aware of the alarm that has been presented. A number of considerations are involved here. Acknowledging alarms add to workload. One principle that can be applied is whether or not to require individual or group acknowledgment of alarms. There is a long-standing debate about the benefits and risks of group acknowledgment whilst it benefits control of workload, it risks missing single critical alarms and the latter are usually considered more important.
- *Analysis*: This refers to the process of analyzing the priority of the alarm. This is separate to the information delivered by the alarm as the operator needs to make judgments about what to do with the alarm. For example, if faced with five high-priority alarms the operator would need to decide if any of them require immediate action, or if they require monitoring or further investigation. Good design of alarm lists helps; options for the designer include grouping of alarms in priority order or using separate alarm lists for different levels of severity. Care needs to be taken if using separate alarm lists, to ensure that the number of lists does not impair performance but instead supports decision-making around what action is required. Design of the alarm name and associated help text is important to provide guidance to the operator.
- *Investigation*: This refers to the process of identifying the underlying cause of the alarm. The alarm system and associated systems need to assist the investigation process by ensuring information is provided at a usable rate and that the data being presented is current (e.g., trend displays need to show the most up-to-date data with any time lag minimized). Other design solutions may include schematic diagrams for subsystems on the plant, and effective use of color-coding that does not detract from other uses of the system.
- *Correction*: This refers to the use of the results of the investigation phase to address the cause of the alarm. Where faults can be fixed directly from the HMI the interface needs to support the best means of doing this. Options include command line entry (in which case help may be necessary for command syntax, and error-trapping should also be built in), or a graphical interface for actions that can be taken such as command buttons to reboot systems and sensors.
- *Monitoring*: This stage of the process involves monitoring the results of the corrective action(s). The design of the system needs to reduce the memory demands on the operator. For example, a system that requires the user to remember to keep switching pages to see certain information is less effective than a system that allows a window to be opened to monitor a specific part of the system to check that the fault has been corrected.

Over time, and with changes to plant and sensor systems, alarms need to be modified. The control system software should automatically save a log of all alarms that are generated so that they can be audited at any time. Ongoing alarm rationalization uses this log to identify nuisance alarms and alarms that are obsolete or require no user intervention. Obsolete alarms should be removed and various options identified for dealing with nuisance alarms. These may include modifying the set point or not raising an alarm until the alarm condition has been detected on two or more sensors.

CONTROLS

Controls provide the means for the operator to interact with the system. They can be hardwired controls such as keyboard, mouse, switches, and rotary control knobs as well as software controls such as drop-down lists and dialogue boxes. In most modern control rooms, a significant amount of the interaction is via software controls combined with hardwired input devices. However, there is also still a need to use hardwired controls, typically housed in control panels such as push-buttons and switches in order to have an independent means of controlling critical plant parameters in the event of a software failure (such as emergency shutdown). In considering interface design for control systems it is necessary to consider both classes of control.

Some of the key principles for interface design that apply to controls are:

- *Compatibility with the task*—ensuring that the control used is well-suited to the type of changes that the operator needs to make to the system.
- *Compatibility with expectations and mental models*—ensuring that the control is compatible with the user's model of how the control is expected to work, for example an "up" arrow means increase, a rocker switch is on when it is pressed down or to the right. There are population stereotypes that drive the users' expectations and not all populations are the same. So in the United States, for example, pressing a rocker switch up provides the expectation that it will turn the function on.
- *Consistency*—ensuring that the controls used are consistent with the expectations of the user and other controls used for similar tasks within the control room. This does not mean that only one type of control should be used as there are benefits to be gained from using different types of controls for different functions. This provides redundant cues to the user that the functions are different.
- *Grouping*—ensuring that controls are grouped together, typically by function, and with the displays that they affect. This includes grouping of groups of controls relating to the same part of the system.

COMPATIBILITY WITH THE TASK

The task analyses conducted in the early stages of the design life cycle provide the basis for considering the compatibility of control devices with the task. A task that requires the user to enter commands and numerical data (e.g., valve percentage open) requires an alphanumeric keyboard as it will be the most effective means of completing such a task. The physical environment at the workstation also needs to be considered to avoid difficulties in operation (Box 13.2).

COMPATIBILITY WITH DISPLAYS

Error rates tend to increase when the design of controls on associated equipment does not match the user's expectations. This principle applies to both hardwired and

> **BOX 13.2 AN EXAMPLE OF THE IMPACT OF WORKSTATION DESIGN ON THE CHOICE OF CONTROL DEVICES**
>
> A modern control system for an on-board military aviation sensor system was replacing an older system. The software was mouse-driven and desk space on the workstation was minimal. A mouse would have required the user to constantly lift and reposition the mouse, which would have become frustrating and inefficient. Poor flying conditions would also likely have caused the mouse to move unintentionally. So the design was changed to accommodate a track ball which required a smaller footprint to operate and was not susceptible to the same issues as the mouse.

software controls. Displays also have population stereotypes relating to how end users expect the display to move. It is important to ensure that the movement of controls and displays is compatible with each other. So turning a rotary knob clockwise creates an expectation that a value will increase. The display also needs to move in the correct direction, in this case upwards, to be compatible with the control movement. Failure to recognize the requirement for control/display compatibility can cause confusion and induce human error.

Early research into control and display relationships (e.g., Fitts and Seeger, 1953) indicated that when the control and the associated display have something in common, then the effectiveness and accuracy of human performance using the system are both improved. This is based on a theory of information processing known as Stimulus-Response (S-R) compatibility. A common example of what happens when a stimulus and a response are not compatible with each other is the Stroop Test (Stroop, 1935). The person taking the test is presented with a series of names of colors but the names are written in different colored inks. The object of the test is to state the color of the writing. When the name of the color and the color of the ink are the same, reaction times are faster (e.g., "BLACK" written in black ink). If the person taking the test sees the word "BLACK" written in red ink, the reaction time tends to be slower. This is because the stimulus and the response required interfere with each other, so they are not compatible.

When this theory is applied to the design of controls the same effect is seen. If the stimulus (the display) and the response (the action of the control) have something in common, then performance improves. An example involving a rotary control is presented in Fig. 13.3.

In each case the control and display have something in common: the movement of the scale pointer is compatible with the movement of the control. If the rotary controls worked in the opposite direction, users commonly make errors in adjusting the scale.

CONSISTENCY

Consistency of the location of controls in relation to associated information displays is important both in hardware and software controls. In Fig. 13.4 the two vertical scales have the rotary control on opposite sides, so if they were together on the same

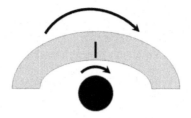

Rotate right, pointer moves right
on scale

FIGURE 13.3

S-R compatibility of controls and horizontal displays.

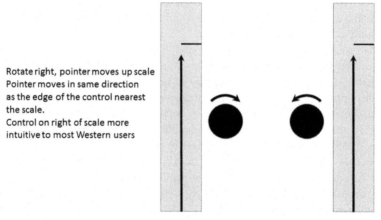

Rotate right, pointer moves up scale
Pointer moves in same direction
as the edge of the control nearest
the scale.
Control on right of scale more
intuitive to most Western users

FIGURE 13.4

S-R compatibility of controls and vertical displays.

workstation it is likely that the user would make a mistake selecting the control for the system they want to manipulate.

The rationale for the arrangement of controls on a software interface or a hard-wired control panel can be based on priority (importance and/or frequency of use), sequence (the order in which they are used in a task), function, or spatial location (such as geographical location). There is no single best way of grouping controls, the choice is based on the task analysis and the user population to ensure that it is logical and avoids errors. Whatever the rationale, consistency needs to be applied so that the same grouping strategy is used between control panels and their respective screens in the user interface.

Consistency also needs to be applied to controls used for the same function in different parts of the interface. For example, if a rotary dial is used to increase temperature on one panel in the control room, then the same type of control should be used for the same hardwired function elsewhere. This also applies to software controls such as radio buttons, drop-down lists, command buttons, and checkboxes.

MANAGEMENT OF INTERACTION

The management of interactions in process control systems is a shared activity between the operator and the system. The operator typically needs to arrange information in the most effective way for task completion, and the system should be designed to support this. The design solutions range from highly complex (e.g., displaying only the information required for a specific type of task such as start-up or shut-down) to simple (e.g., providing sufficient display screens and flexibility to rearrange data in multiple ways).

Navigating through the system to allow system reconfiguration needs to adhere to specific key principles:

- *Minimize the memory load on the operator*: For example, do not expect the operator to have to remember a series of abbreviations for the different screens that they can access.
- *Make navigation through the system intuitive*: For example, provide command buttons to take the user from the current part of the process to the next or previous stage in the process.
- *Provide flexibility*: For example, by combining the previous two points, allow the user to take shortcuts from the current screen to other parts of the system. Being able to call up the overview of the process from a single command button is desirable, as is providing a command line interface for users to be able to skip to a specific part of the process.
- *Ease of change*: Users may need to reconfigure the layout of information rapidly and the system should support this. In windows-based systems this can be undertaken via direct manipulation of the windows between screens. However, care needs to be taken that important windows (such as the alarm list) cannot be accidentally hidden. Solutions include a dedicated position for important information or software that does not allow such information to be overlaid by another window. Preset configurations are another solution where the user can select a layout from a menu depending on the task, or series of linked tasks being performed. User preferences can also be accommodated within controlled boundaries, but care needs to be taken that this does not mean that following handover there is an additional task required to reconfigure the displays. Within windows-based systems preferences can be stored according to login name to avoid this issue. This does potentially present issues if the control screens are shared between operators or different roles at the same time, and so the degree of flexibility does need to be traded off against the principle of consistency.

The process control system is the operator's window on the system and with any interaction there is a need to retain situational awareness. This is a function of the degree to which the operator is immersed within the system, so that if the operator becomes a "machine minder," the risk is that he or she loses awareness of what is happening with the system. There is a balance to be gained from careful allocation of functions between the operator and the process system. Traditionally, this has been a fixed allocation, for example, specific tasks are performed by the process system and others are performed by the operator. As technology becomes more sophisticated, there is the option to consider dynamic functional allocation, such that in, times of high workload, specific tasks can be reassigned to the control system to enable the operator to focus on the most important task in hand. The dynamic allocation could be manual, such as engaging auto-pilot, or automatic, based on operator or process parameters.

MANAGEMENT OF CHANGE

As with any project, changes to the HMI need to be managed. Most users should not have the ability to make changes to system parameters such as alarm set points for example. Control against such changes is normally managed by having administrator or engineer user accounts separate to the operator accounts.

Major changes to the HMI design need to be covered by a formal Management of Change (MoC) process, where they are risk-assessed for their potential to affect operator performance. Due to the nature of an HMI, some changes are easily identified as major changes (such as adding or removing display screens or graphics). Other changes may seem minor, but in terms of the HMI design they are major changes (such as changing the color coding). Care needs to be taken to ensure that it is understood what type of HMI changes need to be the subject of formal MoC.

MoC for elements of an HMI requires human factors analysis of the change as well as engineering appraisal of the proposed change. Considerations include the extent to which the proposed change is likely to affect user performance, including potential errors. Depending on the scope of the change this may require formal human reliability assessment (see chapter: Human Factors in Risk Management). Human factors advice is best sought when deciding whether or not formal analysis is required.

ERROR TRAPPING

The system should be designed to detect operator errors and when appropriate, to check with the operator that their commanded setting was intentional. Designing an error trapping algorithm requires consideration of the nature of the task and how the operator is interacting with the system. Potential errors need to be analyzed to enable error trapping functions to be determined (Box 13.3).

Numerical data entry fields typically require error trapping. The potential to accidentally press the "0" key twice and enter 100 instead of 10 can have significant consequences if the parameter being operated is critical to the process.

BOX 13.3 AN EXAMPLE OF ERROR TRAPPING

If a valve is to be gradually opened or closed the HMI typically presents the operator with "up" and "down" arrows to increase and decrease the percentage open. Error trapping in this instance would be implemented by detecting significant changes in the commanded position, so if the operator repeatedly pressed the "up" button they would be asked to confirm that they wanted the valve to go from 5% open to 50% open. For small changes (e.g., from 5% to 10%) confirmation would not normally be requested. However, there are exceptions. In offshore applications where valves control oil wells under high pressure for example, a change of a few percent on a valve position could have a significant impact on the system, and so smaller changes may also result in requests for confirmation.

However implemented, error checking needs to avoid being seen as a nuisance otherwise users could become desensitized and it will become habit to simply click to indicate confirmation to every system confirmation request. Error trapping should also not slow the operator down in case of emergency response. For example, slamming a critical valve shut may result in a confirmation being required, but this confirmation should not increase the time required for the valve to close.

Where data are invalid (outside the possible range or incorrect format) the system should be designed to detect this. Good feedback to the user is required indicating why their input has not been accepted.

DATA SECURITY

The system should be designed to maintain the integrity of data stored on it, and preserve critical data in the event of system failure or human error. For example, certain critical data such as set points for critical systems should be protected from being altered by any user.

Dual-redundant data storage can also be used for the storage of critical data to protect against a hardware or software failure. The architecture of such a system needs to take into account the system recovery time necessary to maintain safety and the impact on system usability.

MODES OF OPERATION

Many modern HMI systems are designed for several different modes of operation. They can be used to operate the plant, provide training, or for maintenance activities. The current mode of operation needs to be clear to the operator as their ability to control the plant will be affected. Making the interface design significantly different under different modes of operation helps to distinguish modes. For example, a different screen border color can be used to indicate the current system mode.

THE INTERFACE AS A CONTROL ROOM COMPONENT

The HMI is an integral aspect of the whole control operation, and should be considered in conjunction with other aspects of the control room design. For example, the design of the control room console affects the workspace for peripherals, the viewing distances and visual angles, as well as the ability for operators to share critical information.

SUMMARY: CONTROL SYSTEM INTERFACE DESIGN

This chapter has focused on the key principles of HMI design rather than providing the level of detail included in HMI design standards and guidance. However, the interface is an integral part of the control room design and the detail needs to be considered. For example, by determining the viewing distance, it is possible to calculate character size on the screen.

An HMI design that is designed in accordance with best practice interface design principles may still have usability issues if it is not considered as a control room component and designed in isolation. The final key principle is to ensure that all aspects of the control room are considered together in terms of human factors design.

The next chapter "Plant and Equipment Design" discusses some of the other ergonomic design principles which should be taken into account alongside the information discussed in this chapter for projects such as control room design or modification to an existing control room.

KEY POINTS

- Human factors in HMI design is best achieved by being an integral part of the project rather than a separate activity.
- Key human factors design principles need to be applied to the key components of the HMI: the information display, controls, and the management of interactions with the system.
- HMI design is about the display and control of information by the user, and it needs to be approached with a clear understanding of the information requirements of the end user.
- Considerations such as consistency, grouping, and simplicity of information and the way it is controlled by the user need to be made throughout the design life cycle.
- Alarm management should be designed to support the user's task of observing, accepting, analyzing, investigating, correcting, and monitoring the system.
- Controls need to be designed to be compatible with the task, but also with the displays associated with them. They need to do what the user expects them to

do, and operate in a consistent way with other controls that form part of the interface.

- HMI design is not effectively achieved if it does not take into account other human factors design related to the physical arrangement of the control room and control workstations.

REFERENCES

ASM, 2008. ASM Consortium Guidelines Effective Operator Display Design.

EEMUA, 2013. Alarm Systems: A Guide to Design, Management and Procurement. EEMUA Publication No. 191, second ed. The Engineering Equipment and Materials Users Association, London.

Fitts, P.M., Seeger, C.M., 1953. S-R compatibility: spatial characteristics of stimulus and response codes. J. Exp. Psychol. 46 (3), 199–210.

NUREG, (2002). Human-System Interface Design Review Guidelines. U.S. Nuclear Regulatory Commission, NUREG 0700.

Shorrock, S.T., Scaife, R., 2001. Evaluation of an alarm management system for an ATC centre. In: Harris, D. (Ed.), Engineering Psychology and Cognitive Ergonomics: Volume Five—Aerospace and Transportation Systems Ashgate, Aldershot.

Shorrock, S.T., Scaife, R., Cousins, A., 2001. Model-based principles for human-centred alarm systems from theory and practice. In: Johnson, C.W. (Ed.), 21st European Conference on Human Decision Making and Control.

Stanton, N., 1994. Alarm initiated activities. In: Stanton, N. (Ed.), Human Factors in Alarm Design Taylor & Francis, London, pp. 93–118.

Stroop, J.R., 1935. Studies of interference in serial verbal reactions. J. Exp. Psychol. 18, 643–662.

Plant and equipment design

14

C. Munro and J. Edmonds

LIST OF ABBREVIATIONS

ABS	American Bureau of Shipping
ASTM	American Society for Testing and Materials
BDV	Blow Down Valve
C&I	Controls and Instrumentation
HFE	Human Factors Engineering
HMI	Human–Machine Interface
HRA	Human Reliability Assessment
ICC	Isolation Control Certification
LOTO	Lock Out Tag Out
N	Newton
NORSOK	Norsk Sokkels Konkuranseposisjon
NUREG	US Nuclear Regulatory Commission
OGP	International Oil and Gas Producers
P&ID	Piping and Instrumentation Diagram
VCA	Valve Criticality Analysis

The scope of Human Factors Engineering (HFE) was introduced in Chapter 9, Overview of Human Factors Engineering, and a number of HFE areas are described in more detail within Chapter 12, Workstation, Work Area, and Console Design, Chapter 13, Control System Interface Design, and Chapter 16, Environmental Ergonomics. The importance of considering the organizational design during the design stages was raised in Chapter 10, Human Factors Integration within Design/ Engineering Programs, and Section IV, Understanding and Improving Organizational Performance, presents the detailed considerations for relevant "organizational" topics. This section specifically focuses on the application of HFE to plant and equipment design. Although relevant to plant side activities, a separate chapter is presented on materials handling (see Chapter 15: Human Factors in Materials Handling).

The primary end users of process plant interfaces (other than the digital process control interfaces) are the plant operators and maintainers (including vendor specialists), and there are subdivisions within these groups such as for maintenance: electrical, instrument, and mechanical maintainers. The users have distinctly different tasks from each other but they may interact with the same or different aspects of the plant and equipment. The requirements for these interactions need to be understood before applying the relevant human factors criteria to the design of these interfaces. There will inevitably be design trade-offs, but the closer the compliance with HFE requirements, the less likely that the design will induce error, task difficulty, or other risks.

The chapter outlines key plant side roles and discusses specific design issues, including instrumentation design, valve design, access and clearance, design for maintenance, and equipment positioning and layout.

The chapter provides an overview of these topic areas to illustrate relevant considerations during design. However, the reader is directed to HFE standards for detailed information about these topics.

PLANT SIDE ROLES AND ACTIVITIES

There are many activities that are performed in external plant areas, including materials handling, facilities management, possibly helideck operations, fire and emergency response, and security. However, this chapter specifically focuses on the roles related to the operation and maintenance of the plant side process-related equipment. The key roles are described in Table 14.1.

The human interactions with the plant equipment may be simple one-to-one relationships, such as a field operator taking an instrument reading, or a C&I maintenance technician calibrating that same instrument. They both interact with the same equipment, but in a different way.

There is a need to understand the different roles and activities performed on the same item of equipment if the equipment is to meet the needs of these different roles. These may differ by operational state, such as normal (steady state), plant upset, start-up, emergency, or maintenance shutdown. It is not acceptable to perfectly position and design the display of the level gauge in one operational state for one role and then make it near impossible to access it for use by a different role or in a different operational state. It is often the case that it is the maintenance tasks, in particular, that are insufficiently considered during design. It can also be argued that installation and commissioning tasks should also be included. Information therefore needs to be gathered for all end users in support of the design of the plant equipment.

A key method for gathering and understanding these requirements is by using the techniques known collectively as task analysis, although other methods may also be useful to help the understanding of the concept of use. Human Reliability Assessment (HRA) can be used to analyze the vulnerabilities of the equipment to human failure, for both operations and maintenance. Errors relating to maintenance are often overlooked but they present good examples of latent failures where the consequences are realized in a different time and/or different space.

Table 14.1 Typical Plant Side Process-Related Roles and Responsibilities

Typical Roles	Typical Team Allocation	Typical Responsibilities
Field operator or plant operator	Operations	The field/plant operator is responsible for monitoring the plant, performing line walks to ensure steady-state functioning, sampling, manual valve line ups, checking instrument readings, manually adding chemicals to the process, and undertaking preparations for maintenance. It may include the processing of samples in a laboratory. The operations supervisor oversees these activities.
Maintenance supervisor	Maintenance	The maintenance supervisor supervises and organizes plant maintenance. Maintenance includes planned preventative maintenance, inspections, tests, and repairs.
Maintenance technicians	Maintenance	Maintenance technicians directly perform the maintenance on plant equipment including: • Mechanical maintenance, such as detailed inspection and/or the repair or maintenance of pumps, motors, valves, pipework, and vessels. This includes equipment or components being removed and replaced or overhauled and can often include mechanical handling of heavy items. • Electrical maintenance includes the maintenance of switchboards, generators, switches, function testing of motors, and other equipment, integrity checks, power sockets and supplies, appliance testing, and lighting. • Controls and Instrumentation (C&I) maintenance includes the inspection, test, repair, and calibration of field instruments and controllers, and maintaining control loops and control system administration.
Specialist contractors	Maintenance	Specialist contractors often perform specific technical maintenance and are included within the work team at relevant times. Examples include vessel entry and inspection, and maintenance of nucleonics.
Permit coordinator	Maintenance	The permit coordinator manages the permits for plant side activities. This role is generally not plant side, but plant side activity may be undertaken as part of the permit process.

INSTRUMENTATION DESIGN

Operating and maintaining the process facility requires interaction between human and machine elements via the Human–Machine Interface (HMI). The HMI is the window of the system (or equipment) and enables the end user to determine what the system is doing and provides the means to make changes to control its operation. Plant side process-related HMIs vary from simple process variable gauges to more complex local control panels. (Process control system HMIs are discussed separately in Chapter 13: Control System Interface Design.)

The design of displays and controls needs to be commensurate with human capabilities and limitations. Poor design of the displays and/or controls can lead to inaccurate perception of information, incorrect decisions, initiating the wrong control, or not using the control in the manner for which it was intended. Depending on the system being controlled, human error can lead to significant adverse consequences.

HMIs which are designed or selected for use (given that many designs are off the shelf and predesigned), should be usable and reduce the likelihood of error. This means that human factors principles need to be applied in either case: in the design or the assessment of the equipment. The key human factors considerations are:

- understanding who the end users are (this could be different roles performing distinctly different tasks);
- how the HMI will be used and how other HMIs will be used in conjunction with the HMI in question;
- the context of use and issues that might affect the interaction, such as the lighting, orientation of the display, background noise;
- the type of information and level of precision required;
- the level and type of control; and
- the positioning and workstation arrangement within which the HMI is placed.

The following issues are discussed in relation to plant HMIs:

- design of displays;
- design of controls; and
- considerations for control panel design.

DESIGN OF DISPLAYS

Plant side displays are designed to communicate information about a state, location, or measure of a situation or function. They are typically visual or auditory (albeit they can also include tactile). There are four categories of visual information display; quantitative, qualitative, check reading, or representational.

Quantitative displays present precise numerical data, such as a temperature (for example 25°C). This enables the user to record, check, or verify specific parameters. This type of display lacks contextual information such as the relevance of the measure within a range, that is, is 25°C too hot or too cold? Is this within safe operating limits? This relies on the user's knowledge of what the parameter means.

Qualitative displays present information within a context. Within the temperature example, the temperature could be displayed as a dial. If the pointer is at 25°C within a predetermined range from 0°C to 30°C, this indicates to the user that the temperature is closer to the upper limit of that range. The dial may also use additional markings, such as bands of color (e.g., red) to indicate a danger zone, which makes the judgment of the temperature being out of safe operating limits easier to identify and prompts appropriate action.

A lamp indicator is known as a "check reading" display to indicate two (or more) states and is routinely used on control panels to indicate process or equipment status. For example, for a particular process, if a temperature from 0°C to 24°C is safe and 25°C and above is considered unsafe, a simple illuminated red light could be used to represent the unsafe state (i.e., too hot). This offers the user basic information on the status, either safe or not, but it does not communicate the precise temperature or the context of that temperature other than it is too hot.

Representational displays, such as maps or process schematics, are simplified models of the site or process. These may be used plant side on local control panels in software form or they may be less dynamic and presented as a static "picture," using other forms of information display (such as indicator lamps) to add the dynamic element.

Auditory displays are used to provide feedback, to alert or warn the user. Various aspects of the acoustic design can be manipulated, such as tone, pattern, or pitch, to code a message or to indicate a state or a requirement for action. This type of display is effective where action is required quickly such as for an upset condition or emergency or where the operator is moving around the workplace and not necessarily in direct contact with the visual display. There are limits to human discrimination of sounds and the number that can be responded to effectively. Alarm design is discussed in Chapter 13, Control System Interface Design.

There are several human factors standards and guidance documents available to provide relevant criteria for the design of displays, including:

- US Nuclear Regulatory Commission (NUREG)-0700, Rev.2—Human System Interface Design Review Guidelines. US Nuclear Regulatory Commission (2002);
- American Society for Testing and Materials (ASTM, 2007), Re-approved 2013. Standard Practice for Human Engineering Design for Marine Systems, Equipment and Facilities Update. ASTM—F1166-07;
- American Bureau of Shipping (ABS, 2014). Guidance Notes on the Application of Ergonomics to Marine Systems.

DESIGN OF CONTROLS

Controls enable the user to start, stop, or adjust a machine, system, or process. This is achieved by making a control movement. There is a huge variety of control types such as push buttons, switches, levers, rotary knobs, sliders, joysticks, steering wheels, and foot pedals to name but a few. Different control types and specific characteristics are selected dependent on the function and context of use. For example, a control prone to accidental activation is not appropriate where this is likely or where the control is safety critical.

Controls can be classified as discrete (i.e., used for selecting different conditions, such as toggle switches or push buttons) or continuous, such as a steering wheel or pedal.

There are specific characteristics of controls that can be designed to support the end user. Human factors research underpins HFE standards for different control characteristics, including as the appropriate direction of movement, the force required to activate the control, the method and type of feedback (such as auditory, tactile, visual, or a combination), the size, displacement, and spacing of controls. HFE standards, such as ABS (2014), ASTM (2013), and NUREG (2002) can be used to support the design or assessment of controls.

CONSIDERATIONS FOR CONTROL PANEL DESIGN

The location and positioning of displays and controls has a direct impact on their usability, for example:

- there should be a clear line of sight to displays from the viewing position;
- display size should be commensurate with the viewing distance;
- displays should be positioned to avoid reflections and glare which can interfere with clear viewing;
- primary viewing zones should be determined to position the most frequently used and/or important displays in the optimal position;
- secondary and tertiary positions should be used for less important or frequently used displays;
- control panel location should be considered on the basis of the frequency of use, function, and the context of use, that is, sitting or standing, proximity, or orientation to the process.

The displays and controls need to be compatible with each other and with the users' expectations (referred to as their mental model). Grouping HMI elements by function can help to match the mental model of the user and reduce the potential for error.

Displays and controls should also be marked logically. Labeling of displays and controls should clearly indicate their function and incorporate a tag identification system to facilitate finding the correct display and/or control. The orientation and placement of labels for displays and controls should remain consistent throughout the site.

VALVE DESIGN

Valves are a key aspect of human interactions with the process from the plant side. There are various forms of erroneous interactions with valves such as opening or closing the wrong valve. This may be inconsequential, but dependent on the nature and configuration of the valve line-up, this can lead to the incorrect routing of product. It also has the potential to lead to a loss of containment, particularly if the valve is inadvertently opened when a blanking plate has been left off, or when the line is attached to something that is not rated for pipeline pressure such as a tanker. It is not just errors in operation; maintenance errors can include: failing to gain a correct seal on a flange plate or damage during invasive maintenance meaning that the valve passes, fails to function correctly or loses integrity. The error potential is generally well recognized within the industry hence the systems in place for administrative control of valve operation, such as Lock Out Tag Out (LOTO) systems, critical valve registers, and Isolation Control Certification (ICC). It is important though to facilitate the ease of use, ease of maintenance, correct operation, and minimize the potential damage through design as far as possible.

There are limited HFE standards related to valve design across the chemical and process industries. However, the marine industry has compiled useful guidance, namely the ABS (2014) and American Society for Testing and Materials (ASTM, 2013). There is also a petroleum specific standard (Norsk Sokkels Konkuranseposisjon (NORSOK), 2004). These standards cover specific HFE requirements relating to the following and are discussed in the following subsections.

- positioning of valves according to their criticality and frequency of interaction;
- valve size, type, orientation, and torque for human operation; and
- placement of valves for operational and maintenance access, which may include the requirement for mechanical handling.

VALVE POSITIONING

According to ABS (2014), ASTM (2013), and the International Oil and Gas Producers (OGP) Report 454 (2011), the positioning of valves should be based on valve criticality. The technique, referred to as Valve Criticality Analysis (VCA) is an assessment of the criticality of each valve within a process. There may be other valve criticality analyses undertaken on a project, but these may be focused specifically on the safety criticality only. The VCA referred to in this section also takes into account the frequency of interaction. A typical rating scale used is shown in Table 14.2 (based on

Table 14.2 Example Valve Criticality Ratings

VCA Rating	Description
Category 1	Highly critical or frequently operated. Valves essential to normal operation (including start up and shutdown), maintenance, and emergency operations where rapid and unencumbered access is required.
Category 2	Valves that are not critical for normal or emergency operation or maintenance but are used during routine maintenance.
Category 3	Valves which are usually operated rarely and not during critical situations.

ASTM, 2013). The VCA assessment criteria may be specified in more detail by the project so that it is representative of the nature of the process under scrutiny and the level of manual interactions with valves on site.

Each valve on the Piping and Instrumentation Diagrams (P&IDs) is assessed. It is necessary to review each valve in turn, not just the manually operated valves. Control valves, although operated remotely, could be critical to production and would require prompt plant side operator interaction if they malfunctioned. Routine testing, maintenance, and inspection for actuated valves, such as Blow Down Valves (BDVs), require access because of the need to maintain their safety integrity levels. Sample valves are likely to be interacted with regularly, possibly daily or once a shift.

The purpose of the VCA is to:

- define requirements for access and positioning of valves based on relative importance and frequency of use; and
- ensure that the relative importance of valves is considered when compromises need to be made.

In terms of gross valve positioning:

- category 1 valves should be accessed via a permanent access at deck level or via a permanent standing surface;
- category 2 valves should be located with permanent access at deck level, accessed via stairway or other access aid with a purpose-built standing surface or landing;
- category 3 valves do not have specific access requirements imposed (ABS, 2014).

In terms of valve handle height, the recommended positioning is based on the valve type, for example lever-operated valve or a hand wheel, and also whether the valve-stem is vertical, horizontal, or angled. The recommended dimensional ranges can be found in ASTM (2013), but are based on the maritime population so may need to be recalculated for different populations in different parts of the world.

SIZE, TYPE, ORIENTATION, AND TORQUE OF VALVE HANDLES

There are guidelines for the detailed aspects of valves including valve handle sizes, torque limits for manual operation, valve orientation, valve direction of movement, and the valve identification through labeling.

The reader is directed to the ASTM (2013) and ABS (2014) standards for more detail, but two examples are presented as follows:

- 'the force required to initially crack open a manual valve is set on the basis of the expected operator population, valve actuator height, and orientation in relation to the operator. The force requirements vary between 12 and 144 Newton (N) (9–106 pound feet), and should never exceed 450 N (100 pounds)' and
- 'valve handles should close with a clockwise motion of the hand wheel or lever when facing the end of the valve stem'.

VALVE AND PIPEWORK ACCESS

The space around a valve or a pipe should allow for operational or maintenance access. The type of access required would normally be informed by the task analysis, and may include access with or without tools and/or mechanical handling equipment.

There are several guidelines available which mostly concur with each other. Some general requirements and their sources are presented, as follows to illustrate the type of guidance provided:

- a minimum clearance around the outside rim of a valve hand wheel or the end of a lever of 76 mm should be provided (including clearance from obstacles as the valve handle is operated) (ASTM, 2013);
- emergency valves should not be located below deck gratings or behind covers (ASTM, 2013);
- a minimum clearance around a valve for maintenance access of 750 mm should be provided (NORSOK, 2004);
- a minimum clearance of 250 mm between the external diameter of a flange and a fixed obstruction should be provided (NORSOK, 2004);
- a minimum height of 150 mm separation from the bottom of a pipe to the deck/floor should be provided (NORSOK, 2004).

DESIGN FOR MAINTENANCE

Maintenance (including planned and reactive maintenance/repair, inspections, and testing) is a key plant side activity. Many tasks undertaken are safety critical and vulnerable to error. However, designing for maintenance tasks is often overlooked.

Maintenance does not only occur in a planned steady state where personnel have time to plan and proceed at a comfortable controlled pace. It may be that emergency maintenance has to be undertaken in a heighted situation which may involve more pressure on the maintainers. Therefore, equipment should be designed to make the task easier and reduce the potential for human error.

There are a number of keys areas which need to be considered when designing for maintenance including: equipment access, equipment labeling, equipment design, maintenance tools, equipment handling, and equipment safety features. These areas are briefly discussed below, but the reader is directed to ABS (2014) and ASTM (2013) for a more comprehensive overview of the topic areas.

EQUIPMENT ACCESS

Maintainers conduct a wide range of tasks in various different postures and consequently require sufficient access space to the equipment. Access and clearance is discussed in more detail later in the chapter as it is not just specific to maintenance. However, it is useful to consider that access is not only required for the maintainer, but perhaps other team members and the equipment and tools they will use for the task.

EQUIPMENT LABELING

It is important that equipment is labeled correctly in terms of function and identification. A common error in maintenance is to perform isolations or other tasks on similar but incorrect equipment items. Labeling enables the correct item to be identified.

EQUIPMENT DESIGN

The ease of maintenance can be aided by following the simple principles outlined in the human factors standards. Examples of these principles include:

- consider purchasing equipment with self-checking, calibration, and fault finding systems depending on the context of use. This may allow for maintenance to be prioritized to those pieces of equipment that require attention in comparison to checking every item on a rolling program;
- select equipment so that it can be easily accessed from the outside or above, rather than inside or underneath, as these offer easier working postures, and improve visual and physical access. Maintenance that occurs inside and/or underneath a piece of equipment may lead to poorer working positions, injury, and greater exposure to inherent hazards;
- equipment that is frequently replaced or maintained should be positioned so that is has easy access and does not interfere with other pieces of equipment, personnel, or processes;
- frequently replaced component parts or pieces of equipment should be prioritized for easy removal and replacement and consideration should be given to disposable modules, components, and parts to facilitate this process.

MAINTENANCE TOOLS

Equipment selection and design should incorporate considerations for the tools and test equipment required by maintainers. This process should aim to limit the number and complexity of the different types of tools and test equipment a maintainer may require for access as well as checking, repairing, and replacement tasks. Overall, this will reduce the amount and complexity of the tools and testing equipment required to be transported by the maintainer and potentially reduce the scope for human error.

EQUIPMENT HANDLING

Handling of equipment, tools, and test equipment should be considered within the design for maintenance: particularly the aspects of how they will be transported and utilized; whether by an individual, by a team, or via handling equipment.

Equipment that is heavier to lift and impacts on the maintainers' ability to conduct their tasks should be supported to reduce the weight and the subsequent impact on the user. This may involve suspending items of equipment on a lifting arm or winch, or even mounting items on sliders or counterbalancing them. Further detail on material handling can be found in Chapter 15, Human Factors in Materials Handling.

EQUIPMENT SAFETY FEATURES

The safety of personnel is a priority for any process and designing for maintenance equally prioritizes safety. This includes the use of physical guards for protection such as barriers and electrical lock off switches and controls that can be locked.

Organizational processes for safety, such as the use of risk assessment, permit to work systems, LOTO systems, hazard markings, and signs are generally well embedded within the chemical and process industries.

WORKSPACE ACCESS AND CLEARANCE

Within the plant side areas there are many different types of workspace. These include the workspaces within plant buildings such as equipment rooms, laboratories, or storage spaces, as well as the open spaces and spaces around equipment items. The term workspace is applied to any area where work is conducted, even below ground and confined spaces within equipment.

Workspace access and clearance applies to the space envelope required to move around and gain access to plant areas as well as the space within which someone will work. Dimensional space envelopes are defined in standards, such as to ABS (2014), ASTM (2013), and NORSOK (2004). The standards specify the minimum dimensions for walkways, stairways, ramps, ladders, platforms, doorways, manways, and other access routes and methods. They also specify the recommended dimensions of workspaces for users in different postures to work at the workstation or piece of equipment, such as standing, kneeling or working prone. As well as access and clearance, there is a need for the user to be able to reach everything they need to access and see everything they need to look at. The emphasis is on ensuring that users can work easily, achieve comfortable working postures, and avoid compromised execution of the task.

In designing access spaces and workspaces, the dimensions, layout, and orientation of those spaces need to be based on the context of use, including consideration of different roles using the space, the tasks they will be doing, the number of people needing to be accommodated at once, and other aspects such as environmental conditions. The tasks are important as they will influence whether the task requires the transportation of large items of equipment through it and the types of postures required. The type of equipment being brought to a space, used, removed, or replaced as part of maintenance all have to be considered. Allowances should be made for the largest part of equipment expected to be maintained in that space.

In general, it is acceptable to apply the dimensions specified in the standards. However, there are occasions where design compromises have to be made and this is when there is a need to revert back to the core ergonomics discipline to calculate the space envelope. Anthropometry is the measurement of body dimensions and this is used to calculate minimum acceptable levels for safe working in workspaces. All body dimensions have a normal distribution in any population with few people at the extreme of the body dimension size range. There are times where there is a need to design for the largest or the smallest dimension. For example, if reach is determined by the smallest persons reach dimension (e.g., 5th percentile person forward reach), then anyone larger will also be able to reach. Clearance dimensions are normally based on the dimensions of the largest person (e.g., 95th percentile stature, shoulder breadth or girth) so that all people this size or smaller can fit. There are other times when there is a need to accommodate a range, such as 5th–95th percentile body dimension, and if this is difficult to achieve, an adjustable feature to the design may be appropriate.

A safe working envelope needs to be determined by considering the three-dimensional space in which the user will conduct their tasks. In terms of the user and the task requirements, designers may need to consider their height, width, girth, leg length, arm length, hand size, and any other measurements that impact on the user and the task. Additional allowances for personal protective equipment as well as population sizing of the intended user group; including gender and nationality should be applied. International anthropometric data are available for use in design, and the most relevant data set to the user population should be selected.

Consideration should also be given to evacuation routes and muster spaces. Moving quickly and safely into a safe space is of prime importance in an emergency situation when personnel may be in an elevated state of stress or moving more quickly than normal.

Handling of materials is covered in Chapter 15, Human Factors in Materials Handling. However it is important to reflect on this topic in terms of access and clearance. Handling tasks have to be accommodated in terms of height clearance, widths, and lengths for maneuvering, transporting by hand, or using handling equipment or delivery vehicles. Adequate space for turning circles also has to be considered.

EQUIPMENT POSITIONING AND LAYOUT
INTERNAL PLANT SIDE WORK AREAS

The typical plant side internal work areas might include storage areas, laboratories, equipment rooms, or other buildings. As for any workspace there is a need to understand the tasks that are undertaken in different operational scenarios, the equipment requirements for the different roles, and the nature of the interaction, before determining the optimal position and access for the equipment required in the different types of work area. This is discussed in Chapter 12, Workstation, Work Area, and Console Design.

The general principle is to colocate equipment and workspaces that are used together on the basis of the relative frequency or importance of the interaction.

A simple technique known as link analysis can be used to good effect to determine the general arrangement of a work area.

EXTERNAL PLANT AREAS

The external areas of the plant are typically comprised of skid packages or major items of equipment and pipe runs which may be on a single level or may be accessed via decks at different levels. Aside from ensuring adequate workspace access and positioning of the human interfaces, there is a need to consider the logical layout of a plant at a macrolevel. This is rarely discussed in the HFE guidelines and standards. Simple HFE principles can be used to good effect at this macrolevel of design:

- if there is a sequence of interaction, follow it;
- if equipment is functionally linked, group it; and
- if the interaction is frequent or critical, keep it easily accessible or central.

Within building design, the functional work zones are determined and then simple techniques, such as adjacency matrices, are used for identifying the need for colocation or separation of work zones. It is possible to apply a similar technique to plant layout, albeit there will be trade-offs against fundamental engineering requirements such as keeping vulnerable equipment away from critical blast pressure wave sources or flare radiation, achieving the right gradient on a pipe run or maintaining equipment in adequately vented spaces. Despite the possibility of usability and logical arrangement of the work area being subservient to some of these requirements on the basis of safety, there is value in giving them consideration. This does not just aid efficiency of human performance but it can also avoid erroneous operation of the wrong equipment because the person has misidentified their location.

The size of the plant spaces may be constrained by a number of factors: landscape or terrain; the available space; the cost of land (or in the case of offshore assets, the cost, weight, or other constraints of the platform or vessel structure). The site real estate is therefore at a premium and certain equipment items or packages are critical to position first. There may also be the logistics of getting the item into position and the need to consider handling requirements, both during installation and during operations.

SIGNAGE AND LABELING

HFE can be used to optimize signage, labeling, and marking to enhance human performance and safety. Well-designed signage helps to maintain spatial awareness, direct individuals to the right location, keep them away from or protected from hazardous areas, and aid escape, rescue, or evacuation in an emergency. Labeling and markings help the individual to identify the right equipment or material contents and avoid erroneous behaviors. Conversely, if these are poorly designed, presented, or positioned they have the potential to induce an error or impinge on the performance of a task. Different categories of signage and labeling are presented in Table 14.3.

Table 14.3 Different Categories of Signage and Labeling

Term	Definition
Label	Used to identify an item of equipment or component of equipment.
Marking	Used to identify contents and flow direction for pipes.
Sign	Used to identify and provide information to personnel. Informational signs present general nonprocedural information. Instructional signs present instructions for performing a task.
Hazard sign	Used to identify and provide information about hazardous situations. They include three types of hazards: danger, warning, and caution.
Danger	Indicates an imminent hazardous situation which, if not avoided, could lead to death or serious injury (Used for extreme situations).
Warning	Indicates a potentially hazardous situation which, if not avoided, could lead to death or serious injury.
Caution	Indicates a potentially hazardous situation which, if not avoided, could lead to minor or major injury.

Signage and labeling is a means of communicating and portraying information to an individual or group of individuals. There are several considerations to any information transaction. The information provided needs to:

- be visible and readable in different environmental contexts (e.g., day and night, foggy, clear);
- enable the message to be understood; and
- enable the messages to provoke the intended behavior of the recipient.

There are several barriers to achieving this aim, for example, individuals may be overloaded with information, confused or disoriented by the information or fail to understand or recognize the intended behavior.

The reader is directed to ASTM (2013), ABS (2014), and NUREG (2002) for detailed guidance in this area. The HFE design guidance includes:

- how to design a signage or labeling scheme;
- orientation and placement of signs so they are readable and usable at the point of use;
- visibility and legibility, taking account of visual distances and conditions;
- terminology and language to incite the right behavior;
- shapes and sizes of signs;
- use of symbols and pictograms to correctly portray meaning;
- labeling of pipes, flow directions, equipment items, valves, and control panels;
- specific guidance for hazard signs, information signs, and instructional signs; and
- specific guidance for orientation plans.

This chapter specifically considers the application of HFE to improve the health, safety, and usability of plant and equipment. The next chapter provides an outline of the key human factors issues related to materials handling. This is a key component of plant side activities, although the scope is wider than the processing plant.

KEY POINTS

- The primary end users of plant side spaces are field operators and maintainers, but they have different task needs which need to be understood if the design is to meet these needs;
- Instrument design and selection is imperative to enable effective interaction between the users and plant side process interfaces and reduce the potential for error;
- Valve design and placement needs to facilitate ease of use and prevent error during interaction. The HFE standards cover various aspects of their design;
- Design for maintenance is an important consideration to reduce the risk of active or latent error or injury by considering access, labeling, equipment design, and tool use;
- Access and clearance need to be considered to ensure the safe usability of site spaces and accommodate different tasks;
- External plant areas often do not receive adequate HFE attention, but there are some simple principles that can facilitate efficient and safer performance, in much the same way as for internal plant areas;
- Signage and labeling is important for enhancing spatial awareness, accessing the right equipment and aiding evacuation. HFE principles can be used to make improvements across the site.

REFERENCES

ABS, 2014. Guidance Notes on the Application of Ergonomics to Marine Systems.

ASTM, 2007, Re-approved 2013. Standard Practice for Human Engineering Design for Marine Systems, Equipment and Facilities Update. ASTM—F1166-07.

International Oil and Gas Producers (OGP), 2011. Report 454—Human Factors Engineering in Projects. OGP454.

Norsk Sokkels Konkuranseposisjon (NORSOK), S-002, 2004. Working Environment.

NUREG-0700, 2002. Rev. 2—Human System Interface Design Review Guidelines. US Nuclear Regulatory Commission.

Human factors in materials handling

15

E.J. Skilling and C. Munro

LIST OF ABBREVIATIONS

ABS   American Bureau of Shipping
HSE   Health and Safety Executive
IMIRP Industrial Musculoskeletal Injury Reduction Program
LFS   Labour Force Survey
MAC Manual Handling Assessment Chart
MHI   Materials Handling Institute
NIOSH National Institute for Occupational Safety and Health
PPE   Personal Protective Equipment
REBA Rapid Entire Body Assessment
RULA Rapid Upper Limb Assessment
SWL   Safe Working Load

The chemical and process industries produce a multitude of products including plastics, textiles, fuels, and lubricants. A wide variety of materials are also used in the processes themselves. Whether in the form of raw materials or end products, materials handling will likely be involved. Materials handling is defined as:

the movement, protection, storage and control of materials and products throughout manufacturing, warehousing, distribution, consumption and disposal.
Material Handling Institute (MHI, 2016)

Materials handling incorporates a wide range of manual, semi-automated, and automated systems. All forms of materials handling involve an operator at some point, either directly if the materials require manual handling, through interaction with the mechanical equipment used to handle loads, such as pushing or pulling a trolley, or in the use and maintenance of semi-automated or automated equipment.

Human factors can be applied to the design or modification of any materials handling system. The integration of human factors principles aims to ensure that the capabilities and limitations of the operators are taken into consideration. Human factors in materials handling should be considered alongside engineering, purchasing, and product management. This combination helps to ensure that the systems provided are effective for the organization and the operators without compromising productivity or causing injury. The overriding principles when considering material handling tasks are to:

- eliminate *manual* materials handling as far as possible;
- assess the operators, load, task, work area, and environment; and
- (re)design these factors to reduce risk;

The chapter considers handling injuries, the analysis and planning of materials handling tasks, and key considerations for mechanical and manual handling. Automated and highly mechanized machinery, such as bulk carriers and feeders, where there is no direct operational involvement of humans is not covered in this chapter, other than discussing maintenance.

HANDLING INJURIES

Injuries, including fatalities and multiple fatalities from handling activities, are particularly relevant from a human error perspective, particularly in relation to collision and dropped objects which could involve handling heavy loads in critical areas of the process plant. Human error can also occur during the maintenance of mechanical handling systems. This means that there is a need to analyze and reduce the potential for human error, as well as considering specific aspects of the equipment and work context.

Many musculoskeletal injuries are attributed to handling in the workplace. In 2013–14 an estimated 909,000 working days were lost in the United Kingdom due to handling injuries (Health and Safety Executive (HSE), 2014a). The Labour Force Survey (LFS), 2014 reports:

- an average of 6.6 days were lost for each handling injury;
- handling injuries made up nearly one quarter of all reported injuries;
- eight percent of major/specified injuries involved handling;
- more than one quarter of over-7-day injuries involved handling.

PLANNING AND ASSESSMENT OF MATERIALS HANDLING

One of the principles of material movement is *"planning of all material movement activities for maximum efficiency and safety"* (Kroemer, 1997).

This principle is supported by incorporating a variety of methods including tabular and graphic aids, assembly charts, operation process charts, travel charts, activity relationship charts, and critical-path diagrams. These techniques can be used to plan the layout of work areas, handling activities and events, the human locations, the equipment required, and communications. The first assessment is whether the handling can be mechanical or whether manual handling is required for all or part of the task.

ASSESSMENT OF THE LOAD

The assessment of the load should identify if it is safe for an individual to handle or whether there are other hazards relating to the load which mean that it should be mechanically handled.

Assessing the weight of a load is an important factor but it is not the only factor affecting the safe execution of materials handling or potential injury risk. The following factors should also be considered (adapted from HSE, 2012):

- load size and shape;
- texture of the load; sharp or soft, malleable or solid;
- the rigidity, flexibility, or stability of the load;
- resistance or friction associated with moving the load;
- hazard identification; including weight (Safe Working Load, SWL), temperature, radioactivity, chemical composition, and occupational health hazards;
- coupling or "the quality and ease of the grasp or connection" to the load by an operator's piece of equipment;
- location of the load.

Guidance figures for lifting and lowering are shown in Figs. 15.1 and 15.2.

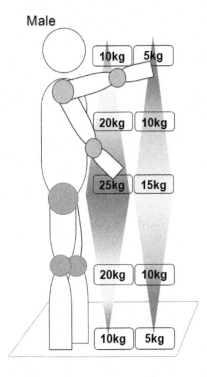

FIGURE 15.1

Male lifting and lowering load guidance.

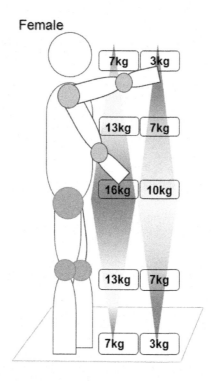

FIGURE 15.2

Female lifting and lowering load guidance.

Guidance values for carrying, pushing, and pulling as recommended by the HSE and the American Bureau of Shipping (ABS) are listed in Table 15.1.

It is not the case that two people can handle twice the load or that three people can handle triple the load during team handling. The following guidance is provided:

- for a two-person lift, reduce the load limit by 2/3 of the sum of the individual load limit;
- for a three-person lift, reduce the load limit by 1/2 of the sum of the individual load limit;
- avoid a four or more person lift as higher levels of communication and coordination are required.

There are other factors that should be assessed for manual handling, including the task, work area, and operator factors. These are discussed later in the chapter.

Every opportunity should be taken to eliminate or reduce manual handling. Where the manual handling limits are exceeded or other factors dictate the need for mechanical handling, the selection of mechanical assistance should be appropriate to the load.

Table 15.1 Examples of Pushing/Pulling and Carrying Guidance

	HSE (1992, as amended)		ABS (2013)	
	Male limit	Female limit	All male	Male and female
Push and pulling—force to stop or start load	20 kg	15 kg	Not specified	Not specified
Push and pulling—sustained force to keep the load moving	10 kg	7 kg	Not specified	Not specified
Carrying an object up to 10 m with two hands	25 kg	17 kg	20 kg	19 kg
Carrying an object more than 10 m with two hands	Risk assess	Risk assess	20 kg	14 kg
Carrying an object more than 10 m with one hand, at the side of the body—for example, tool box	Risk assess	Risk asses	10 kg	9.5 kg

MECHANICAL HANDLING

MATERIAL HANDLING STUDY

A method for completing a materials handling study is presented by the Energy Institute (2016). This is intended for designers, site managers, operators, or the competent responsible person involved in a transfer. It includes three stages:

- Stage 1—Record the details of any item that may need to be moved that is greater than 20 kg or irregular in size, including item description, weight, identification (label, location, name, or drawing), and the item's dimensions;
- Stage 2—For each item recorded in Stage 1 identify the following:
  - the most suitable materials handling equipment to remove and replace the item;
  - the most suitable materials handling equipment to bring maintenance equipment to the item;
  - the costs required to lease, hire, or purchase the handling equipment;
- Stage 3—Assess the work area layout considering the following:
  - adequacy of the space for equipment and people to move in and out of the area;
  - adequacy of the clearance for any movements required in the area, including removing, replacing, repositioning, and loading/unloading;
  - adequacy of the movement routes being free from obstruction;
  - adequacy of the storage for materials handling equipment.

EQUIPMENT SELECTION

Within the chemical and process industries there are two general categories of materials that need to be moved by equipment and/or people: unit and bulk materials. The type of materials handling equipment is dependent on the following:

- the load size and weight;
- the terrain and anticipated travel path;
- environmental conditions;
- how frequently the unit will be used;
- the people who will use the device.

Unit handling

There is a wide range of materials handling equipment designed for mechanical assistance. They have different functions, features, and lifting limits and should be selected based on meeting the functional needs and ergonomics requirements. Table 15.2 presents typical equipment.

There are a number of human factors considerations, such as:

- Operator/maintainer: strength and endurance (where relevant) and load limits for pushing/pulling;
- Equipment: mobility and maneuverability, load stability and security, visibility, equipment control panel design, handle design;
- Work area/environment: body motions and postures, space requirements, weather conditions, ability to hear and see signals, flooring, pinch points.

Bulk handling

Equipment used mostly for in-process movement of large volumes of powders or liquids such as conveyors, cranes, or industrial hoists require less direct operator intervention are described in Table 15.3.

There are several factors to consider for crane operations which can vary depending if the crane is fixed on site or required temporarily. Restricted access to site can complicate the lift. The crane may be stationary with the load being manipulated around the crane, or alternatively, the crane may be mobile and maneuver the load and the body of the crane together to reach a set destination, such as ship to structure handling.

The Crane Operator Toolkit (Industrial Musculoskeletal Injury Reduction Program (IMIRP Society), 1999) outlines typical human factors assessment issues for crane operations including:

- connecting the crane to the load;
- control of the load through the lifting process to its lift destination. This may require a complex series of controls and adjustments that can be physically and mentally demanding. A crane operator must be aware of load, the crane structures (e.g., the lifting boom, the supporting line, and coupling equipment) and how and where these are positioned and interact with the immediate environment. The controls and the cockpit arrangement are critical;

Table 15.2 Typical Equipment Used for Mechanical Assistance

Type of Equipment	Typical Uses	Description
Carts	Transporting of loads from the starting point to end point of a task.	– four wheels and a supporting shelf for the load; – available in a range of sizes, heights, widths, and load limits; – additional features may include brakes, directional or steering wheels, and handles for maneuvering;
Dollies	Transporting of loads from the starting point to end point of a task.	– wheeled low level supporting structures; – placed under materials or loads which allow them to be easily transported on a flat surface; – normally a minimum of four wheels, with brakes; – handles are not normally part of this piece of equipment;
Scissor lift tables	Transporting, accessing, loading, or unloading where there is a variation in heights incorporated into the task.	– similar to carts, however, they have the added feature of a lowering and raising platform;
Mobile hoists	Used when loads exceed the physical lifting capability of an individual or team.	– mobile lifting device; – often a variety of coupling options to secure the materials to be maneuvered;
Fixed lifts/ hoists	Used when loads exceed the physical lifting capability of an individual or team. Hoists may have an increased SWL capability if they are secured to a supporting structure.	– similar to mobile hoists, except that they are secured into the ceiling or wall; – ceiling hoists may be fixed, have limited movement or move in an "H" frame pattern to cover the majority of the floor space for maximum lifting potential; – wall-mounted hoists typically have extending arms, with a degree of rotation;
Vacuum or magnetic lifts	Used in production lines for the assembly of goods and frequent handling of loads.	– offer a flexibility in lifting, in which suction or a magnetic connection is used to secure the load; – once activated or engaged, the operator can move the materials with ease on a supported mechanical arm;
Telehandlers	Typically used to maneuver or manipulate materials or objects on site.	– four-wheeled driven vehicles; – incorporating a drivers cabin and the required controls for the operation; – a variety of adaptations are available and examples include, lifting arms, pallet forks, fork mounted lifting hooks and material buckets; – safety requirements include operator certification; – typically a banksman, ground person, or supervisor is present for the task to aid the telehandler driver and act as a second pair of eyes for safety.

Table 15.3 General Principles for In-Process Equipment Selection

	Conveyors	Cranes	Trucks
General usage	Moving uniform loads continuously from point to point	Moving a variety of loads to any point within an area	Moving loads over a variety of suitable surfaces
Material volume	High	Low to medium	Low to high
Material shape	Regular or irregular	Irregular	Regular
Material weight	Light to heavy	Heavy	Light to heavy
Support	None or in containers	Suspended on a pallet, skid, or container	From below by a pallet, skid, or container
Operator required to be present	No	Yes	Yes

- communication between team members;
- location of the lift and path of travel. The line of sight of the load should be considered. Clearly this is a major accident risk if there is a dropped or uncontrolled movement and zoning of dangerous/vulnerable areas should be identified. There should also be consideration of the equipment itself introducing an ignition source in areas at risk of fire and explosion;
- weather, particularly in relation to wind speeds and lifting heights.

The issues are not specific to human factors, but given the safety critical nature of the task, there is a need to consider the potential for human error. The risk of human error during an operational task may be reduced through engineering controls, but the potential for these to be compromised, for example, as a result of not completing a maintenance task, may need to be reviewed. Other human factors issues, such as having effective control panels, good sightlines, suitable training, and good team communication, also reduce the potential for error.

FACILITY LAYOUT

Often the facility layout cannot easily be changed but in some circumstances the design can be planned to better suit the materials handling process. There is also potential to put the lifting team/handler in a vulnerable position where pinch points can occur or where loads may be dropped. Restricted space may affect access and maneuverability of mechanical lifting devices such as hoists.

Where a *process* layout exists, that is, all processes of the same type are grouped together; the advantage is that products may flow through the same workstations. This type of layout requires a relatively large number of transportation lines with the potential for increased materials handling.

Where there is a *product* layout, that is, where all the processes, machines, and activities needed for the product are kept together, materials handling routes can be planned and floor space reduced.

Flow diagrams and flow charts can be used to analyze material flow to calculate material handling travel distances, time, and cost. They can assist by minimizing the number of transportation lines and making transport routes as short as possible (Kroemer, 1997). They also assist with identifying potential bottle necks, areas of reduced efficiency, and increased/unnecessary manual handling.

ENVIRONMENT

When working outdoors, sudden air movements, for example, from gusts of wind can make large loads more difficult to handle safely, as already discussed in relation to crane operations.

A large contrast in brightness between areas may also make it difficult for the handler, such as a fork lift truck driver to adapt and may cause a temporary reduction in visibility. Lighting should be well directed and sufficient for the material handling tasks.

MAINTENANCE OF EQUIPMENT

Designing for maintenance should include the following considerations:

- the location of loads: including access for maintainers, clearance, and space to accommodate the maintainers and maintenance equipment;
- use of modular components to facilitate easier maintenance;
- consideration of the tools and equipment required for maintenance, that is, fewer tools, easier to carry and less opportunity for human error using the wrong tool;
- fitting pad eyes and other facilities to facilitate easier and safer maintenance.

REDUCING THE RISK OF MANUAL HANDLING

As well as assessing the load (discussed earlier) there are other factors which can increase the risk of injury: the task, the work area, and the operator. These factors overlap and should not be considered in isolation of each other.

ASSESSMENT OF THE TASK

It is important to consider factors such as the following:

- the start and end points of the task and the variety of movements required by individuals' and/or equipment;
- the frequency of the handling, for example, a load of 13kg lifted once or twice a day from the floor and transported less than 10m for storage may be within guidelines for an individual and not necessarily require equipment (unless other factors prevail). However, if the same load is transported the same distance multiple times an hour, then arguably mechanical equipment would be a good investment;
- the use of equipment when personal limits are reached.

ASSESSMENT OF THE WORK AREA

The assessment of the work area includes space and flooring.

Space

Operators require sufficient space to be able to undertake manual handling activities without having to adopt awkward postures. Lack of space may prevent materials being reached easily or prevent them being lifted correctly.

Walkways and doorways should be designed to take account of handling requirements. Automatic doors should be considered where loads are moved frequently either manually or using equipment such as trolleys/carts or sack trucks, thus avoiding the need to stop and/or adopt awkward postures to reach handles.

Overhead clearance should be considered, including in storage areas, to remove the need for operators to stoop when accessing, moving, or handling materials to those areas.

Sufficient space should be available for suitable rests or stands within the work area to allow equipment to be placed on them for inspection or maintenance or to readjust the lift team positions safely. Space should also be available around the stands for work to be undertaken without the need for awkward postures or for the equipment to be balanced or partially supported by the operator.

Access to materials can be improved by using devices such as pull out racks. This allows greater storage but without the need for the operator to over reach and adopt awkward postures. Powered platforms can allow materials to be stored at higher levels without the need to use ladders to access them. Recommended storage heights are presented in Table 15.4 (adapted from Pheasant and Haslegrave, 2006).

Flooring

Floor surfaces where materials handling is required, either indoors or outdoors, should be level, free from obstructions, and in good repair. Poor flooring or lack

Table 15.4 Recommendations for the Design of Storage

Height (mm)	Application
<600	Suitable zone for storage of rarely required lighter items.
600–800	This zone is OK for heavier items and good for lighter items.
800–1100	This is the BEST zone for storage.
1100–1400	OK for light items but poor for heavy items. Visibility is likely to be limited for some people.
1400–1700	This zone will be limited for visibility and accessibility.
1700–2000	This zone will be very limited for access and will be beyond the reach of some people.
>2000	This zone will be out of the reach of most people.

of maintenance can increase the likelihood of slips, trips, and falls and tension and effort required by the operator to either manually transfer a load or use a mechanical handling device. Pushing or pulling a trolley is significantly more difficult if the floor is uneven or cracked and can make the load unstable.

When possible, handling activities should be carried out on one level, avoiding gradients or changes of level (HSE, 1992 as amended). Where ramps or slopes are necessary, they should have a low gradient. ABS (2013) recommends that ramps used where materials handling occurs should have 4° of inclination, with a maximum inclination of 7°. The force required to push a load is approximately 2% of the load weight, for example, with a slope gradient of 1°, the push force for a 100 kg laden trolley is 2 kg. However, if the ramp gradient increases to 5°, the push force is increased to 9 kg and a slope gradient of 10° requires a push force of 17.5 kg. Ramps should also have a nonskid surface and have level landings at the top and bottom.

ENVIRONMENTAL FACTORS

Thermal conditions may make materials handling more difficult. High temperatures increase the likelihood of exhaustion and may reduce the operator's capacity for manual handling. If the conditions make the hands sweat, this may reduce grip, either when manually handling a load or when using a steering wheel or lever. Conditions which create the need for thick/padded gloves also reduce grip.

Rain, ice, and snow can increase the risk of slipping if on foot or of collision if using equipment such as a pallet truck outdoors. Specific assessments should be undertaken for operators working at heights, for example, on a tank. Conditions may appear satisfactory at ground level but may differ when above ground level on a low friction surface.

When working outdoors, sudden air movements, for example, from gusts of wind, can make large loads more difficult to handle safely. Insufficient lighting may mean that an operator fails to see a change of surface level.

OPERATOR FACTORS

Individual capability

Individuals vary in their capability for manual handling on the basis of age, gender, stature, fitness, health, training, and experience. Gender differences were discussed earlier. Additional analysis is required if the task involves young operators (under 18 years of age), pregnant operators, or those with a health condition or injury.

Personal protective equipment

Many materials handling tasks require the use of personal protective equipment (PPE) such as protective footwear, glasses, and gloves due to the nature of the loads being transferred. Operators may also be wearing PPE relating to other tasks that they perform as part of their role. PPE should not make the task more difficult as this

Table 15.5 Manual Handling Analysis Tools

Tool	Use
Manual Handling Assessment Chart (MAC) (HSE, 2014b)	Lifting, carrying, and team handling
Job Stress Index (Mital et al., 1993)	Analysis of lifting
National Institute for Occupational Safety and Health (NIOSH) Lifting Equation (Waters et al., 1994)	Analysis of lifting
Rapid Entire Body Assessment (REBA) (Hignett and McAtamney, 2000)	For dynamic work. Provides scoring for trunk loading, postures, forces, and movement.
Rapid Upper Limb Assessment (RULA) (McAtamney and Corlett, 1993)	For seated work. Provides scoring for neck and upper limb loading, postures, forces, and movement.

encourages operators to deviate from procedures by not using the PPE. PPE should be well-fitting and not restrict movement. It should be designed to minimize the likelihood of catching on loads during transfer. Gloves should be close-fitting so that they allow dexterity whilst maintaining gripping. They should be made of materials suitable for the task at hand, such as cut resistant or heat-retardant.

Training

Manual handling training needs to be specific and include topics such as posture, movement, and lifting techniques. Trainees should be taught "dynamic risk assessment" so they are able to assess risks before performing any material handling task. Systems should be in place to allow operators to report when they identify risk factors as changes occur subsequent to the original assessment.

MANUAL HANDLING ANALYSIS TOOLS

There are different analysis tools to support the evaluation of manual handling tasks, such as those presented in Table 15.5.

SUMMARY: HUMAN FACTORS IN MATERIALS HANDLING

Human factors applies to the whole materials handling system (manual, semi-automated, or automated) and should be included in the analysis and planning of handling tasks to reduce the potential for injury. The next chapter provides a review of environmental factors relevant to human factors.

KEY POINTS

- All forms of materials handling involve an operator at some point, either directly, if manual handling is required, through the interaction with mechanical equipment used to handle loads, or in the maintenance of semi-automated or automated equipment;
- In 2013–14 an estimated 909,000 working days were lost in the UK due to handling injuries;
- Designing a materials handling system should involve the analysis of the task, the load, the work area, and the operator with the aim of eliminating manual handling as far as possible;
- Mechanical handling equipment should be selected only after a thorough analysis has been conducted to ensure the most appropriate equipment is selected;
- Storage of materials for access at any stage of the process should be designed taking human factors into consideration;
- Designing for maintenance is essential to ensure the safe and efficient ongoing usage of equipment.

REFERENCES

American Bureau of Shipping, 2013. The Application of Ergonomics to Marine Systems. ABS, Houston, TX.

Energy Institute, 2016. Human Factors Safety Information Bulletin (No. 1): Manual and Mechanical Handling.

Health and Safety Executive, (1992). Manual Handling Operations Regulations (1992) (as amended). Guidance on Regulations L23 (third edition). HSE Books 2004 ISBN 9780717628230.

Health and Safety Executive, (2012). Manual Handling at Work—A Brief Guide(pdf). Available from: <http://www.hse.gov.uk/pubns/indg143.pdf> (accessed 28.01.2016.).

Health and Safety Executive, (2014a). Handling Injuries in Great Britain (pdf). Health and Safety Executive. Available from: <http://www.hse.gov.uk/statistics/causinj/handling-injuries.pdf> (accessed 21.12.2015.).

Health and Safety Executive, 2014b. Manual Handling Assessment Charts (the MAC Tool). HSE Books., ISBN 9780717666423.

Hignett, S., McAtamney, L., 2000. Rapid entire body assessment. Appl. Ergon. 31 (2), 201–205.

Industrial Musculoskeletal Injury Reduction Program (IMIRP Society), 1999. Crane Operator Toolkit. Common Injury Jobs.

Kroemer, K.H.E., 1997. Ergonomic Design for Material Handling Systems. CRC Press, Boca Raton, FL.

Labour Force Survey, (2014) (online). Available from: <http://www.ons.gov.uk> (accessed on 21.12.2015.).

Material Handling Institute, (2016). Material Handling Overview (online). Available from: <http://www.mhi.org/> (accessed 6.01.2016.).

McAtamney, L., Corlett, N., 1993. Rapid upper limb assessment: a survey method for the investigation of work related upper limb disorders. Appl. Ergon. 24 (2), 91–99.

Mital, A., Nicholson, A., Ayoub, M., 1993.). A Guide to Manual Materials Handling. Taylor & Francis, Washington, DC.

Pheasant, S., Haslegrave, C.M., 2006. Bodyspace, second ed. CRC Press, Boca Raton, FL.

Waters, T., Putz-Anderson, V., Garg, A., 1994. Applications Manual for the Revised NIOSH Lifting Equation. US Department of Health and Human Services., (NIOSH) Publication No. 94-110.

Environmental ergonomics 16

E.J. Skilling and C. Munro

LIST OF ABBREVIATIONS

A(8)	Average Vibration Magnitude Values over a 8-hour Workday
CAD	Computer-Aided Design
CIBSE	Chartered Institute of Building Services Engineers
CCOHS	Canadian Centre for Occupational Health and Safety
CLO	Measurement of Dry Clothing Insulation
dB(A)	Measurement Low Intensity A-Weighting
dB(B)	Measurement Moderate Intensity B-Weighting
dB(C)	Measurement High Intensity C-Weighting
EAV	Exposure Action Value
ELV	Exposure Limit Value
HAVS	Hand Arm Vibration Syndrome
HSE	Health and Safety Executive
Leq	Sustained Noise Level Exposure
OSHA	Occupational Safety & Health Administration
PPD	Predicted Percentage of Dissatisfied
PMV	Predicted Mean Vote
PPE	Personal Protective Equipment
t_a	air temperature
t_g	150 mm diameter black globe temperature
t_{nw}	natural wet bulb temperature
TOG	Measurement of Dry Clothing Insulation
WBGT	Wet Bulb Globe Temperature
WBV	Whole Body Vibration
%RH	Percentage of Relative Humidity

Environmental ergonomics concentrates on the interaction between people and their physical environment with specific emphasis on thermal comfort, lighting, noise, and vibration. These factors are relevant to engineering projects as well as the operational stage of a system life cycle, albeit that the key opportunities for effective solutions are most likely to be during design. Poorly designed working environments or insufficient management of environmental factors can have a significant adverse effect on health, well-being, and performance. The adverse effects can be both psychological and physical and it is widely recognized that poor design can lead to lower productivity, increased potential for human error and physical discomfort, injury or death in extreme conditions of cold or heat.

This chapter discusses the measurement of environmental factors and reflects on the guidance available in relation to the physiological tolerances for thermal factors; lighting requirements for different types of work; noise in relation to noise-induced hearing loss, interference of communications, and noise as a stressor; and the adverse effects of vibration on health and performance.

MEASURING ENVIRONMENTAL FACTORS

During engineering design projects or operations, a review of environmental (and other) factors is often undertaken using a Work Environment Health Risk Assessment (WEHRA). Typical topics are shown in Table 16.1 with possible causes. This is broader than the typical environmental ergonomics scope of noise, vibration, thermal comfort, and lighting, but draws the topics together under several physical elements which can have a detrimental effect on human performance, health, and safety.

Environmental factors may be assessed quantitatively through taking measurements or using prediction modeling and then assessing these findings against relevant criteria from regulations, guidelines, or standards.

There is benefit in gaining subjective, qualitative feedback from surveying the opinions of the workforce for existing workplaces. This can be done through questionnaires and/or informal interviews.

THERMAL COMFORT

Thermoregulation is the physiological process within the human body, which regulates and maintains the internal core temperature at approximately 37°C (Wilson and Corlett, 2005). Deviation from 37°C can cause ill health and any deviation of 2°C can cause death.

Thermal comfort describes an individual's perception of their immediate environment, ranging from feeling too hot, moderate or too cold. Thermal comfort is defined as "that condition of mind which expresses satisfaction with the thermal environment" (Wilson and Corlett, 2005). It becomes apparent to the individual that they are uncomfortable if there are changes in the environment or extremes of thermal conditions. Thermal comfort is not simply related to temperature alone; it is a combination of six environmental and personal factors; air temperature, radiant temperature, humidity, air flow, metabolic rate, and the clothing worn by individuals. The combination of these factors can be complicated, and due to the nature of individuals experiencing and expressing their own perceptions of thermal comfort, there can be a wide range of opinions and levels of satisfaction within a group of users sharing

Table 16.1 Work Environment Health Risk Assessment

Work Environment Factor	Possible Causes
Noise	• Presence of noisy equipment • Use of noisy portable tools • Impact noise
Vibration	• Whole body vibration • Hand arm vibration
Thermal comfort	• Extremes of temperature • Draughts • Humidity • Poor air quality • Heavy manual work in hot areas • Sedentary work in cold areas • Exposure to rain, ice, high winds
Lighting	• Lighting not compatible with the task • Glare and reflective sources
Work arrangement/layout	• Confined/restricted space • Difficult access/reach • Static or fixed posture working • Kneeling, squatting of lying positions
Chemical hazards	• Products, intermediates, by-products, and waste • Proprietary chemicals • Emission points to atmosphere for chemical agents • Building construction materials • Fumes/gases emitted from welding, cutting
Biological hazards	• Water-borne bacteria • Bacterial growth in air conditioning
Radiation hazards	• Infrared/ultraviolet light radiation • Ionizing radiation
Mechanical hazards	• Collision with moving vehicles • Sharp objects • Moving parts

the same work environment. Establishing the ideal thermal environment to suit everyone can be difficult. Using room temperature as an example, it is unlikely that it will be considered "ideal or comfortable" by more than 60% of people (McKeown and Twiss, 2004). Despite the difficulties in satisfying individuals, it is important to assess the thermal environment and thermal comfort factors.

AIR TEMPERATURE

Air temperature is the temperature of the air surrounding an individual and is typically measured in degrees Celsius (°C) or degrees Fahrenheit (°F). It is traditionally measured using mercury in glass thermometer, with thermocouples and thermistors being used more recently (Wilson and Corlett, 2005).

RADIANT TEMPERATURE

Radiant temperature is the heat radiated by a heat source such as a furnace, a fire, or the sun. The temperature is measured with a black globe thermometer and recorded in Celsius (°C) or degrees Fahrenheit (°F). The impact of radiant temperature is larger than that of air temperature, when considering the heat absorbed or lost by an individual to their environment.

HUMIDITY

Humidity is the amount of moisture in the air and is expressed as a percentage of Relative Humidity (%RH). Hygrometers containing wet and dry bulbs are used to calculate humidity; alternative methods include hair hygrometers and capacitance devices.

AIR VELOCITY

Air velocity describes the speed at which air is moving over an individual and is measured in meters per second (m/s). Kata thermometers or hot wire anemometers are used to measure air velocity.

METABOLIC RATE

The type of work undertaken has an impact on the perception of temperature. Tasks that are more physical in nature raise the metabolic rate, the body produces heat, and the individual feels hotter. Tasks that are less physically demanding are associated with a lower metabolic rate, less body heat is produced, and individuals feel cooler than someone performing more physical tasks. Metabolic rate is measured in kilocalories. Figures for common work tasks have been established using laboratory and mobile data collection.

CLOTHING

The clothing worn by individuals has a significant effect on the thermal comfort; too many items of clothing may lead to the individual becoming too hot or to be too cold if they are not wearing adequate layers of insulation. Clothing can be used to self-regulate thermal comfort, with individuals adding or removing layers of clothing to adapt to their immediate environment. Personal Protective Equipment (PPE) can be a source of thermal discomfort for individuals, especially if it is not carefully selected as it can lead to the individual being too hot or too cold and affects their ability to conduct their tasks. Measurements of clothing and insulation are measured in CLO and TOG respectively and used to calculate thermal comfort (Wilson and Corlett, 2005).

PSYCHOLOGICAL AND PHYSICAL EFFECTS OF THERMAL COMFORT

Thermal discomfort caused by the internal or external environment being too hot or too cold may lead to a variety of psychological or physical effects.

Mild thermal discomfort in hot environment can affect cognitive capability and concentration, and can lead to irritability and exhaustion. The physical effects associated with mild "hot" thermal discomfort include sweating and dehydration.

Extremely hot environments may lead to heat stress and cause heat stroke, delirium, and an inability to process information or communicate coherently. The physical symptoms range from fainting, progressive dehydration to systemic organ failure, and eventually death.

Mild thermal discomfort in cold environments may lead to reduced motivation and reduced ability to process information. Cold environments can cause physical reactions, including shivering and reduced blood flow peripherally to the hands and feet; causing issues with dexterity. Subsequently, this can have an effect on the ability to manipulate controls. Cold environments, including cold draughts can cause an increased risk of musculoskeletal injury.

Extreme cold environments can cause individuals to experience a loss of situational awareness and cause irrational behavior, as well as cardiac and respiratory failure, and eventually death.

The physiological, physical, and psychological impacts affect work capability and productivity. Thermal discomfort can cause staff to be less efficient and more prone to error, which increases the risk of injury or accidents.

ASSESSMENT OF THERMAL COMFORT

Thermal comfort is assessed using different approaches including thermal indices, thermal scales, and comfort surveys.

THERMAL INDICES

Wet Bulb Globe Temperature (WBGT) can be used to calculate a composite temperature figure which accounts for the measured air temperature, radiant temperature, and humidity. The formula is presented in Box 16.1.

BOX 16.1 WET BULB GLOBE EQUATION

Indoor environments or outdoors with minimal sun exposure (solar load):

$$WBGT = 0.7t_{nw} + 0.3t_g$$

Outdoor environments, including sun exposure (solar load):

$$WBGT = 0.7t_{nw} + 0.2t_g + 0.1t_a$$

where

t_{nw} = natural wet bulb temperature,
t_g = 150 mm diameter black globe temperature,
t_a = air temperature.

Taken from Wilson and Corlett (2005)

The composite temperature can then be reviewed in conjunction with measurements for likely metabolic work rate (based on tasks and rest) and the clothing being worn. This is used to define an exposure limit for users and judge the suitability of the environment using the guidance in Table 16.2 and the WBGT correction values in Table 16.3.

PMV AND PPD

The Predicted Mean Vote (PMV) is the most established thermal comfort scale, shown in Fig. 16.1 (adapted from ANSI/ASHRAE Standard 55, 2004). The scale can be used to survey a group within a working population and be used in conjunction with the six environmental and personal factors as previously discussed by applying equations. The result provides a measure of PMV for that population within that environment. In an internal environment, an acceptable range should be from −0.5 PMV to +0.5 PMV with a zero PMV being the ideal result.

The "Predicted Percentage of Dissatisfied" people (PPD) is related to the PMV (ANSI/ASHRAE Standard 55, 2004). As the PMV score moves away from zero, either towards −3 or +3, the PPD increases as more people are dissatisfied in an environment where they feel colder or hotter than they would prefer.

COMFORT SURVEYS

Comfort surveys are used to subjectively measure the work environment. The results summarize the subjective perception of thermal comfort in a specific work environment, and are used to derive control measures or interventions. It is recognized that it is difficult to satisfy all users sharing a work environment.

CONTROL MEASURES

A sample of acceptable work area temperature limits is presented in Table 16.4, taken from NORSOK (2004). These limits can be used to assess and monitor the environment. If temperatures fall out with these limits, control measures should be implemented by an organization.

Control measures to improve thermal comfort include administrative controls, engineering controls, acclimatization, and PPE/clothing:

- administrative controls include: rest periods and breaks, limits of exposure to an uncomfortable environment;
- engineering controls include: heaters and air conditioning;
- acclimatization controls relate to planning and monitoring acclimatization programs;
- PPE controls include: selecting clothing that is suitable for the thermal environment and using specialist PPE for extreme environments, such as gloves, hats, shoes, and jackets.

Table 16.2 WBGT Exposure Limits for Various Levels of Work and Workload

Work/Rest cycle: allocation of work	United States Department for Labor Occupational Health and Safety Administration (OSHA, 2015) Physical workload			Canadian Centre for Occupational Health and Safety (CCOHS, 2011) Physical workload			Work/Rest cycle: allocation of work
	Light	Moderate	Heavy	Light	Moderate	Heavy	
Continuous work	30.0°C (86°F)	26.7°C (80°F)	25.0°C (77°F)	31.0°C	28.0°C	Not specified	75–100% work
75% work (25% rest) each hour	30.6°C (87°F)	28.0°C (82°F)	25.9°C (78°F)	31.0°C	29.0°C	27.5°C	50–75% work
50% work (50% rest) each hour	31.4°C (89°F)	29.4°C (85°F)	27.9°C (82°F)	32.0°C	30.0°C	29.0°C	25–50% work
25% work (75% rest) each hour	32.2°C (90°F)	31.1°C (88°F)	30.0°C (86°F)	32.5°C	31.5°C	30.5°C	0–25% work

Table 16.3 Clothing Worn and WBGT Correction Values

Clothing Type	CLO Value	WBGT Correction Value (°C)
Cotton trousers and shirt (Hanson and Graveling, 1999)	Not specified	+3.6
Lightweight clothing, appropriate for summer (OSHA, 2015)	0.6	0
Light work clothing (Mital et al., 2000)	Not specified	0
Cotton overalls (OSHA, 2015)	1.0	−2
Cotton overall, jacket (Mital et al., 2000)	Not specified	−2
Work clothing, appropriate for winter (OSHA, 2015)	1.4	−4
Winter work clothing, double cloth coveralls, water barrier (Mital et al., 2000)	Not specified	−4
Permeable, water barrier (OSHA, 2015)	1.2	−6
Light weight vapor barrier suits (Mital et al., 2000)	Not specified	−6
Vapor-barrier suit, hood, gloves, boots (Hanson and Graveling, 1999)	Not specified	−7
Fully enclosed suit with hood and gloves (Mital et al., 2000)	Not specified	−10

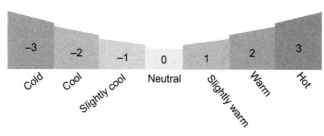

FIGURE 16.1

PMV scale.

Table 16.4 Recommendations for Work Area and Temperatures

Work Area	Temperature Min/Max (°C)
Utility and general process areas	Outdoor temperature or 5–35
Meeting rooms and offices	20–24
Stores (storage of large parts)	16–26
Machinery room (unmanned)	5–35

LIGHTING

For most tasks, vision is the main sense for receiving information and lighting is a critical element for achieving good performance. Visual comfort and performance is achieved through appropriate work area lighting and through avoiding glare, reflection, and shadows. Poor lighting and poor visibility can contribute to accidents and poor performance as well as "momentary blindness" (momentary low field vision due to the eyes adjusting from brighter to darker, or vice-versa, surroundings).

When assessing lighting in the workplace, the following aspects should be taken into account:

- the adequacy of the lighting in the different work areas;
- the presence of glare, poorly distributed light or flicker;
- maintenance arrangements for lighting;
- arrangements for emergency lighting.

LIGHTING OF WORK AREAS

Illuminance is the amount of light emitted from a light source falling on to a surface. Illuminance is measured in lux (lx). In the open during daylight hours illumination levels vary between 2000 and 100,000lx. At night, artificial light levels are typically between 50 and 500lx.

The required level of illuminance should be determined by the nature of the work being undertaken and in particular, the level of detail of visual information. The age of the operators, their quality of vision, and the speed and accuracy required by the task also need to be considered.

In the United Kingdom, the Health and Safety Executive (HSE) provide guidance in for lighting levels for different types of work HSG38 (1997), presented in Table 16.5 The Chartered Institute of Building Services Engineers (CIBSE) (2004) and BS EN 12464-1 (2002) also provide lighting levels for indoor work places.

Work area lighting levels recommended by CIBSE (2004) relevant to the chemical and process industries are presented in Table 16.6.

Large differences in illuminance between adjacent areas can reduce visual comfort and visibility leading to potential safety issues; this is typical when going between indoor and outdoor working areas. Guidance for the maximum ratios of illuminance between adjacent areas is presented in Table 16.7 (adapted from HSG38, 1997).

Where there is a difference between the recommended average illuminance level and the maximum illuminance ratios, the higher value should be used, at least where the areas are closest, such as doorways.

Lighting sources within the workplace are usually made up of variable proportions of natural daylight and artificial light. The amount of daylight depends on the time of day and year, global location, and the design and layout of the building. Artificial light is required to supplement natural light.

Table 16.5 Lighting Levels

| Activity | Typical Location or Work | HSE Guidance | | CIBSE Guidance |
		Average illuminance (lx)	Minimum illuminance (lx)	Standard maintained illuminance (lx)
Movement of people, machine, and vehicles	Parking area, corridors, bulk stores	20	5	100
Movement of people, machine, and vehicles in hazardous areas	Construction site, loading bays, plant rooms	50[a]	20[a]	150
Working requiring limited perception of detail	Large components assembly, storage rooms	100[b]	50[b]	200
Work requiring perception of detail	Offices	200	100	500
Work requiring perception of fine detail	Laboratories, control rooms	500	200	750–1000

[a]Guidance given taking only safety into account, where perception of detail is required or the need to recognize a hazard, the figure should be increased for work requiring perception of detail.
[b]Guidance given with the purpose of avoiding visual fatigue.

Artificial lighting can be divided into three main categories:

- general lighting—used for general illumination of an area, for example, by a ceiling fixture in a room;
- localized general lighting—provided by overhead or wall fixtures in addition to general lighting to increase the lighting levels at a particular work area, for example, at a work bench or console;
- task lighting—provided for increased illumination over for example a computer workstation, typically controlled by the user.

The type of light fixtures need to be considered in relation to the work area. These are classified as:

- direct—these project almost all of the light downward and can create shadows or glare so should be used with care, especially with low ceilings;
- direct-indirect—these distribute light equally upwards and downwards with only a little light being transmitted horizontally which reduces glare;
- indirect—these project almost all of the light upward so ceilings and upper walls must be clean and reflective to allow the light to project back to the

Table 16.6 Recommended Lighting Levels (CIBSE)

Area	Recommended Lighting Level (lx)
General office	500
Computer workstations	300–500
Computer-aided design (CAD) areas	300–500
Tool shops	300–750
Spot welding	500–1000
Inspection and testing	500–2000
Rest rooms	150
Entrance halls	200
Gatehouses	200
Corridors and stairs	100
Entrances and exits	200
Boiler house	100
Control room	300
Mechanical plant room	150
Electrical plant room	100
Loading bays	150
Warehouse/bulk stores	100

Table 16.7 Maximum Ratios of Illuminance for Adjacent Areas

Situation	Typical Location	Maximum Ratio
Where each task is individually lit and the surrounding area is lit to a lower illuminance	Local lighting in a control room	5:1
Where two working areas are adjacent but one is lit to a lower illuminance	Localized lighting in a materials store	5:1
Where two work areas are lit to different illuminances, and are separated but with frequent movement between them	Materials store area inside the warehouse and a loading bay outside	10:1

work area. They cause the least direct glare and are usually used in office type environments;

- shielded—these use diffusers, lenses, and louvers to cover bulbs from direct view so reducing glare but distributing light.

EFFECTS OF INAPPROPRIATE LIGHTING

The following effects can be the result of unsuitable lighting for the work being undertaken.

Glare

Glare is experienced when parts of the visual field are excessively bright compared with the general surroundings, where a bright light source or reflection interferes with how a person sees an object. It causes the eyes to adapt to the brighter light making the detail on the darker work area more difficult to see.

Direct glare occurs when the eyes have to look directly into a light source, such as the headlights of an oncoming car. Indirect glare is reflected from a surface into the eyes, such as the headlights of a car reflected in the rear view mirror or the sun coming through a window reflecting on to a monitor. Both types of glare need to be avoided as they make visual work more difficult, increasing the risk of error or injury. Some ways to avoid or correct glare include:

- use of indirect or shielded lighting fixtures;
- use of several low intensity fixtures rather than one large intensity fixture;
- increasing the brightness of the area around the glare source as this reduces the contrast between the source and the background;
- use of matt work surfaces rather than shiny or polished surfaces;
- positioning workstations so that light fixtures or windows are parallel to the operators line of sight, rather than directly in front or behind them;
- avoiding workstations being positioned directly below light fixtures;
- providing adjustable window coverings.

Poorly distributed light

When light is poorly distributed, part of the work area will be dark and dull and this reduces the ability to detect detail. This is due to insufficient light intensity or excessive differences in illumination making adaptation of the eyes slower. This can be overcome by applying the illuminance ratios mentioned previously in Table 16.7 and by using light colors on ceilings and walls to enable a more even distribution of light and softening of shadowed areas.

Flicker

Light flicker is rapid repeated changes in light intensity so that the light appears to flutter and be unsteady (CCOHS, 2013). At certain frequencies the eye is sensitive to flicker and it can cause visual discomfort, visual fatigue, and distraction. Some individuals are particularly sensitive to flicker, such as people with epilepsy, where flicker at certain frequencies can lead to seizures. The following interventions are recommended to resolve issues related to flicker (HSG38, 1997):

- change lamps (bulbs or tubes) near the end of their lives according to a planned maintenance program;
- check electrical circuits for faults;
- use high frequency control gear.

LIGHTING MAINTENANCE

A planned maintenance program should include cleaning and replacement of lamps, maintenance of fittings, and an assessment of illuminance. Illuminance levels of lamps decline with age and can be assessed using a standard photometer (light meter). CIBSE (2004) provides guidance on the maintenance of illuminance levels.

EMERGENCY LIGHTING

Emergency lighting is required in all areas where operators may be at risk if the standard lighting fails. There are two forms:

1. Standby lighting—this enables operators to continue with essential work. The illuminance required is dependent on the work taking place but can range from 5% to 100% of the normal levels;
2. Escape lighting—this enables safe evacuation of a building. BS 5266 (1999) recommends that escape lighting should reach necessary illuminance levels within 5 seconds of failure. This can be extended to 15 seconds if operators are familiar with the work area. Illuminance levels for escape routes should follow the levels for corridors (100 lx) as shown in Table 16.7. Battery or generator powered escape lighting should be designed to operate for 1–3 hours to ensure everyone is able to leave the building safely.

NOISE

High noise levels can cause hearing damage and adversely affect concentration, communication, and cognitive capability. This may affect an individual's ability to safely execute a task and may lead to an increase in error.

ASSESSING NOISE EXPOSURE

Noise exposure should be monitored and mitigated if they exceed regulatory standards or cause task issues.

Noise exposure levels are typically measured in decibels (dB) using a sound level meter. The human hearing response is nonlinear so an increase of 3 dB actually represents a doubling of the sound intensity. Sound level meters use scales of sound as follows (Wilson and Corlett, 2005):

- A-weighting—measuring low intensity;
- B-weighting—measuring moderate intensity;
- C-weighting—measuring high intensity.

These are measured and recorded as dB(A), dB(B), and dB(C), with dB(A) reflecting the human range of sound detection.

The sustained noise level that personnel are exposed to can be represented as *Leq* which describes the average sound level over a set period of time. Other measurements include:

- peak noise levels;
- background noise levels;
- median noise levels.

Multiple noise measurements are typically presented on a map or plan of the workplace. This provides an overview of the working area and locations of the higher noise sources. Measurements are compared with international standards for exposure limits relevant to the type of work environment giving consideration to the variety of tasks being conducted and the equipment being used. Control measures should be implemented to reduce noise exposure as necessary.

SHORT-TERM NOISE EXPOSURE EFFECTS

Short-term exposure to high noise levels can lead to:

- temporary hearing loss; full loss or reduced ability to hear within a person's normal range;
- tinnitus; which is described as ringing in the ears and can affect sleep;
- inability to communicate;
- distraction and reduced concentration.

LONG-TERM NOISE EXPOSURE EFFECTS

Prolonged exposure to high noise levels can lead to the same effects except that the damage may be permanent.

NOISE EXPOSURE REQUIREMENTS

Noise exposure limits are provided by regulatory bodies and standards organizations such as ISO. In the United Kingdom, the limits are as follows (HSE, 2012):

- daily or weekly personal noise exposure
  lower exposure action value—80 dB, based on an exposure duration of 8 hours;
  upper exposure action value—85 dB, based on an exposure duration of 8 hours;
- peak sound pressure
  lower exposure action value—135 dB;
  upper exposure action value—137 dB.

It also states that prolonged very low or very high frequency noises should be avoided, and noise levels should not interfere with communications, warning signals, or mental performance.

NOISE DISTRACTION

Noise distraction is relative to the context of the task. Individuals undertaking tasks that require greater concentration may be distracted by much lower noise levels, whereas he or she may tolerate higher levels if little concentration is required.

The tasks conducted by operators in control rooms are typically safety critical and require a high level of concentration. However, control rooms are often used as a central hub for handovers, meetings, issuing permits, and as traffic routes to other areas. This can increase the noise levels in the control room and cause distraction or prevent important communications being heard. A background level of 55–60 dBA is suggested by Invergård and Hunt (2009) as this is less likely to interfere with communication between operators, which would likely be between 60 and 65 dBA at a distance of 1 m. Control measures should be implemented to ensure the background noise does not distract operators. This may require consideration of traffic routes, the separation of tasks/roles, and noise attenuation measures.

NOISE CONTROLS

There are a range of controls that can be implemented to reduce the noise levels in a workplace. These should be considered in accordance with the hierarchy of controls shown in Fig. 16.2 (HSE, 2015).

The first consideration should be the elimination of the noise source followed by substitution (such as alternative equipment) to reduce noise levels.

Engineering controls include:

- sound proofing of the workspace or work area;
- sound proofing noise sources, for example, by encasing equipment or machinery;

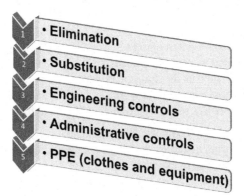

FIGURE 16.2

Hierarchy of controls.

- regular maintenance to reduce the peak noises levels;
- using more reliable equipment to reduce exposure times;
- using noise attenuation measures, such as rubber mounts.

Organizational controls include;

- health surveillance;
- task rotation to reduce the duration of exposure;
- hearing protection should be suitable for the noise environment and provide a good fit.

VIBRATION

Prolonged exposure to vibration can lead to musculoskeletal and body part discomfort and may result in permanent injury.

A commonly known health effect of exposure to vibration is Hand Arm Vibration Syndrome (HAVS) which is caused by vibration transmitted to the hand and arm during the operation of handheld power tools, such as grinders, cutting equipment, or powered hammers. It can also be caused by the operation of hand guided equipment or by holding materials being processed by machines. Symptoms typically include:

- tingling and numbness in the fingers which can result in an inability to do dexterous work (such as, assembling small components) or everyday tasks (such as, fastening buttons);
- loss of strength in the hands;
- fingers going white (blanching) and becoming red and painful on recovery.

Factors which may increase the risk of HAVS include:

- the characteristics of the vibration (vibration magnitude (expressed in m/s^2));
- tool design including weight and center of gravity;
- tool maintenance;
- tool handle design;
- type of material being worked upon;
- awkward posture and working overhead;
- work organization, such as the duration of exposure;
- environmental factors such as cold work;
- individual characteristics such as smoking and medical history.

Whole Body Vibration (WBV) also has several health effects such as spinal, analrectal, and gastrointestinal disorders, but they have been difficult to verify in research as being conclusively related to exposure to vibration. Exposure to WBV does induce physiological responses. The most basic is an increase in heart rate; about 10–15 beats/minute above resting levels. Blood pressure can also increase and some studies have revealed a slight increase in breathing rate and oxygen consumption that may be related to increased muscular activity.

At vibrations of about 10–25 Hz visual acuity decreases which can affect visual tasks. Workers typically affected include drivers, particularly drivers of heavy plant vehicles and especially off-road use. Vibrations are transmitted via the seat to the buttocks, from the vehicle floor to the feet, and from the vehicle headrest to the head. Vibration through the feet can also be a problem for staff standing on platforms of machinery that vibrate.

WBV is relevant for all platforms subject to motion, but particularly to vehicles and may be affected by factors such as:

- road surface condition;
- vehicle design and age;
- vehicle suspension system and maintenance;
- seat design, suspension level, and condition;
- vehicle cab layout that causes the adoption of poor postures;
- work organization such as spending long periods in the vehicle without breaks;
- an individual's skill and technique for driving, such as driving too fast or not avoiding uneven surfaces and pot holes;
- lighting, for example, poorly lit areas may make it harder for drivers to avoid uneven surfaces;
- Cold temperatures which may encourage staff to spend longer in the vehicles.

Standards for assessing vibration such as ISO 5349-1 (2001a,b), ISO 5349-2 (2001a,b), and EN 14253 (2003) differentiate between HAVS and WBV.

The key factors are:

- the level of vibration;
- the duration of exposure;
- the design of the equipment used, including seating and handle types;
- the postures adopted to use the equipment;
- other environmental factors such as lighting and temperature.

MEASURING VIBRATION LEVELS

Exposure Action Values (EAVs) and Exposure Limit Values (ELVs) are set in the United Kingdom for HAVS and WBV to protect employees at risk from vibration (HSE, 2005).

EAV defines the level of vibration at which action must be taken to reduce exposure. ELV sets the maximum level of vibration that should not be exceeded in any single day. Both values are calculated based on the average exposure experienced over an 8-hour period.

For HAV:

- the EAV is 2.5 m/s² A(8);
- the ELV is 5.0 m/s² A(8). *This means that a tool with a vibration value of 5.0 m/s² should be used for no more than 8 hours in a single day. A tool with a vibration value of 10.0 m/s² should be used for a maximum of 4 hours in a single day.*

The HSE provides an online HAV exposure calculator to support the assessment of the Partial Exposure score for each individual tool and a Total Exposure score (shown in m/s^2 A(8)). The calculator also calculates the time to reach the EAV and ELV.

For WBV:

- the EAV is 0.5 m/s^2 A(8);
- the ELV is 1.15 m/s^2 A(8).

Manufacturers of plant and machinery must prove that their products do not exceed the maximum levels and organizations should be given access this information. The HSE provides a similar exposure calculator for WBV and the manufacturer's data are used with exposure duration to produce Partial Exposure and Total Exposure scores.

CONTROLLING THE RISKS OF VIBRATION

Should an assessment indicate that vibration is a problem, the first action is to determine whether the source can be eliminated. Where this is not the case the hierarchy of control should be followed (see Table 16.8).

Table 16.8 Controlling the Risks from Vibration

Hierarchy	HAV	WBV
Substitution	Replace the tool with one producing a lower vibration magnitude.	Replace the machinery with those designed and constructed to reduce WBV and state vibration emissions.
Engineering controls	Ensure tools are well balanced and avoid tools which exceed 2 kg if they are handheld without additional support.	Select seating to minimize vibration, for example, suspension seating.
	Ensure handles are designed to absorb vibration, for example, rubber not wood/metal.	
Administrative controls	Reduce exposure through work schedules.	Reduce exposure through work schedules.
	Train operators in the use of tools and best postures to adopt.	Train operators to adjust seating and controls and to drive at speeds suitable for the road conditions.
	Ensure the workstation layout is assessed.	Educate operators in WBV and ensure they are aware of the reporting system in place.
	Educate operators in HAVS and ensure they are aware of the reporting system in place.	Maintain machinery and road surfaces.
	Maintain tools.	Provide health surveillance.
	Provide health surveillance.	
PPE	Provide suitable PPE for the job and the work environment, for example, gloves and warm clothing.	Provide suitable PPE for the job and the work environment, for example, warm clothing.

SUMMARY: ENVIRONMENTAL ERGONOMICS

The design and modification of the work environment should incorporate human factors principles to prevent adverse physiological and psychological ill health. The standards and guidance referred to within this chapter can support organizations in managing environmental factors which may otherwise have a detrimental effect on safety and/or productivity.

KEY POINTS

- Poorly designed work environments or insufficient management of environmental factors can have a significant effect on health, well-being, and performance.
- Thermal comfort should be assessed to reduce the risk of adverse physical and psychological effects.
- The lighting should be appropriate for the tasks being undertaken and the workspaces.
- Noise exposure must be within safe limits to reduce the risk of hearing damage.
- Noise can also cause distractions and communication difficulties.
- Prolonged exposure to vibration can lead to discomfort, ill health, and, if prolonged, potential injury.
- Factors which can or may affect exposure to vibration must be assessed and suitable control measures put in place.

REFERENCES

ANSI/ASHRAE Standard 55, 2004. American Society of Heating, Refrigerating and Air-Conditioning Engineers, Inc. ISSN 1041-2336.

BS 5266-1, 1999. Emergency Lighting: Code of Practice for the Emergency Lighting of Premises Other Than Cinemas and Certain Other Specified Premises Used for Entertainment. British Standards Institute 1999, ISBN 9780580330445.

BS EN 12464-1:2002 Light and lighting-Lighting of work places-Part 1: Indoor work.

Canadian Centre for Occupational Health and Safety, 2013. Lighting Ergonomics—Light Flicker. Available from: <http://www.ccohs.ca/oshanswers/ergonomics/lighting_flicker.html>.

Code for Lighting CD-ROM CIBSE (Chartered Institute of Building Services Engineers), 2004. ISBN 9781903287224.

European Committee for Standardization (CEN), 2003. Mechanical Vibration Measurement and Calculation of Occupational Exposure to Whole-Body Vibration with Reference to Health Practical Guidance (Standard No. EN 14253:2003). CEN, Brussels.

Griffin, M.J., 1996. Handbook of Human Vibration. Academic Press, London.

Hanson, M.A., Graveling, R.A., 1999. Development of a Draft British Standard: The Assessment of Heat Strain for Workers Wearing Protective Equipment. Historical Research Report, Research Report TM/99/03 1999. Institute of Occupational Medicine (IOM), Edinburgh, (PDF; Online) Available from: <http://iom-world.org/pubs/IOM_TM9903.pdf> (accessed 15.02.16.).

Health and Safety Executive (2005) Control of Vibration at Work Regulations 2005. Guidance on Regulations L140 HSE Books 2005 ISBN 978 0 7176 6125 1.

Health and Safety Executive (HSE), 2015. Construction Physical Ill Health Risks: Noise (Online). Available from: <http://www.hse.gov.uk/construction/healthrisks/physical-ill-health-risks/noise.htm> (accessed 22.12. 15.).

Health and Safety Executive (HSE), Control Room Design (Online). Available from: <http://www.hse.gov.uk/comah/sragtech/techmeascontrol.htm> (accessed 22.12.15.).

Health and Safety Executive (HSE), 2011 Leadership and Worker Involvement Toolkit, Management of Risk While Planning Work: The Right Priorities (Online). Available from: <http://www.hse.gov.uk/construction/lwit/assets/downloads/hierarchy-risk-controls.pdf> (accessed 22.12.15.).

Health and Safety Executive (HSE), 2012. Noise at Work. A Brief Guide to Controlling the Risks. Published 11/12 (PDF; Online). Available from: <http://www.hse.gov.uk/pubns/indg362.pdf> (accessed 22.12.15.).

Hot Environments—Control Measures, OHS Answers Factsheets, Canadian Centre for Occupational Health and Safety (CCOHS), (2011) (Document last updated on August 22, 2011) (Online). Available from: <https://www.ccohs.ca/oshanswers/phys_agents/heat_control.html> (accessed 15.02.16.).

International Organization for Standardization (ISO), 2001a. Mechanical Vibration Measurement and Evaluation of Human Exposure to Hand-Transmitted Vibration—Part 1: General Requirements (Standard No. ISO 5349-1:2001). ISO, Geneva.

International Organization for Standardization (ISO), 2001b. Mechanical Vibration Measurement and Evaluation of Human Exposure to Hand-Transmitted Vibration—Part 2: Practical Guidance for Measurement at the Workplace (Standard No. ISO 5349-2:2001). ISO, Geneva.

Invergård, T., Hunt, B., 2009. Handbook of Control Room Design and Ergonomics. A Perspective for the Future, second ed. CRC Press, Boca Raton, FL.

Lighting at Work HSG38, second ed. 1997, HSE Books ISBN 9780717612321.

McKeown Dr., C., Twiss, M., 2004. Workplace Ergonomics: A Practical Guide. IOSH Services Limited.

Mital, A., Kilbom, Å., Kumar, S., 2000. Ergonomics Guidelines and Problem Solving, Vol. 1. Elsevier Science, Oxford.

Norsk Sokkels Konkuranseposisjon (NORSOK), S-002, 2004. Working Environment Thermal Environmental Conditions for Human Occupancy.

United States Department of Labour, Occupational Safety & Health Administration, OSHA, 2015. Technical Manual, Section III, Chapter 4 (Online). Available from: <https://www.osha.gov/dts/osta/otm/otm_iii/otm_iii_4.html#3> (accessed 22.12.15.).

Wilson, J.R., Corlett, N., 2005. Evaluation of Human Work. Taylor & Francis, Boca Raton, FL.

Human factors in the design of procedures

17

E. Novatsis and E.J. Skilling

LIST OF ABBREVIATIONS

COMAH	Control of Major Accident Hazards
CSB	Chemical Safety Board
HSE	Health and Safety Executive
HTA	Hierarchical Task Analysis
IT	Information Technology
LAN	Local Area Network
NOPSEMA	National Offshore Petroleum Safety and Environmental Management Authority
P&IDs	Piping and Instrumentation Diagrams
PPE	Personal Protective Equipment
SME	Subject-Matter Expert
TTA	Tabular Task Analysis

This chapter discusses the critical role of procedures within a management system that helps ensure safe and reliable operations. The inadequate management of procedures is one of the main contributing factors to incidents, evidenced through investigation reports of major accidents and summary reports of the underlying human factors causes of industry events.

Organizations therefore have a responsibility to ensure that procedures are available and easily accessible, are accurate and presented in a user-friendly way, and are followed by people. Doing so has many benefits for both individuals and the organization, such as enabling a repeatable standard of performance, and helping mitigate risk, particularly in relation to human failure.

A definition for procedures is provided, along with how procedures differ from, and link to, other types of documents in a management system. Procedures can be presented in different formats and guidance on how to determine an appropriate format is outlined. This is important to ensure that safety-critical procedures are recognized and to avoid overload of step-by-step procedures which may discourage use.

A process for procedure development is presented, with a focus on tools and methods used at the different stages. These methods include task analysis to generate an accurate picture of the task as well as how to engage end users in the development process to ensure compliance.

Principles for the presentation of procedures are described, and a list of items is outlined to help users design procedure templates and write procedures. Finally, this chapter provides tips and considerations for implementing programs to improve the usability of existing procedures.

DEFINING PROCEDURES

Documented procedures are an essential part of an organization's management system. Within this book, procedures are defined as "instructions for how to perform jobs." Organizations typically use procedures to define a range of activities, including operations, maintenance, emergency response, and support function tasks, some of which may be safety-critical. Procedures can be used to ensure that the right information is available to train and support personnel to perform tasks correctly and to mitigate risk, particularly in relation to human failure. Procedures therefore have a key part to play in reducing the potential for major accidents.

There are different types of procedures, which vary in their format and level of detail. The appropriate format depends on several factors: whether the task is safety-critical; the complexity of the task; how often the task is performed; and the competence of the user (UK Health and Safety Executive (HSE), 2015). A flowchart based on this guidance is presented in Fig. 17.1 to help users determine the most appropriate format of procedure to use. The guidance aims to prevent an overload of step-by-step procedures.

Step-by-step procedures are the main type of procedure discussed in this chapter. In some circumstances, job aids may be sufficient and are an important tool to guide and assure performance. Common types of job aid include:

- *Checklists*—for example, to systematically check all important elements have been considered;
- *Diagrams*—for example, to illustrate the layout of a unit;
- *Flowcharts*—for example, to show how a control of work system operates.

It is important that organizations distinguish between procedures and other types of document but show how they interlink with each other. Often organizations blend "how to" procedural documents with guidance or standards, making it difficult for end users to follow the procedural component. The common types of documents within management systems, aside from procedures, are defined as follows:

- *Policies*—statements about strategies for realizing business objectives. These documents are usually corporately owned, limited to the number of business processes within the organization, and are typically one page in length;
- *Standards*—minimum mandatory legal and/or business requirements to support the policies. Organizations typically aim to limit the number of standards by

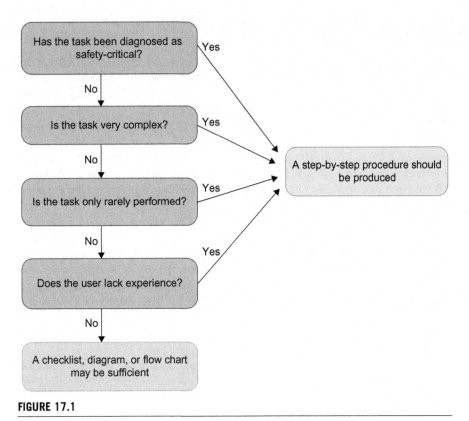

FIGURE 17.1

Flow chart for deciding on the format of procedures.

defining them at a corporate level. Sometimes local standards are required when the context of the operations differs substantially, such as operating in another country where legislation may differ;

- *Guidelines*—supporting documents to procedures and standards, providing more detail and references, including educational information on a topic.

Organizations usually distinguish between documents that indicate business requirements ("what" is required), and performance expectations ("how" it is required). Policies, standards, and guidelines define the "what" and "why," whereas procedures define the "how" and "when." Models like this explain the relationships between documents, in particular where standards support the policies, procedures support the standards, and guidelines support procedures. Such a structure can provide the confidence for end users that following procedures also ensures compliance with higher level standards and policies.

LESSONS FROM PAST INCIDENTS

The International Association of Oil and Gas Producers Incident Analysis (2013) has shown inadequate work standards/procedures to be one of the top 10 causal factors of fatal and high potential incidents each year from 2010 to 2013. Similar trends have been reported by the National Offshore Petroleum Safety and Environmental Management Authority (NOPSEMA, 2015), where issues with procedures were identified as one of the five top root causes for offshore incidents in Australia each year from 2005 to 2014. Moreover, analysis by Hare et al. (2015) on dangerous occurrences reported to the UK HSE from the chemical industries between 2007 and 2013 showed that operations and maintenance procedures were two of the common underlying causes of these incidents.

The inadequate management of procedures has contributed to well-known incidents. For example, the investigation into Piper Alpha found that the maintenance procedures had not been followed prior to the incident (Cullen, 1990). The US Chemical Safety Board (CSB) investigation into the Texas City Refinery explosion found that underlying conditions encouraged operators to consciously deviate from written operating procedures and that procedures were out of date and no longer relevant (CSB, 2007). It is clear why the subject of usable procedures is recognized by the UK HSE as a key human factors topic that needs to be managed by organizations working in hazardous industries.

Managing procedures well has a number of individual and organizational benefits, as shown in Box 17.1.

WHY PROCEDURES FAIL

Although procedures are a key element of a safety management system and an important training tool (HSE, 1999b), organizations may rely too heavily on procedures as their primary means of controlling risk, and fail to apply an appropriate hierarchy

---

BOX 17.1 BENEFITS OF USABLE PROCEDURES

✓ Help achieve a repeatable and recognized standard of performance;
✓ Enable training and competency measurements to be more consistent;
✓ Enable consistency in reviewing standards of performance;
✓ Support optimal performance, for example, personnel do not have to rely on memory of how to perform tasks;
✓ Provide a record for the organization on how to perform tasks;
✓ Achieve better compliance with procedures where end users have been involved in their development;
✓ Mitigate risk, particularly in relation to human failure;
✓ Meet legislative requirements, where procedures support a standard that defines minimum mandatory legal requirements.

of control. Other reasons for failure may be that personnel do not follow the procedures or, in some cases, the procedures may not be adequate. One frequently seen example is when experienced personnel become complacent and no longer use the procedure to guide them on task steps. For safety-critical tasks, the procedure is (or should be) designed to support and assure error-free performance. If the procedure is not used, it cannot help to prevent errors. For instance, a checklist or step-by-step procedure is there to make sure each step is carried out and in the right order. If a task is performed from memory, the chances of error are increased and the control measure is no longer functioning as intended.

Common types of procedural failures are presented in Fig. 17.2. The failures are divided into unintentional behaviors (errors) and intentional behaviors (noncompliance), with examples provided for each. These examples highlight that there are many ways people can fail to follow procedures. The underlying causes of these failures need to be thoroughly investigated.

The common underlying reasons why procedures fail are presented in Box 17.2.

It is important that the behaviors demonstrated by people within the organization support a culture of following procedures. This applies to all levels within an organization, and behavioral expectations should be made clear. For example:

- *All individuals need to:* learn the procedures that apply to them in their job; follow procedures; and identify and raise issues with procedures to their supervisor;
- *Supervisors need to:* ensure that their teams comply with procedures, and support them to address issues with procedures;

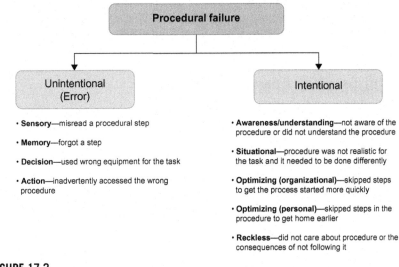

FIGURE 17.2

Common types of procedural failures.

BOX 17.2 COMMON REASONS WHY PROCEDURES FAIL

- *Missing or not accessible*—a procedure for the task may not be available, it may be missing or difficult to find or access in the hard copy or electronic storage system.
- *Inaccurate or incomplete*—the procedure may not be an accurate or a complete representation of the task.
- *Unrealistic or unfeasible*—it may not be possible to practically complete the task in the way described in the procedure.
- *Poorly written*—the procedure may be overly long, use complex language, or provide ambiguous instructions, making it difficult to read.
- *Poorly presented*—the layout of the procedure may be difficult to follow because it is not presented in sequential order, may combine task steps, or use a small font.
- *Too many procedures, overload*—providing too many procedures, for example, providing procedures for all simple and frequent tasks, can overwhelm people and discourage use.
- *Incorrect versions*—when versions of the procedure are not controlled effectively, for example, when multiple versions are uploaded to an electronic system and are not well labeled, people may select and use an old version.
- *Following procedures is not managed and reinforced*—the purpose and importance of procedures and following them may not be actively supervised and reinforced as part of the safety culture, making the use of procedures appear "optional."

- *Managers need to:* explain the expectation for following procedures to the workforce, and not tolerate shortcuts to procedures.

Applying a procedure can be likened to completing a jigsaw. A jigsaw is easier to complete if there is an accompanying picture and all the pieces needed are available and correct. People are more likely to comply with a procedure if they have a clear understanding of the purpose of the task, if all the information is available and there is no unnecessary information to confuse or distract them. The remainder of this chapter explains how to ensure the jigsaw is easy for people to complete.

PROCEDURE DEVELOPMENT PROCESS

The HSE (2015) recommends that all Control of Major Accident Hazards (COMAH) sites should have a procedure for managing procedures. This procedure should:

- Identify which tasks need procedures and the detail that is required;
- Identify the procedure owner;
- Define how procedures will be kept up to date, managed for change and distributed;
- Define how to ensure the involvement of end users, consistency with other information such as signage and labeling, adequate training in procedures, competence of personnel, and the compliance of personnel with procedures.

An overview of the process for developing procedures is presented in Fig. 17.3. This process and the methods within it are crucial for ensuring that the content of

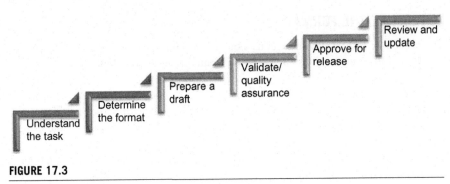

FIGURE 17.3

Procedure development process.

procedures is accurate. These development steps are described in the following subsections.

UNDERSTAND THE TASK

Organizations need to be clear about which of their procedures are safety-critical. A safety-critical task can be defined as a task which, if carried out incorrectly or not at all, could lead to serious plant damage, loss of containment, injury, or fatality (Institute of Petroleum, 2003). Other definitions and alternative methods are used to identify safety-critical tasks. The Energy Institute uses a rating scale taking account of (1) the consequences of human failure and (2) the level of human involvement (Energy Institute, 2011). Another method is presented by the UK HSE (1999a) and is discussed in Chapter 6, Human Factors in Risk Management.

It is important to ensure the task is fully understood regardless of whether it is safety-critical or not. The main method used in support of this step is task analysis (Kirwan and Ainsworth, 1992), which is a systematic method for analyzing a task in terms of its goals, operations, and plans. First, a Hierarchical Task Analysis (HTA) is often performed to define the task structure and the associated actions. In a HTA, the task is split into a number of subtasks that can then be further divided to produce a tree structure (Kirwan and Ainsworth, 1992).

A Tabular Task Analysis (TTA) is then performed to document further detail about each step within the task, including, for example, who does what, necessary equipment, hazards which may be present, and potential errors. If it has not already been undertaken, Human Reliability Analysis (HRA) should be undertaken for safety-critical tasks. Practical instructions for task analysis methods are provided by the Energy Institute (2011).

Walking and talking through the task with experienced users or Subject-Matter Experts (SME) is a key part of the analysis. This helps to ensure that the information collected is correct and provides an accurate representation of the task. The information gathered using the task analysis can be used to develop a draft of the procedure as it provides the basis for the step-by-step procedure.

DETERMINE THE FORMAT

The procedure format should be consistent across all procedures or classes of procedure. The following sections should be built into procedure templates.

- *purpose statement*—a brief high-level goal or intent of the procedure;
- *precautions and hazards, with explanations of why they are issues*—include only the key health, safety, environmental, and equipment hazards for the task, not all issues from the risk assessment;
- *equipment, tools, and Personal Protective Equipment (PPE) required*—include any special operating, maintenance equipment and PPE, or materials that must be in place, available or obtained before starting the procedure. Standard PPE does not need to be listed;
- *initial conditions to be satisfied before starting the activity*—include any conditions that must be met or actions which must be performed before starting the procedure such as permit preparation or isolations, and key competency requirements needed by personnel conducting the procedure;
- *references to relevant documents*—include any references, such as Piping and Instrumentation Diagrams (P&IDs) essential to implementing the procedure;
- *procedure overview*—include a high level overview of the task steps, using this structure to present detailed task steps;
- *sequential procedure steps*—present the steps of the task in the order they are to be performed.

Sometimes more than one format is required. For example, a safety-critical procedure may require each task step to be signed off. A different template may be required for standard operating procedures which may limit sign-off to critical steps or "hold points." An emergency response procedure may include decision-action diagrams to aid decision-making (e.g., if condition "X," perform action "Y").

PREPARE A DRAFT

Procedures should be drafted by individuals who are familiar with the task, ideally the SMEs who actually perform the work. A team approach can be used, especially for tasks that involve expertise from different disciplines or end users from different areas. Teams may include operators, engineers who design the equipment, safety personnel, contractors, and technical writers. This approach promotes "buy-in" from potential users of the procedure and improves compliance with use.

The level of detail that the procedure provides depends on the decisions made by the organization regarding who the procedures should be designed for, and their level of experience. However, a general rule of thumb is that procedures should be written with sufficient detail so that someone with appropriate technical competence but limited experience or knowledge of the procedure can successfully implement the procedure. Organizations may need to consider the suitability of the procedure for use by contractors or those unfamiliar with local terminology and work practices. Less experienced users may require a more detailed version of a procedure, whereas experienced users may be able to use shorter job aids for guidance and task/error assurance.

VALIDATE/QUALITY ASSURANCE

On completing a draft of the procedure, it should be validated prior to obtaining approval for release to ensure that it meets the end user needs. This should be done by a walk through or, where possible, a test of the procedure within the workplace. The validation process helps to identify potential errors with the new procedure.

APPROVAL FOR RELEASE

Approval needs to be obtained from relevant parties, including end users. The timing for providing staff with training on new procedures needs to be carefully considered and rolled out in a coordinated manner. It is important to communicate the reasons for new or changed procedures to encourage buy-in and compliance. It is not usually enough just to circulate revised procedures to "read and sign off" particularly if the procedure is safety-critical or if there are significant changes.

REVIEW AND UPDATE

Once a procedure is in use, it is also important that it is reviewed and kept up to date. Procedures should be systematically reviewed on a periodic basis to ensure that they remain appropriate and relevant (US Environmental Protection Agency, 2007). A review date should be added to each procedure when it has been reviewed. It is necessary to control "procedural drift" by ensuring that the procedure is followed consistently over time. Feedback from personnel should be sought and, if the procedure is not being used, investigations need to be undertaken to find out why. For key procedures, ongoing audit arrangements can check task performance against the procedures concerned so that any drift is identified early. Supervisors can also do this less formally as part of their normal supervisory activities.

Applying these human factors principles for procedure development helps to reduce the likelihood of organizational incidents occurring as a result of how procedures are managed.

PRESENTATION OF PROCEDURES

There are a number of factors to consider when presenting procedures to help reduce the risk of error in use. Eight presentation topics are described and should be taken into account when writing procedures.

FORMAT

Procedures that are easy to navigate encourage personnel to use them. Whether procedures are provided as hard or soft copies, the following points should be applied:

- Index procedures for easy reference—the grouping of procedures should be logical and intuitive for end users;
- Ensure the hard copy or electronic system for packaging/storing procedures is easy to find, access, and navigate—for example, hard copy procedures should be provided in a hard-wearing ring binder, with dividers to organize procedures into sections which are logical for end users. The binder should be kept close to where end users will use it;
- Use a consistent format for all procedures, as previously explained.

Many organizations make procedures available on a Local Area Network (LAN) and/or providing procedures on handheld electronic devices (e.g., iPads). Critical procedures may also need to be available as hard copies, such as for emergency scenarios or scenarios where the LAN is unavailable. In this case laminated checklists or pocket cards should be made available to meet the needs of end users.

Table 17.1 presents the advantages and disadvantages of making procedures available on handheld electronic devices.

Table 17.1 Advantages and Disadvantages of Making Procedures Available on Handheld Electronic Devices

Advantages	Disadvantages
• Potential to use additional applications on the same device. For example, take photos, call up P&IDs and photos, access the LAN/document management system and intranet, video call in real time for technical help at the point of use • Access to related documents at the point of use (End users should be involved in organization of the documents to ensure they can easily find procedures, and not just make sense to a document controller or Information Technology (IT) professional) • Potential for the use of electronic forms and checklists • Easy to track who has accessed a document and when	• The cost involved in design, supply, and maintenance of the devices • Document updates require configuration management • Lack of industry evidence on the operational benefits of tablets or similar handheld devices • There may be a limited range of intrinsically safe devices for use in potentially explosive atmospheres • Potential failure of the battery or other technical issues during critical tasks • Questionable benefit for office-based functions. If they are only introduced for field-based roles, electronic and paper versions need to be managed

PRECAUTIONS

Procedures need to highlight key hazards and precautions in a simple and consistent manner. This section should be kept concise, and not include detailed lower risk hazard information that would be captured in a risk assessment. Procedures need to:

- include a separate section at the start of the document to highlight hazards and precautions;
- use consistent colors and everyday symbols to highlight hazards (e.g., ⚠);
- embed the hazard or precaution at the relevant task step, in addition to listing it at the start;
- include a brief description of what can happen, why, and the consequences of ignoring the precaution. This encourages compliance.

TASK STEPS

The task analysis helps determine the task structure and task steps in their order of sequence. The following guidance is presented regarding the task steps:

- limit task steps to one action per procedural step—for example, "insert the new gasket in the flange joint" is one action. "Insert the new gasket in the flange joint, then insert stud bolts and nuts, and bring flanges together" is three actions and not appropriate as a single procedural step;
- present the task steps sequentially; in the order that they should be performed;
- break down long procedures into smaller parts—a procedure overview section can be used;
- provide a means to keep track of task steps by using a reference for each step, for example, 1, 2, 3;
- highlight important safety-critical tasks and task steps.

LAYOUT

Procedures that are laid out with clear headings and open white space are easier to read. Apply the following points to ensure the layout is clear:

- use headings and subheadings from the task analysis to break a procedure down;
- make important headings larger instead of <u>underlining</u>;
- use open space in the printed text to avoid clutter;
- justify (align) text to the left—this makes the spaces between words equal distance and therefore easy to read;
- use one blank line between paragraphs of printed text;
- avoid indenting more than three adjacent blocks of text.

GRAMMAR AND LANGUAGE

The language used for procedures should be simple, precise, and action-oriented. Apply the following points to ensure that grammar and language in procedures is as clear as possible and not confusing or ambiguous:

- use simple, short, and concise sentences;
- use an active voice starting the statement with a verb so that it is interpreted as an action, for example, "remove the lid," as opposed to "the lid should be removed";
- avoid negative statements, such as "do not start opening if pipe is not fully depressurized." If a negative statement is needed, use only one, avoid double negatives;
- avoid nested statements as they are difficult to read, for example, "the operator, who must get authorization from the shift supervisor, will need to obtain the appropriate safety interlocks";
- describe precise actions with quantitative units, if relevant, for example, "hold the button for 10 seconds";
- present actions in the correct order, for example, "wait until the valve is fully open before adding liquid," as opposed to "do not start adding liquid until the valve is fully open";
- spell out numbers before a unit of measure—for example, "collect one 10 kg container," rather than "collect 1 10 kg container" which may be misread as a 110 kg container.

TERMINOLOGY

The terminology used in procedures needs to be clear, simple, and importantly, understood by end users. Apply the following points to ensure terminology is clear and suitable for end users:

- avoid complex language. It is better to use a simpler version of a word, for example, "start" rather than "commence";
- avoid ambiguity, such as "the valve on the right" which is orientation dependent;
- use terminology understood by end users;
- define abbreviations in brackets following the first use;
- provide a glossary of terms and abbreviations at the beginning of the procedure;
- avoid using technical jargon and language unless the expression is well understood; and
- use terminology consistently, that is, use the same term for the same meaning.

FONT

Features of the font type used can affect how easy the procedure is to read. The following points should be used when writing procedures:

- use a simple font of sufficient size, for example, Arial font at a minimum of 12 point;
- avoid the use of ALL CAPITALS unless highlighting information and do not overuse them. Capitals take longer to read than sentence case;
- use bold, italics, and underlining for highlighting only and ensure that the use is consistent.

VISUAL AIDS

Visual aids can be helpful to convey large amounts of information more succinctly and clearly than text. When using visual aids, the following points should be considered:

- use color sparingly and consistently and comply with color conventions familiar to end users. The document should never rely on color due to the use of black and white printing and because 10% of the population are color blind;
- diagrams or flow charts can be used to illustrate the process being described;
- photographs can be useful but care should be taken with regards to picture quality and ensuring that it presents sufficient detail;
- locate diagrams alongside the related text so they can be easily associated;
- use maps for emergency procedures, locations, and layout of areas with a consistent orientation;
- use tables, charts, and graphs for statistics, production data, and figures. The principles in Table 17.2 can be used to indicate the scale of the numbers:

Table 17.2 Best Practice for the Presentation of Numbers within a Procedure

Nondecimal Right-align numbers that do not contain decimals or the same number of decimal places	Decimals Align decimal points	Leading Zeros Provide a leading zero before the decimal point
600	28.56	0.125
5790	628.45	0.345
567,234	4567.890	0.222

An organization will benefit from using the presentation principles explained in this section to support procedure writers. Procedure development is a specific skill which requires training but, once accomplished, can make a significant improvement to the quality and usability of procedures.

PRACTICAL TIPS FOR IMPLEMENTING PROCEDURE IMPROVEMENT PROGRAMS

GAP ANALYSIS

It is important to conduct a thorough gap analysis of current practice in procedure development. The aim is to identify strengths and weaknesses across all relevant factors, including procedure access, content, presentation, and the use of procedures. Data should be collected across all functions (including operational, maintenance, and support functions) and different sites, as there may be different practices. A gap analysis should do the following:

- assess a range of procedures against the organization's principles for the presentation of procedures;
- review existing systems for managing procedures;
- review training materials for procedure development and use;
- conduct focus groups or interviews with a sample of end users to understand strengths and problems with procedures from their experience;
- review other available information that indicates the level of compliance with procedures (e.g., investigation reports, audit reports, survey results).

The gap analysis will provide useful information in identifying the extent of the improvement needed and areas of good practice to be retained and extended, as well as key development areas. The gap analysis process can also help the project team gain buy-in to the project.

EVALUATION CRITERIA

Evaluation criteria appropriate for the project objectives should be identified while the project is being designed. For example, if an objective is for the organization to improve the readability of procedures (see Box 17.3), a suitable set of criteria could include document length, number of words in the document, Flesch Reading Ease, and Flesch-Kincaid Grade Level. "Before" and "after" comparisons can be made on these measures to assess the level of improvement.

SELECTION AND TRAINING OF PROCEDURE WRITERS

Often a technical person, administrator, document writer, or person with spare time on their hands is given the job of rewriting procedures. They may not always be the

---

BOX 17.3 ASSESSING THE READABILITY OF PROCEDURES

Procedures and other documents can be tested for how difficult they are to read. Rudolf Flesch emphasized the importance of readability, and wrote many articles and books on the subject, including The Art of Readable Writing (1974). Flesch's Reading Ease formula has become a popular measure. It uses two variables, the number of syllables and the number of sentences for each 100-word sample, and predicts reading ease on a scale from 1 to 100. A score of 30 indicates "very difficult" and 70 "easy" (Dubay, 2004). The reading ease formula was later modified to develop another test called the Flesch-Kincaid Grade Scale, which produces a grade-level score (Dubay, 2004). The score indicates the US educational grade level required to read a document.

The Flesch Reading Ease Score and the Flesch-Kincaid Grade Level are built into the *Readability Statistics* in the *Proofing* function within Microsoft Word. When writing procedures, it can be useful to test documents using these measures to obtain an indication of their readability. As a general guide, aim for a reading ease score of 60–70 for procedures. Remember that the tool is simply measuring sentence and word length. Improving readability to get a better score means shortening sentences and minimizing the use of unnecessarily long words. However, it is important to ensure that the procedure still maintains logical flow and accuracy.

---

most suitable person for the job. It is recommended an appropriate set of selection criteria be developed, against which to select procedure writers. Once the writers have been selected, they should be provided with training on the human factors principles and methods covered in this chapter, along with project-specific information.

PILOTING THE APPROACH

As with many projects, it is useful to pilot the selected procedure improvement approach in one area prior to extending it to others. This allows changes to the approach and training to be made prior to wider implementation.

COACHING AND SUPPORT TO PROCEDURE WRITERS (POST-TRAINING)

Training for procedure writers needs to include coaching immediately after the training. This helps procedure writers to practice their skills and be provided with feedback as they get started. Spot checking the work of procedure writers periodically after the initial training and coaching can help to "catch" issues before they develop. Often people fall back into poor habits or start introducing new approaches that do not fit with the agreed and appropriate human factors approach. Procedure writers also need to be given sufficient time to do their work well.

SUMMARY: HUMAN FACTORS IN THE DESIGN OF PROCEDURES

Procedures guide people to perform tasks correctly and mitigate error, particularly in relation to human failure. They therefore have an important role in reducing the potential for major accidents. To ensure that procedures are usable and used, a clear process is needed for their development. This process includes using task analysis to develop accurate content, and following presentation principles to ensure usability and reduce the risk of error in use. Involving end users in drafting and validating procedures is critical to encouraging compliance with them. This development process needs to occur within an organizational culture where the importance of procedures and following them is actively managed and reinforced.

The next chapter discusses various aspects of organizational performance, including safety culture, management of change, training and competence, and effective supervision. All of these topics either have a bearing on procedures or are supported by having well-developed procedures.

KEY POINTS

- Improving the usability of procedures can help to improve human performance during their application, thus reducing the likelihood of errors and incidents. They act to guide the user and assure error-free performance.
- Following the key steps in the procedure development process can help to improve the usability of procedures.
- Task analysis is a key method for developing procedures to provide an accurate representation of the task.
- End users must be involved in the development of procedures to achieve compliance with them.
- Following the key principles for the presentation of procedures can help improve their usability.
- Organizations should have in place a procedure for managing procedures, and provide human factors guidance and training to procedure writers.

REFERENCES

Chemical Safety Board, 2007. Investigation Report BP Texas City, Texas Report No. 2005-04-I-TX.

Cullen The Hon. Lord Williams Douglas, 1990. Public Inquiry into the Piper Alpha Disaster, Vols. 1 and 2. HMSO, London (Reprinted 1993).

Dubay, W., 2004. The Principles of Readability. Impact Information, Costa Mesa, CA.

Energy Institute, 2011. Guidance on Human Factors Safety Critical Task Analysis. Energy Institute: London.

Hare, J.A., Goff, R.J., Holroyd, J. 2015. Learning from Dangerous Occurrences in the Chemical Industries. Symposium Series No. 60, Hazards 25.

Health and Safety Executive, 1999a. Human Factors Assessment of Safety Critical Tasks. Offshore Technology Report OTO 1999 092.

Health and Safety Executive, 1999b. Reducing Error and Influencing Behaviour HSG48, second ed.. HSE Books., ISBN 9780717624522.

Health and Safety Executive, 2015. Human Factors Briefing Note No. 4: Procedures. Available from: <http://www.hse.gov.uk/humanfactors/topics/procedures.htm> (accessed 18.09.15.).

Institute of Petroleum, 2003. Human Factors Briefing Notes No. 6: Safety Critical Procedures. London.

International Association of Oil and Gas Producers, 2013. OGP Safety Performance Indicators 2013 data, Report No. 2013s.

Kirwan, B., Ainsworth, L.K., 1992. A Guide to Task Analysis. Burgess Science Press, Basingstoke.

NOPSEMA, 2015. Annual Offshore Performance Report. Regulatory Information about the Australian Offshore Petroleum Industry. <www.nopsema.gov.au/resources/data-reports-and-statistics/>.

US Environmental Protection Agency, 2007. Guidance for Preparing Standard Operating Procedures (SOPs): EPA QA/G-6. Office of Environmental Information, Washington, DC.

FURTHER READING

Energy Institute, 2011. Guidance on Human Factors Safety Critical Task Analysis. Energy Institute: London.

Health and Safety Executive, 1999a. Human Factors Assessment of Safety Critical Tasks. Offshore Technology Report OTO 1999 092.

Health and Safety Executive, 1999b. Reducing Error and Influencing Behaviour HSG48, second ed.. HSE Books. ISBN 9780717624522.

Health and Safety Executive, 2015. Human Factors Briefing Note No. 4: Procedures. Available from: <http://www.hse.gov.uk/humanfactors/topics/procedures.htm> (accessed 18.09.15.).

Kirwan, B., Ainsworth, L.K., 1992. A Guide to Task Analysis. Burgess Science Press, Basingstoke.

Understanding and improving organizational performance

Safety culture and behavior

18

E. Novatsis

LIST OF ABBREVIATIONS

ACSNI   Advisory Committee on the Safety of Nuclear Installations
EHS   Environment Health and Safety
HRO   High Reliability Organization
HSE   Health and Safety Executive
INSAG   International Nuclear Safety Advisory Group
PPE   Personal Protective Equipment
RAF   Royal Air Force

This chapter discusses the importance of safety culture, its link to safety performance, and how it can be defined, assessed, and developed. Since the inception of the term "safety culture" after the Chernobyl nuclear disaster in 1986, researchers and practitioners have used different methods to define the concept. Narrative descriptions, organizational culture models, safety climate scales, cultural maturity models, and behavior standards are discussed as key ways to describe this concept. Emphasis is placed on practical descriptions that help employees at different levels to understand their day-to-day role in keeping the organization focused on safety. Importantly though, safety culture cannot be defined and viewed in isolation from the wider organizational culture. Organizations need to provide the right structures, processes, and practices to enable people to stay focused on safety.

Once an organization has a clear definition of the safety culture it is working towards, a range of methods can be used to assess strengths and gaps to determine development priorities. Measurement options include perception surveys, focus groups, interviews, observation, behavioral gap analysis, and incident investigation reviews. These methods each have advantages and disadvantages; therefore, organizations are encouraged to use a structured multi-method approach to gather a deep and comprehensive assessment of the safety culture.

After this understanding is gained, organizations can determine the most appropriate ways of developing the gaps identified and sustaining improvements. Examples of practical and engaging activities are provided for developing specific behaviors, safety culture themes, and for organization-wide safety culture reinforcement.

BOX 18.1 CASE STUDY: LOSS OF ROYAL AIR FORCE (RAF) NIMROD XV230

On 2 September 2006, the RAF Nimrod XV230 experienced a mid-air fire and crashed during a routine mission in Afghanistan. The incident resulted in the deaths of the 14 personnel on board and total loss of the aircraft. The most probable immediate cause of the event was a release of fuel during a mid-air refueling operation or a leak from the fuel system, which ignited when it contacted an ignition source.

The independent review into the incident (Haddon-Cave, 2009) revealed several organizational causes, including a safety culture within the Ministry of Defence that "allowed 'business' to eclipse airworthiness" (p. 13). Some key examples of the weaknesses in the safety culture revealed in the review include:

- Warning signs were not acted upon—significant earlier incidents with other Nimrod aircraft were not used as opportunities to identify problems that were potentially also relevant in the Nimrod XV230, including issues with the same aircraft elements and systems.
- Risk identification activities lacked rigor and independent oversight—three inherent design flaws introduced by modifications to the aircraft over several years were not identified in the subsequently developed safety case. The review describes the production of the safety case as "a story of incompetence, complacency and cynicism" (p. 10); it contained obvious errors of fact, analysis, and risk categorization and did not identify risks that led to the loss of the aircraft.
- Continuous organizational changes and cuts were imposed during the 7 years prior to the incident—the severe resource cuts, budget cuts, and ongoing changes against increasing operational demand were distracting, and diluted the focus on safety. Concerns about the impact of these financially driven actions on safety and airworthiness were raised but not acted on sufficiently.
- Senior managers failed to set high safety and ethical standards—leaders within the organizations involved in producing and checking the safety case did not provide appropriate resources, supervision, or time for this important task. The leaders involved in decisions regarding the ongoing organizational changes and cost cutting also failed to assure adequate focus on safety.

Finally, tips from wide and varied experiences of helping companies develop safety culture are provided, including what works well and common challenges that need to be mitigated.

Accordingly, this chapter is divided into three sections: defining safety culture, assessing safety culture, and developing and sustaining safety culture.

DEFINING SAFETY CULTURE

The term "safety culture" was introduced into the industrial safety lexicon in 1986 by the International Nuclear Safety Advisory Group (INSAG) in its summary report on the Post-Accident Review Meeting on the Chernobyl Accident (1992). Deficiencies in design, safety analysis, learning from previous events, and operating practices that contributed to the explosion at the nuclear power plant reflected a poor safety culture. The importance of safety culture to both operating and regulatory regimes, and throughout all phases of the asset life cycle was also emphasized in the investigation report.

Subsequently, investigations into other major accident events in various industries have explained some of their findings using the term "safety culture," including Piper Alpha (Cullen, 1990), the train crash at Clapham Junction (Hidden, 1989), Texas City, (Chemical Safety Board, 2007) and the loss of the RAF Nimrod XV230 (Haddon-Cave, 2009) to name a few. With this trend came an interest from academics and practitioners to describe what is meant by "safety culture." This section outlines some of the main ways that safety culture has been defined and comments on their utility for organizations.

NARRATIVE DESCRIPTIONS

Narrative descriptions are one way that organizations, including regulators, have described safety culture. For example, the Advisory Committee on the Safety of Nuclear Installations (ACSNI, 1993) following the Chernobyl accident event described safety culture as:

> the product of individual and group values, attitudes, perceptions, competencies and patterns of behaviour that determine commitment to, and the style and proficiency of an organisation's health and safety management. Organisations with a positive safety culture are characterized by communications founded on mutual trust, by shared perceptions of the importance of safety and by the efficacy of preventive measures (ACSNI, 1993, p. 23).

Simplified narrative definitions such as "the way we think and behave in relation to safety" might be useful in the branding of safety culture programs. For organizations that wish to develop their own narrative description, Guldenmund (2000) summarizes the narrative definitions of safety culture proposed by researchers from 1980 to 1997. Such definitions must however be accompanied by a more practical description of what individuals need to do to create a culture that keeps a healthy focus on safety. Otherwise, people might be left wondering what kind of thinking and behavior is required from them.

USE OF ORGANIZATIONAL CULTURE MODELS

Predating the term safety culture was research into organizational culture and climate, and this work has been used to help define safety culture. Guldenmund (2010) summarizes several models of organizational culture, explaining that culture is commonly described as having multiple layers. For example, Schein's (1992) model of organizational culture makes the distinction between three levels or layers at which culture can be studied and analyzed: (1) outside layer of artifacts; (2) middle layer of espoused (adopted) values, and (3) inner core of basic underlying assumptions. Some examples of this model from a safety perceptive include:

- *Artifacts*—the visible and verbally identifiable elements in an organization. For example, posters on golden safety rules, procedures, incentives, Personal Protective Equipment (PPE), and safety-oriented behaviors that are demonstrated (or not) by personnel;

- *Espoused values*—written or spoken statements made by people in the organization. For example, the company policy "everyone safe, everyday"; senior managers opening meetings by stating the importance of safety; or attitudes towards behaviors associated with safety communications, procedures, or training;
- *Basic underlying assumptions*—underlying shared convictions regarding safety. For example, an underlying assumption in a poor safety culture might be that people are careless, and unsafe acts are always deliberate and deserve discipline. Alternatively, in a more positive safety culture an underlying assumption might be that people try to do the best they can but occasionally they get things wrong, and the human and organizational influences on their behavior need to be understood.

This layered approach to describing safety culture taken from organizational culture models can be helpful when it comes to measuring safety culture. For instance, a survey of people's perceptions of behavior might help determine some of the outer layers. However, it will require in-depth interviews and focus group discussions to uncover explanations underpinning the perceptions that might also reveal deeper underlying assumptions.

FACTORS MEASURED IN SAFETY CLIMATE PERCEPTION SCALES

Notably, a similar term "safety climate" is used in the safety literature and preceded safety culture, being first used by Dov Zohar in a study on perceptions of organizational climate for safety in industrial organizations in Israel (Zohar, 1980). Safety climate has been regarded as the surface features of safety culture determined from perceptions of the workforce (Cox and Flin, 1998), or similarly as a "snapshot" of the state of safety providing an indicator of the underlying safety culture of a work group, plant, or organization (Flin et al., 2000, p. 178).

Another way to help define safety culture is to consider what factors are measured in safety climate scales. Researchers have developed many different scales to measure perceptions of safety in organizations. These safety climate scales vary in the factors they measure. It is therefore important to look for common factors measured across different scales as this can help organizations know what factors to include in their descriptions of safety culture.

Two factors are highly consistent in safety climate scales: manager commitment to safety and supervisor commitment to safety. The high representation of these two factors is consistent with Zohar's multilevel model of safety climate (Zohar, 2000; Zohar and Luria, 2005), which proposes that climate is measured at an organization level (i.e., management safety-related actions) and at a group level (i.e., supervisor safety-related actions), as managers and supervisors influence safety outcomes in different ways. Researchers have also identified a coworker level in safety climate scales (e.g., Seo, 2005; Seo et al., 2004; Lu and Tsai, 2008). Such evidence for the multilevel nature of safety climate suggests that when organizations are developing a description of safety culture, safety-related factors pertaining to management, supervision, and other members of the workforce must be represented.

Moreover, Flin and colleagues (2000) conducted a review of 18 scales used to assess safety climate in energy, chemical, transport, construction, and manufacturing sectors. They found that in addition to management as the most common factor, perceptions of safety management systems, risk, work pressure, and competence were commonly assessed. Rules and procedures were also identified in several of the studies reviewed. Another review of 16 safety climate scales by Seo and colleagues (2004), which included some of the same scales as the aforementioned study, similarly identified management commitment, competence, work pressure, and risk as common factors measured. In addition, this study found supervisory support, coworker support, employee participation, perception of hazard level in the work environment, and perceived barriers to safety as dimensions frequently assessed in the scales. The common factors identified in these reviews can assist organizations in developing their own safety culture description by ensuring these dimensions are represented.

HIGH RELIABILITY ORGANIZATIONS

Another body of literature relevant to the discussion of safety culture definition is that of High Reliability Organizations (HROs) (Weick et al., 1999; Weick and Sutcliffe, 2007). HROs are organizations that operate for long periods under difficult conditions and have few major incidents. As defined by Weick and colleagues (1999, p. 81) "the processes found in the best HROs provide the cognitive infrastructure that enables simultaneous adaptive learning and reliable performance." The five key elements discussed in this literature are summarized below.

- *preoccupation with failures rather than successes*—being wary of long periods of success and encouraging reporting and identification of errors as they are recognized as warning signs to larger failures;
- *reluctance to simplify interpretations*—steps are taken to create more complete and detailed understanding of what is going on. Signals that something might be wrong are analyzed and different points of view and discussion of tough issues are encouraged;
- *Sensitivity to operations*—frontline operators are set up to enable them to maintain situation awareness. Frontline operators are highly informed about operations; people are familiar with operations beyond their own job. Managers are sensitive to the experiences of their frontline operators and discuss their experiences of the operation with them;
- *Commitment to resilience*—being ready to tackle events and are not disheartened when errors and events occur. Assume errors will occur and have systems to identify, correct, and learn from errors, and being focused on continually developing people's skills and knowledge;

- *Deference to expertise*—decisions are taken by the people with the greatest expertise about the events in question, even if they are lower in the organizational hierarchy; decision-making structures are flexible.

These elements of HROs deserve specific mention as they describe practices that need to be established and thinking styles that need to be encouraged for safe and reliable operations. These elements are an important addition to safety culture definitions. For example, the HRO concepts have been included in the development of behavioral standards discussed later in this section.

CULTURAL PROGRESSION MODELS

Safety culture has also been defined using culture progression or maturity models (e.g., Health and Safety Executive (HSE), 2000; Energy Institute, 2015). The maturity model concept was adapted from the Software Engineering Institute (Paulk et al., 1993) where it was used to improve the way software was built and maintained.

This type of maturity model was developed in response to efforts to improve the behavioral and cultural aspects of safety in the offshore oil and gas sector. A report by The Keil Centre for the HSE (2000) reviewed behavioral modification programs and showed that good behavioral programs may work at one location, but fail in another. The explanation for this was that the tools and techniques required for improving safety culture differed depending on the current level of safety culture at each facility. The Safety Culture Maturity® Model (HSE, 2000), which was developed to address the findings of this research, is illustrated in Fig. 18.1. This model

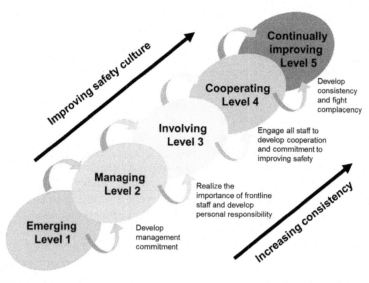

FIGURE 18.1

Safety Culture Maturity® Model (Registered Trademark of The Keil Centre Limited).

was designed to help organizations establish their current level of Safety Culture Maturity® and identify actions required to improve their culture. Five levels of maturity are defined, from Emerging through to Continually Improving. Each level is defined by 10 descriptors which have been identified as the key elements of safety culture according to this model.

Some organizations have since developed their own version of a maturity or progression model. For example, one was developed by Shell, and is now available through The Energy Institute (2015). These models provide a practical method for an organization to define (and assess) its safety culture at various facilities and by different workgroups, which can help identify appropriate improvement actions.

BEHAVIOR STANDARDS FOR HEALTH AND SAFETY

Another practical approach taken to describing safety culture is the use of behavioral standards or behavioral competency frameworks for safety. This work was pioneered in a joint project by the Wood Group and The Keil Centre (Hayes et al., 2007). The framework was designed from academic research which identified leadership behaviors that support workplace safety outcomes (HSE, 2003), an industry report (Step-Change in Safety, 2004), which identified safety behaviors following a review of 11 offshore fatalities and in-company research, which identified positive and negative safety behaviors of relevance to the organization. The internal research applied two job analysis methods: critical incident interviewing (Flanagan, 1954) and the repertory grid technique (Kelly, 1955). The data generated identified positive and negative safety behaviors across three occupational groups: Managers, Supervisors, and Everyone. The "Everyone" occupational group applies to all employees, including managers and supervisors.

An overview of the behavior standards framework is illustrated in Fig. 18.2. The behaviors for the three employee groups are linked by four themes (Standards, Communication, Risk Management, and Involvement) that were identified during the research. Beyond the high level framework, further detail is provided for each behavioral descriptor using positive and negative behavioral indicators. An example for Managers, Set High Standards, is presented in Fig. 18.3.

The advantage of this approach to describing safety culture is that it outlines what different members of the workforce are required to do, from management practices through to frontline behaviors. Many companies in the oil and gas and other high hazard industries have developed their own version of this model. Key advantages are explained by companies that have adopted this approach (e.g., Hayes et al., 2007; Hunter and Lardner, 2008; Hayes et al., 2008), including that it incorporates behaviors for all roles, and shows the linkage between roles. They also favor the approach because it emphasizes mindfulness in the risk management theme consistent with principles of HROs, can be easily integrated into management systems, and is relevant for both personal and process safety.

FIGURE 18.2

Overview of a behavior standards framework.

From Ronny Lardner, (Keil Centre), Paul McCormick and Emily Novatsis (Woodside) - TESTING THE VALIDITY AND RELIABILITY OF A SAFETY CULTURE MODEL USING PROCESS AND OCCUPATIONAL SAFETY PERFORMANCE DATA in Hazards XXII, 2011, SYMPOSIUM SERIES NO. 156.

Managers' Behaviors
To improve our safety culture...

Set high standards	I will...	I will not...
	MP1.1 Explain safety expectations to the workforce	MN1.7 Fail to plan how to achieve desired safety performance
	MP1.2 Verify that the workforce understands and follows safety expectations	MN1.8 Delay following up on agreed safety actions
	MP1.3 Tackle significant safety issues without delay	MN1.9 Tolerate variable and inconsistent safety standards
	MP1.4 Emphasize that production never compromises safety	MN1.10 Allow short-term production pressures to win over safety
	MP1.5 Recognize good safety behaviors and performance	MN1.11 Place undue emphasis on occupational safety to the detriment of process safety
	MP1.6 Address poor safety behaviors and performance	

FIGURE 18.3

Example of a detailed description of a behavioral competency.

From Ronny Lardner, (Keil Centre), Paul McCormick and Emily Novatsis (Woodside) - TESTING THE VALIDITY AND RELIABILITY OF A SAFETY CULTURE MODEL USING PROCESS AND OCCUPATIONAL SAFETY PERFORMANCE DATA in Hazards XXII, 2011, SYMPOSIUM SERIES NO. 157.

AN ORGANIZATIONAL CULTURE FOCUSED ON SAFETY

More recently it has been suggested that the term safety culture and the focus on changing safety culture is not helpful for safety management (The Safety Institute of Australia, 2014). In fact Hopkins (2015) proposed the term should not be used at all. It is argued that the focus should instead be on organizational and management practices and on defining characteristics of an *organizational culture* that supports safety. There is some merit in this approach because it recognizes that what has become known as "safety culture" is only one aspect of the broader context in which personnel operate. From a practical point of view this can help organizations avoid becoming fixated on safety culture without appreciating the importance of management practices and the broader organizational culture in determining safety outcomes.

Whether companies choose to use the term safety culture or not, organizations should ensure they have a clear and practical definition of the behaviors required from all levels of the workforce to help the organization remain focused on safety. The organization then needs to provide the right structures, processes, and practices to support people to demonstrate those behaviors.

DIFFERENT TYPES OF SAFETY

Whichever approach is used to describe safety culture, a key consideration when developing a definition is what types of safety are relevant for the organization. As we have seen from major accident events such as Texas City and Macondo, organizations can focus too much on one type of safety (e.g., personal or occupational safety) to the detriment of another type of safety (e.g., process safety). When an organization sets out to describe its safety culture, the types of safety relevant to the organization (e.g., personal, process, product, or contractor safety) need to be discussed, agreed, and reflected in the definition that is produced.

Another important discussion is whether the definition is just focused on safety or whether health, environment, and quality will also be included. In many organizations, one or more of these disciplines are grouped under the same function. While there is less literature on health and environment culture, The Keil Centre's experience is that when companies conduct internal research on what behaviors are needed to create excellence in safety, health, environment, and quality, similar behaviors emerge. For example, organizations want all employees to speak up about issues, regardless of whether the issue concerns safety, health, environment, or quality.

Finally, organizations must ensure that their definitions of safety culture are relevant to both office-based and site-based personnel. As illustrated in the Nimrod case study in Box 18.1, activities such as the safety case development and organizational changes were conducted by office-based personnel. Their poor execution at an earlier point in time contributed to the incident at a later time. Office-based staff can struggle to grasp the impact that their activities have on safety, therefore definitions of safety culture must include and be relevant to them.

SUMMARY

Since the term safety culture was first used in 1986 by INSAG, researchers and practitioners have used different methods and models to define it. These include narrative descriptions, organizational culture models, safety climate scales, cultural maturity models, and behavior standards. It is important that organizations are clear on what types of safety are important for their operations and ensure that all employees are clear on what is required from them to maintain an organizational culture that is focused on safety or a strong safety culture. The adopted description must include what is required from managers in terms of establishing appropriate management practices and displaying the right behaviors through to the expectations for supervisors and frontline staff.

ASSESSING SAFETY CULTURE
RATIONALE FOR ASSESSING SAFETY CULTURE

Evidence of a link between safety culture and safety performance justifies the need to assess and improve it. Major accident events are one way of establishing a link between safety culture and performance. When major events occur, often aspects of the safety culture are identified in the investigations as being deficient in some way, as illustrated in the example in Box 18.1. This is obvious but demonstrates that aspects of safety culture contribute to safety performance.

Another way of illustrating this relationship is studies measuring safety climate; an indicator of the underlying safety culture and safety performance criteria. These studies provide evidence that more positive safety climates, as measured by the safety climate perception scales, are associated with lower incidents, which is described as a negative relationship. A complete review of the research is not provided here, rather a sample of studies is provided for illustration. In an experimental intervention study, Zohar (2000) found that safety climate predicted minor injuries requiring medical attention in a manufacturing setting 5 months after the assessment. A longitudinal study by Wallace and colleagues (2006) found that foundation climates (management–employee relations and organizational support) and safety climate (measured at time one) had direct and indirect effects on occupational accidents in the transport sector (measured a year later). Furthermore, Clarke's (2006) meta-analysis of 35 studies on safety climate and performance found moderate relationships between safety climate and accidents/injuries.

More recently significant negative relationships were found between a measure of perceived safety behaviors and self-reported incidents (Novatsis et al., 2010) and actual personal and process safety incidents (Lardner et al., 2011). In another study on process safety, safety climate scores correlated negatively with hydrocarbon leaks during the 12 months preceding and the 12 months following the assessment (Kongsvik et al., 2011). In other words, lower safety climate scores were associated with more leaks in these time periods.

Evidence for the criterion validity of safety climate is building in terms of personal and process safety performance. Given safety climate is considered a snapshot of the underlying safety culture, there is support for using such assessments to complement other methods of assessing safety culture, as they provide an indicator of safety performance.

ASSESSMENT PREREQUISITES

There are important considerations to make before a safety culture assessment is undertaken. A safety culture assessment should not be conducted without clear support from senior management. This may be obvious, but it must be stressed that the type of discussion needed with senior managers and the commitment required from them is critical to the success of the assessment. The function or organization leading the assessment should discuss the assessment prerequisites outlined in this section with senior managers and be candid about their role in the assessment process. This includes the role of senior managers in communicating the purpose and importance of the assessment to the workforce in encouraging participation, providing feedback, and acting on the findings. A frank discussion about the readiness of them (potentially) and the organization to change should also be broached.

The purpose and required outcomes of the assessment must be clear. The purpose may be to achieve general improvements in safety management and safety performance such as reducing incident rates, changing behavior or improving compliance. The assessment may be more targeted where the intent is to help solve a particular problem or to understand the culture within a high risk occupational group. The purpose needs to be clear as this must be clearly communicated to participants and may influence organization of the assessment and the methods used.

Timing of the assessment must be considered to help ensure a good response rate and avoid overload or confusion from other assessment activities. It is best to avoid conducting an assessment at the same time as major operational activities such as a shut down or planned start-up. It is also important to ensure the assessment is coordinated with the activities of other support functions; for example, a planned employee engagement survey. School holidays and other common holiday periods should also be avoided. Depending on the purpose of the assessment, organizations may want to time the delivery of the assessment results to coincide with annual activity planning. This way, the recommendations for improvement can inform corporate, department, team, and personal plans. Depending on the roster worked, the length of time over which the assessment is conducted must also be considered. For example, if a survey method is being used and the organization has long swings, the survey may need to remain open for at least 6 weeks to provide all shifts with the opportunity to participate.

Another consideration is the involvement of the workforce in the planning and coordination of the assessment. This can be set up in the way of a working or steering group, with representation from different occupational groups and levels of the workforce. This type of workforce participation can help the assessment team understand practical issues, such as access to people for workshops or timing considerations. They can also have an active role in encouraging participation in their area and working with managers to feedback the results.

Safety culture assessments are usually best delivered as a combined effort between internal resources and external expertise. It is important to determine up front what can be delivered internally and the amount of external support needed.

Expert guidance can add significant value to steer and support the assessment phase, interpret results without bias, decide on appropriate actions, and provide objectivity.

The method or methods that will be used to conduct the assessment should of course also be considered and determined. These approaches are discussed in detail in the following sections.

ASSESSMENT METHODS

There are three main steps in an assessment process: measuring the safety culture, identifying strengths and development areas, and deriving an action plan. A key difference in measurement methods is whether they are quantitative or qualitative. Quantitative methods have their origins in psychology and involve collecting numerical data (e.g., a safety climate survey) and the data are analyzed statistically. Qualitative methods have their origins in sociology and anthropology and involve collecting data from individuals or groups that is not numerical in nature; for instance interviews or observation. Some of the main methods for assessing safety culture are discussed here along with their utility for organizations.

Safety climate surveys

Standardized safety climate surveys are a common assessment method used to provide an indication of the underlying safety culture. Such surveys are typically administered to a large cross section of the workforce, where personnel are asked to provide their perceptions of safety-related dimensions individually and anonymously. Items are often presented as statements; for example, "there is sometimes pressure to put production before safety on this facility." Responses to the items are assessed on a multipoint rating scale; for example, "strongly agree" to "strongly disagree." Where this approach is used, a minimum 50% workforce response rate (preferably higher) should be aimed for, to be considered representative of the wider workforce. Data are typically numerical and analyzed statistically. When open-ended questions are included these are often coded into themes to allow statistical analysis.

Many different safety climate surveys have been developed, applied, and tested in different countries and for different industry sectors (see HSE, 1999, for a review of a sample of tools and European Agency for Health and Safety, 2011 for a summary list). Some are commercial products, some are freely available, and others are provided in academic papers. When selecting a safety climate survey organizations need to determine if the survey is fit for purpose. Important features to consider when selecting an off-the-shelf tool include:

- *Validity*—understanding what the survey measures and how well it does so. There are different types of validity. For example, face validity refers to whether the items appear valid to experienced industry users. Organizations should review the items and language to determine whether it will look valid to end users, and therefore be more likely to be accepted and completed. Content validity refers to whether the content of a questionnaire covers a representative sample of the behavioral domain to be measured; health and safety in this case. As discussed earlier, the dimensions measured in surveys vary, and organizations should check whether the content was developed from research, and whether it covers a comprehensive set of relevant safety-related dimensions. The other type of validity important to describe here is criterion validity, which refers to whether there is a relationship between scores on a measure and scores on relevant performance criteria. Organizations should check whether any research has been conducted to assess whether the safety climate survey relates to health and safety performance criteria. Validity and reliability information is usually included in a technical manual and organizations should review this before selecting an instrument;

- *Reliability*—the survey's consistency. In others words it is the degree to which the survey measures the same way each time it is used under similar conditions. The technical manual should be checked to determine whether the reliability of the survey's dimensions is acceptable;
- *Customization*—surveys differ on the degree to which they can be tailored. Organizations may have requirements to include specific demographic items or custom open-ended questions. It is important to check that the survey design can accommodate these needs. Moreover, the availability of online and hard copy versions of the survey should be determined, based on requirements;
- *Reporting*—the report outputs also differ between standardized surveys. The depth of analysis, types of graphs, narrative summaries, the ability to compare time-points, and comparisons against industry norms are all factors that can vary. The standard output and ability to tailor this should be examined for its suitability;
- *Support*—some survey providers offer support for what to do with the results. Checking that such advice is available may be important for companies that do not have an internal technical expert who can provide such guidance.

Behavioral gap analysis

An alternative to a safety climate survey is a behavioral gap analysis. This method can be used to measure workforce perceptions of how frequently the behaviors in behavior standards (described in the first section of this chapter) are demonstrated. The gap analysis can be conducted as an online or hard copy survey or in a workshop setting. This method differs from climate surveys in that specific *behaviors* are measured, rather than *statements towards safety-related elements* in the company. For example, each of the positive behaviors in Fig. 18.3 would be assessed. People are asked to rate how frequently managers display behaviors, such as "explain safety expectations to the workforce." Responses are provided on a four- or six-point rating scale, ranging from "Seldom/Never" to "Always." This method can provide very specific feedback on improvement areas, especially when administered as part of a workshop, to understand why the behaviors are or are not being displayed and to gather ideas for improvement.

Ewen and Medina Harvie (2011) describe a case study where a behavioral gap analysis was undertaken on a project in 2008 using the behavior standards. The results were used to develop an improvement plan with the client and effort made by the contract manager to set high standards for safety. This assessment initiative contributed to a 42.5% improvement in the lost work cases per million man hours and a 24% improvement on the total number of recordable cases per million man hours. These practitioners also describe where assessment using the behavior standards helped to identify the need for a contract manager to improve their Environment Health and Safety (EHS) communication. Actions were taken to improve this behavior along with reinforcement of the positive communication behaviors needed from supervisors and the workforce. They found a measureable behavioral improvement when the assessment was repeated, together with a significant improvement in EHS performance. Moreover, they report that the ongoing application of the behavior standards over a 4-year period on this contract contributed to a steady decline in total recordable injury frequency rate from 6.70 in 2005 to 0.00 in 2008.

Cultural progression assessment

Cultural progression models were discussed earlier as a way to define safety culture: the Shell model (Energy Institute, 2015) and the Safety Culture Maturity® Model (HSE, 2000) being the better known ones in industry. These models can also be used as assessment tools. The aforementioned models are similar in that both assess a number of safety culture dimensions, each of which have descriptors representing a different level of maturity on that dimension. Participants are required to select the descriptor for each dimension that best describes the situation at their site. A workshop setting is preferred for this type of assessment because after ratings have been made participants can explain their findings, make comparisons, and discuss what is needed to progress to the next level. The facilitator guides and records the discussion. This type of assessment is practical, engaging, and provides both quantitative and qualitative data which can be used to develop the maturity of the organization's safety culture.

Focus groups

A focus group is an open group discussion and in this context can be used to gather people's views on safety culture at a site or organization. They can be used to collect quantitative data, qualitative data, or both, depending on the nature and purpose of the focus group. For example, a behavioral gap analysis can be conducted as a focused discussion where behavioral ratings are collected and then a discussion is facilitated on why those ratings were made. Focus groups must be structured, otherwise valuable discussion time may be spent on irrelevant topics or too long spent on a particular topic. Structure can be achieved by planning the topics that will be covered and the process that will be used. Effective focus groups also require a skilled facilitator to keep the discussion on track. This method can be a useful addition to a safety culture assessment that has relied heavily on survey data. For example, focus groups could be undertaken in addition to the survey to gather qualitative information, or be conducted after the survey to elaborate on the findings and to determine improvement actions.

Interviews

Interviews are another method to gather information about the safety culture. They provide qualitative data and like focus groups, must be planned and use a structured interview guide to improve reliability and validity. When used to gather data on all aspects of the safety culture, the interview guide should include open questions about the interviewee's perceptions about safety culture and management practices and then target specific safety culture dimensions. Interviews can also be used when deeper insight into specific or complex issues is needed, for example, a theme that has emerged from other assessment data, or to understand more sensitive issues that may not be appropriate for discussion in a group setting.

In whichever stage of the assessment process that interviews are used, interviewee selection is important. For instance, a new employee may have a different view to a long-term staff member and a manager may have a different view to someone who is not in a leadership role. Moreover, someone in an operational site–based role may see things differently from someone in an office-based support role. It is therefore important to interview personnel in different roles, disciplines, and levels to pick up potential differences in perception. It can be helpful to interview people who are new to the organization but who have worked in the same industry for another company. These employees see things with "fresh eyes" and can make comparisons readily about the safety culture by making comparisons to what they have observed elsewhere. Finally, interviews are time intensive, so the number that can be practically conducted requires consideration. They are typically used to complement other safety culture assessment methods.

Expert coaches and observation

A person external to the organization can operate as an expert coach to provide an assessment of the safety culture. This person is typically a skilled and knowledgeable health and safety professional who is familiar with safety culture and other human factors issues. This approach can be helpful because the person chosen has no knowledge of the local safety culture and therefore offers a "fresh pair of eyes" on how things are being done. The external expert observes work being done, attends meetings and other activities, conducts interviews, and asks questions to understand how things are done. They spend an agreed length of time at the site or in the organization and on completion of their assignment provide feedback, recommendations, and coaching on how aspects of the safety culture could be improved.

A related assessment method is the use of observation. This tends to be more structured and systematic than the involvement of an expert coach. Managers, supervisors, and other members of the workforce are observed doing their daily tasks so that information can be gathered on working practices and processes. Companies often use behavioral observation programs as part of their safety culture toolkit to reinforce positive practices and identify development opportunities. When observation is used as part of a safety culture assessment activity, they are usually conducted by someone who is external to the organization or from a different part of the business. A standard set of observation prompts can be helpful to such observers. Similar to an expert coach, observation by an impartial "outsider" can provide insights into the safety culture to which local personnel may not be aware.

Outcomes of incident investigations

Investigation reports can also be used to understand safety culture. As illustrated in the Nimrod incident described in Box 18.1, failings in the culture and management practices are often highlighted when incidents occur. As part of a safety culture assessment, the investigation reports of an organization's recent serious or high potential events can be examined. The assessor should look for positive behaviors that were demonstrated and may have helped minimize the consequences of the incident. Negative behaviors and failings in the application of management practices should also be identified along with trends across incidents which may highlight vulnerabilities in the organization. Of course, the quality of information gleaned using this method is limited by the quality of the investigators, investigation, and report. The subject of human factors in incident investigation is addressed in Chapter 8, Human Factors in Incident Investigation.

TRIANGULATION

The methods described in this section each have advantages and disadvantages and these are summarized in Table 18.1. It is good practice to use a multimethod approach called triangulation to achieve a more robust assessment. Data are collected using different methods, are obtained from more than one source, and are collected by multiple people. For example, an organization might review serious incidents from the last year to identify key trends, conduct a safety perception survey, and combine this with a

Table 18.1 Advantages and Disadvantages of Different Safety Culture Assessment Methods

Assessment Method	Advantages	Disadvantages
Safety perception survey	• Confidential • Standardized • Quick to complete • Potential to reach large sample • Produces quantitative data, making it easier to compare results over time • When conducted online, do not have to enter data or text comments manually	• Impersonal • Items left to interpretation, little opportunity for clarification • Potential for "tick and flick" • Difficult to turn results into actions to improve safety • Often necessary to hold workshops or interviews to clarify or explore results • May be subject to social desirability effects (e.g., the tendency to respond to items in a way that will be seen favorably by others)

(Continued)

Table 18.1 Advantages and Disadvantages of Different Safety Culture Assessment Methods (Continued)

Assessment Method	Advantages	Disadvantages
Behavioral gap analysis	• Specific behaviors are identified for improvement • Behaviors assessed are those contained in the behavior standards which are also used for developing safety culture • Shows linkages between different groups on the same theme • When conducted as a workshop, can educate people about safety culture at the same time as collecting data and results can also be discussed	• When conducted as an online or paper gap analysis, it has the same disadvantages as a perception survey • When planning actions, there is a temptation to work on too many behaviors • Needs an experienced facilitator when conducted as a workshop • Time-consuming
Focus groups	• Provide a deep picture of cultural issues • Can obtain a range of views • Can just focus on issues of interest or need • Participants can suggest solutions to issues they identify • Discussion and involving people can create buy-in and engagement • Can educate about safety culture at the same time as collecting data	• Restricted range of issues can be addressed • Lack of a validated structure may mean that issues are missed • Participants may be unwilling to open up if trust and confidence are low or if they are more reserved • Answers can be influenced by group dynamics and dominant personalities • Not easy to make comparisons between sites or over time • Relies on ability of facilitator to extract relevant issues in the time available and keep discussion on track
Interviews	• Provide a deep picture of cultural issues • Helpful to discuss sensitive or complex issues	• Time-consuming • Restricted range of issues can be addressed • Not easy to make comparisons between sites or over time
Coach/ Observation	• Impartial, fresh set of eyes • Data collected in real time • Can target specific areas of concern • Bring expertise	• Observation can be time-consuming • Bias on the observer's observations • Could be seen as "intruder" to group therefore people not open to talk • Time-point comparisons difficult, hard to repeat analysis
Incident investigation outcomes	• Can identify key issues to be included and explored further in other assessment methods • Cost-effective as data are already gathered	• Quality of data obtained depends on the quality of the investigation • Is retrospective and may not be a current reflection of the culture

series of workshops to explore the results more fully. This approach can help assess the outer and deeper layers of the safety culture. Whilst triangulation is accepted as a useful approach, it must be tailored to suit the organization and must be well structured.

SUMMARY

Important considerations must be made prior to undertaking a safety culture assessment. These include being clear on the purpose of the assessment, gaining management support and employee participation, coordinating timing, identifying resources, and determining assessment methods. A range of methods can be used to assess an organization's safety culture, including both qualitative and qualitative techniques. Each method has advantages and disadvantages. It is recommended that a triangulated approach is used, which employs multiple methods to obtain a more robust assessment.

A key objective of conducting a safety culture assessment is to identify strengths in the safety culture and areas that require development. This information can help determine appropriate actions that can be implemented to improve the current state of the safety culture, including the necessary supporting management practices. Such actions are best integrated into the organization's health and safety or organizational development improvement plan.

DEVELOPING AND SUSTAINING SAFETY CULTURE

Once an organization has a robust understanding of its safety culture and improvement areas, it must accept the challenge of safety culture improvement. Although the development process usually starts with a set of improvement actions, organizations must acknowledge and accept that safety culture development is not a one off "tick the box" item in an annual plan. The organization changes, people come and go and new issues emerge, demanding an ongoing effort to stay focused on sustaining safety culture. Otherwise, as illustrated in the Nimrod incident in Box 18.1, an organization can become distracted and fail to maintain its focus on safety.

Even when improvements in the safety culture have been made, periodic reinforcement will be required. Persistence and patience will also be needed as safety culture development requires changes to behavior, thinking patterns, and management practices, which can take time. It can therefore be helpful for organizations to adopt a continuous improvement approach, such as the "Assess, Plan, Do, Monitor" framework (Step-change in Safety, 2000) or a similar quality management model. Using such a framework for safety culture development can reinforce that the focus required is never over and it needs ongoing maintenance.

Importantly, the intervention or development activity selected must suit the current maturity of the safety culture at the location in question. The level of maturity can differ between sites at the same organization. For example, a safety culture assessment may reveal that managers are not demonstrating the right behaviors or establishing organizational practices that support safety, so an intervention aimed at providing them with the knowledge and skills to lead safety may be appropriate. A workforce-designed and -led development activity is unlikely to work in this situation. Even if a specific maturity assessment of the safety culture has not been conducted, an effective assessment should provide enough information to gauge the maturity in order to determine appropriate development activities.

Tools and activities to develop and sustain safety culture are many and varied. This section provides examples of tools that can be used for developing a specific safety culture theme or specific behaviors and examples of organization-wide reinforcement activities.

PRACTICAL EXAMPLES OF TOOLS TO DEVELOP SPECIFIC BEHAVIORS

Leadership development programs or activities are used to help leaders understand their role in influencing safety culture and to develop the skills to do this effectively. Such activities may include:

- *workshops that provide education on specific safety behaviors*. For example, an Australian oil and gas company designed a workshop targeting behaviors that needed improvement as determined from a safety culture assessment. Topics covered included communication with the workforce, verification that the workforce understood the safety management systems, risk assessment of organizational change, recognition of safe behavior, and human factors in incident investigation. The workshop resulted in developmental goals for these behaviors from all managers that could be tracked for improvement at a subsequent assessment;
- *practical tools for leadership skill development*. For example, an American electrical and gas services company developed a set of tools for executives based on findings from a behavioral gap analysis. The tools included how to conduct safety tours and inspections, incident investigation process reviews, project development safety reviews, and safety recognition. Executive safety development plans that made use of these tools were established and tracked.

Non-technical skills programs are another helpful way to develop aspects of the safety culture, targeting skills needed to demonstrate specific safety behaviors. For example, findings from a safety culture assessment may reveal that people are not speaking up about safety issues or intervening on poor behaviors they observe. This may be because they do not have the skills to do so effectively and confidently. A program covering the process and skills for intervening and having conversations about safety issues could be made available to address this gap.

Workplace toolkits containing interactive activities are another method used to develop safety culture. Activities are provided for each theme in the organization's safety culture model and are designed to develop one or more of the behaviors within that theme. For example, under the theme of "Learning," an activity may encourage participants to explore the power of story-telling as a learning method and get them to practice this method. The activities are short, practical, and do not rely on expert knowledge to conduct. They can be used if and when needed to develop or reinforce particular behaviors.

Discussion cards are another flexible and interactive method used to develop safety culture. The Eurocontrol Safety Culture Discussion Cards (2015) are a good example of this method. Each card contains a challenging question that targets a specific behavior within the company's safety culture model. The card also provides an explanation of the importance of that topic. The cards have multiple applications (e.g., safety moments, comparing views or examining influences between topics) and can be used by individuals or groups. Like the workplace toolkit, they do not rely on expert knowledge to use. Such a tool can educate and engage people whilst also reinforcing the behaviors, thinking styles, and management practices needed to strengthen the safety culture.

Engaging learning scenarios that require people to identify and/or demonstrate behaviors that require development are also an effective development method. Lardner and Robertson (2011) provide an example of a management decision-making scenario to help managers understand the organizational failings from a serious incident. Novatsis et al. (2012) describe an engaging method of educating people about isolation incidents. Learning from events within and outside of the organization is a critical element of a strong safety culture. If this aspect of the safety culture is weak, scenario-based learning methods, whether they are written, computer-based, videos, games, or stories, can be powerful learning tools.

PRACTICAL EXAMPLES OF ORGANIZATION-WIDE REINFORCEMENT EFFORTS

An effective way of reinforcing and sustaining safety culture is to integrate the selected approach into management systems rather than keeping it separate to them. For example, behavior standards (described earlier in this chapter) can be integrated into human resource and safety management systems. The encompassing safety behaviors can be used in selection processes to design interview questions that target the behaviors, included in induction processes or be used to develop performance and development goals. The behaviors could also be used to guide prestart discussions or tool-box talks, to conduct behavioral audits, or to undertake postincident reviews. Finding many ways to link the behaviors or themes in an organization's safety culture description to management systems and practices provides reinforcement throughout an employee's life cycle with the company. In this way it becomes part of daily practice and not a stand-alone workshop that someone attended at some point during their employment. Organizations must clearly communicate how their approach to developing safety culture links to and complements other programs and systems, otherwise personnel may get confused and disengage.

Some organizations choose to conduct a company-wide initiative every 1–2 years to refresh and reinforce the importance of safety culture. This has been done in a variety of ways, including through videos, workshops, toolkit activities, discussion cards, or by linking their safety culture approach to a major industry or company event.

BOX 18.2 TIPS AND CHALLENGES IN SAFETY CULTURE DEVELOPMENT

Tips

- ✓ ensure high risk issues/areas are addressed
- ✓ do the right things at the right time (e.g., match to maturity)
- ✓ integrate into management systems and show relationships to other programs
- ✓ be flexible
- ✓ plan for reinforcement and prepare to reinforce and persist
- ✓ use a mix of specialists and trained focal points to implement, support, and drive
- ✓ keep things simple and deliverable
- ✓ involve line managers
- ✓ stop doing something that is not working

Challenges

- getting traction for activities against other business objectives
- getting acceptance that it is long-term work and not "tick-box"
- having enough resources to support teams after initial activity
- maintaining motivation in areas where safety culture is already positive or improved greatly
- building shared understanding across office- and site-based teams
- avoiding overload to the business
- identifying how to best monitor and evaluate
- surviving leadership or organizational changes

SUMMARY: SAFETY CULTURE AND BEHAVIOR

There are many examples of practical and engaging methods to develop safety culture which can be used after an assessment or for reinforcement when required. Importantly, the selected approach must match the maturity of the local safety culture. The behavior or theme that requires development must be clear, along with the group or area that needs improvement. Tips for developing safety culture are provided in Box 18.2, along with common challenges that may need to be planned for and mitigated. A key success factor for sustaining a strong safety culture is to integrate the approach into existing management systems. This encourages personnel to demonstrate the required safety behaviors as part of daily practice.

Maintaining the focus on safety culture is critical during organizational change, a topic covered in the next chapter.

KEY POINTS

- Safety culture has been identified as a contributing factor in major accident events. There is evidence of a link between safety culture and safety performance.
- Safety culture has been defined in different ways, although practical definitions are favored that describe what each level within the organization needs to do to maintain focus on safety.
- There are many methods that can be used to conduct a safety culture assessment, each with advantages and disadvantages.

- Using multiple methods for safety culture assessment can create a more robust, valid, and reliable understanding of the strengths and gaps in the safety culture.
- Safety culture development requires ongoing effort and a continuous improvement mind-set.
- There are many engaging and interactive methods to develop and sustain safety culture after assessment activities or for ongoing reinforcement efforts.
- The development activities selected must suit the current level of maturity of the local safety culture.
- Integrating the chosen safety culture approach into management systems helps sustain safety culture development efforts.

REFERENCES

ACSNI Study Group on Human Factors. 1993. Third Report: Organising for Safety. Advisory Committee on the Safety of Nuclear Installations.

Chemical Safety Board, 2007 Investigation Report BP Texas City, Texas. Report No. 2005-04-I-TX.

Clarke, S., 2006. The relationship between safety climate and safety performance: a meta-analytic review. J. Occupat. Health Psychol. 11 (4), 315–327.

Cox, S., Flin, R., 1998. Safety culture: philosopher's stone or man of straw. Work & Stress. 12 (3), 189–201.

Cullen, 1990. The Hon. Lord W. Douglas. The public inquiry into the Piper Alpha disaster.

Energy Institute, 2015. http://www.eimicrosites.org/heartsandminds/culture.php (accessed 03.12.15.).

Eurocontrol. 2015. Safety Culture Discussion Cards. http://www.skybrary.aero/index.php/Safety_Culture_Discussion_Cards (accessed 10.12.15.).

European Agency for Health and Safety at Work, 2011. Occupational Safety and Health Culture Assessment: A Review of Main Approaches and Selected Tools. European Agency for Health and Safety at Work.

Ewen, J., Medina Harvie, Z., 2011. An Alternative Approach to Safety Culture Development. Woodgroup, Aberdeen.

Flanagan, J.C., 1954. The critical incident technique. Psychol. Bull. 51, 327–358.

Flin, R., Mearns, K., O'Connor, P., Bryden, R., 2000. Safety climate: identifying the common features. Safety Sci. 34, 177–192.

Guldenmund, F.W., 2000. The nature of safety culture: a review of theory and research. Safety Sci. 34, 215–257.

Guldenmund, F.W., 2010. (Mis)understanding safety culture and its relationship to safety management. Risk Anal. 30 (10), 1466–1480.

Haddon-Cave, C., 2009. An Independent Review into the Broader Issues Surrounding the Loss of the Royal Air Force (RAF) Nimrod MR2 Aircraft XV230 in Afghanistan in 2006. The Stationery Office, London.

Hayes, A., Lardner, R., Medina, Z., & Smith, J. 2007. Personalising Safety Culture: What Does it Mean for Me? Loss Prevention 2007, 22–24, May 2007, Edinburgh, UK.

Hayes, A., Novatsis, E., & Lardner, R. 2008. Our Safety Culture: Our Behaviour is the Key. Paper presented at the Society of Petroleum Engineers (SPE) International Conference on Health, Safety, Environment in Oil and Gas Exploration and Production, 15–17 April, 2008, Nice, France.

Health and Safety Executive, 1999. Summary Guide to Safety Climate Tools. Offshore Technology Report 063. HSE Books.

Health and Safety Executive, 2000. Safety Culture Maturity Model. Offshore Technology Report. HSE.

Health and Safety Executive. 2003. The Role of Managerial Leadership in Determining Workplace Safety Outcomes. http://wwwhse.gov.uk/research/rrpdf/rr044.pdf.

Health and Safety Laboratory. 2015. The Safety Climate Tool. http://www.hsl.gov.uk/products/safety-climate-tool (accessed 10.12.15.).

Hidden, A., 1989. Investigation into the Clapham Junction Railway Accident. Department of Transport, HMSO., ISBN 0101082029.

Hopkins, A., 2015. Celebrating the Work of Andrew Hopkins Event. RMIT University, Melbourne, 16 November, 2015.

Hunter, J., Lardner, R. 2008. Unlocking Safety Culture Excellence: Our Behaviour is the Key. IChemE Symposium Series No. 154.

International Nuclear Safety Advisory Group. 1992. INSAG-7 The Chernobyl Accident: Updating of INSAG-1. Safety Series No. 75-INSAG-7.

Kelly, G., 1955. The Psychology of Personal Constructs. Norton, New York, NY.

Kongsvik, T., Johnsen, S.A.K., Sklet, S., 2011. Safety climate and hydrocarbon leaks: an empirical contribution to the leading-lagging indicator discussion. J. Loss Prevent. Process Indust. 24, 405–411.

Lardner, R., & Robertson, I. 2011. Towards a Deeper Level of Understanding from Incidents: Use of Scenarios. Symposium Series No. 156, Hazards XXII.

Lardner, R., McCormick, P., & Novatsis, E. 2011. Testing the Validity and Reliability of a Safety Culture Model using Process and Occupational Safety Performance Data. Paper presented at Hazards XXII—Process Safety and Environmental Protection Conference, 11–14 April, 2011, Liverpool, UK.

Lord Cullen Report, 1990. Public Inquiry into the Piper Alpha Disaster, Vols. 1 and 2. HMSO.

Lu, C.S., Tsai, C.L., 2008. The effects of safety climate on vessel accidents in the container shipping context. Accident Anal. Prevent. 40, 594–601.

Novatsis, E., McCormick, P. & Lardner R. 2010. Testing the Validity and Reliability of a Safety Culture Model: A Practitioner's Perspective. Paper presented at 13th International Symposium on Loss Prevention and Safety Promotion in the Process Industries, June 2010, Brugge, Belgium.

Novatsis, E., McCormick, P., & Lardner, R. 2012. Beyond Bulletins and Presentations: Use of Scenarios to Learn from Incidents. Paper presented at Society of Petroleum Engineers (SPE)/Australian Petroleum Production & Exploration Association (APPEA) International Conference on Health, Safety, and Environment in Oil and Gas Exploration and Production, 11–13 September 2012, Perth, Australia.

Paulk, M.C., Curtis, B., Chrissis, M.B., Weber, C.V., 1993. Capability Maturity Model for software, Version 1.1. Software Engineering Institute: Carnegie Mellon University, Pittsburgh, PA.

Schein, E.H., 1992. Organizational Culture and Leadership, second ed. Jossey-Bass, San Francisco, CA.

Seo, D., 2005. An explicative model of unsafe work behavior. Safety Sci. 43 (3), 187–211.

Seo, D., Torabi, M.R., Blair, E.H., Ellis, N.T., 2004. A cross-validation of safety climate scale using confirmatory factor analytic approach. J. Safety Res. 35, 427–445.

Safety Institute of Australia, 2014. Organisational Culture. Safety Institute of Australia Ltd, Westmeadows.

Step-Change in Safety 2000. Changing Minds. Step-Change in Safety: www.stepchangein-safety.net.

Step-Change in Safety 2004. Fatality Report—Report into offshore fatalities 2000–2002. Step-Change in Safety: www.stepchangeinsafety.net.

Wallace, J.C., Mondore, S., Popp, E., 2006. Safety climate as a mediator between foundation climates and occupational accidents: a group-level investigation. J. Appl. Psychol. 91 (3), 681–688.

Weick, K., Sutcliffe, K., 2007. Managing the Unexpected. Jossey-Bass, San Francisco, CA.

Weick, K.E., Sutcliffe, K.M., Obstfeld, D., 1999. Organizing for high reliability: processes of collective mindfulness In: Sutton, R. Staw, B. (Eds.), Research in Organizational Behavior, Vol. 21 JAI, Greenwich, CT, pp. 81–124.

Zohar, D., 1980. Safety climate in industrial organizations: theoretical and applied implications. J. Appl. Psychol. 65 (1), 96–102.

Zohar, D., 2000. A group-level model of safety climate: testing the effect of group climate on microaccidents in manufacturing jobs. J. Appl. Psychol. 85 (4), 587–596.

Zohar, D., Luria, G., 2005. A multilevel model of safety climate: cross-level relationships between organization and group-level climates. J. Appl. Psychol. 90 (4), 616–628.

Managing organizational change

19

K. Gray and J. Wilkinson

LIST OF ABBREVIATIONS

CCPS   Center for Chemical Process Safety
CSCE   Canadian Society for Chemical Engineering
ECMT   Extended Corporate Management Team
HSE   Health and Safety Executive
IAEA   International Atomic Energy Agency
NASA   National Aeronautics and Space Administration
RAF   Royal Air Force

Change is a fact of business life and requires proactive management for any organization to continue to trade successfully. In simple terms, a business that does not adapt its products or services to what is often a changing market and economic environment will not survive. Even if such factors are stable, organizations need to continue to improve and innovate to remain competitive. Therefore, the process by which organizations manage their business changes is vital, and that includes the way they engage, prepare, and support their people in doing so. Organizational change can be proactive, such as improvements to the business structure, organization and processes, or identifying and seizing business opportunities. It can also be more reactive, such as adapting to changes in the surrounding business, market, or technological environment.

Typically, change management programs have focused on engineering, plant, or technical change (referred to as "engineering change" throughout the rest of this chapter). This type of change management is critical, and is well covered in other literature. Changes related to organizational structures, processes, or people, typically receive less focus and are managed poorly, if at all. In practice, "organizational" change is typically intertwined with those more traditional change management processes. What matters is how significant such organizational change is likely to be, and how well it is identified, assessed, and addressed. The established rigorous arrangements applied to engineering change need to be mirrored in those applied to organizational change, particularly as the likely effects on safety of such change are not as well recognized.

> **BOX 19.1 THE IMPACT OF ORGANIZATIONAL CHANGE AT BP TEXAS CITY IN 2005**
>
> At the BP Texas City Refinery, one of the contributions to the accident in 2005 was the failure to recognize and assess the impact of significant organizational change on the staff. In the 5 years preceding the incident, there were staffing cuts of 25%, including board operators, such that one board operator, rather than two, was left to manage three complex process units and manage the start-up of the isomerization tower.

Many examples of organizations failing to consider the effects of organizational change on the human aspects of the operation can be found. One of those examples is from the BP Texas City Refinery accident in 2005 (Box 19.1).

This chapter discusses the key issues relevant to organizational change, and approaches that consider the human elements. Two perspectives are explored: managing the risks associated with the human aspects of the change and managing the psychological adjustment, although there is overlap between the two elements.

WHAT IS ORGANIZATIONAL CHANGE?

> *"We trained hard, but it seemed that every time we were beginning to form up into teams we would be reorganized. Presumably the plans for our employment were being changed. I was to learn later in life that, perhaps because we are so good at organizing, we tend as a nation to meet any new situation by reorganizing; and a wonderful method it can be for creating the Illusion of progress while producing confusion, inefficiency, and demoralization."* (Ogburn, 1957)

Although often attributed to Petronius Arbiter, a 1st century Roman author and courtier, this quotation is actually from Charlton Ogburn Jr, a 20th-century soldier, referring to a US Army unit's experience in World War 2. Regardless, it illustrates that organizational change is not a new phenomenon and the Roman army, like any modern army, was indeed subject to change, such as sudden strategic shifts, leadership changes, and resource and manpower squeezes.

Organizational change can be simply defined as any change to business processes, organizational structure, staffing levels, or culture within a company. The Health and Safety Executive (HSE) uses an inclusive definition that lists examples of the kinds of changes that typically take place.

> *...business process re-engineering, delayering; introduction of "self-managed" teams; multi-skilling; outsourcing/"contractorisation"; mergers, de-mergers and acquisitions; downsizing; changes to key personnel; centralisation or dispersion of functions; changes to communication systems or reporting relationships.*
>
> **HSE (2003), p. 1**

> **BOX 19.2 "WHAT IS DIFFERENT FOR THE MAJOR HAZARD INDUSTRIES IN MANAGING ORGANIZATIONAL CHANGE"**
>
> *What is the difference for major hazard industries as opposed to those managing more conventional hazards from the viewpoint of organizational change?*
>
> A major hazard business (where major hazards are under control) have different organizational change considerations. Many successful organizations are decentralized with independent business units (see Wilkinson and Rycraft, 2014, for a recent review). However, some organizational arrangements do not generally suit major hazard needs, such as flat management structures. Major hazard businesses need a comparatively higher level of hierarchical structure and stability to establish and maintain control. This means that greater caution is required in making changes.

This is related to major hazard industries but is equally relevant to other operational environments. The key is that any change may qualify where it impacts individuals, teams, or the organization significantly. It requires consideration of peoples' capabilities and limitations, and the right structure for people to deliver the business plan successfully and safely (Box 19.2).

DRIVERS FOR CHANGE

There are many change drivers. They include innovation, maintaining competitiveness, product/service changes, new technology, ongoing profitability, and efficiency improvements, including continuous improvement in health and safety performance.

There are also other apparently more benign drivers such as: harmonizing standards and procedures after mergers, office relocation, distancing of central supporting functions from sites, changes to maintenance and inspection intervals or practices, recruitment, selection and training changes or challenges, and sometimes just new managers with new ideas (list compiled from International Atomic Energy Agency(IAEA), 2003, and HSE, 2003).

Finally, financial considerations are powerful motivational factors for both the implementation of any change itself, and for how the change is implemented. However, organizations are recognizing that short-term views aimed at making savings, or rewarding shareholders, can lead to far greater costs in recovering control at a later stage (and meanwhile introduce significant but hard-to-predict risks to safety (IAEA, 2003)).

MANAGING THE RISKS OF ORGANIZATIONAL CHANGE

The field of human factors involves optimizing the interaction of people with one another, with equipment and facilities, the physical work environment, and with their organizational structure and management. All of these factors are potentially affected by organizational change, and need to be assessed with the same rigor as engineering changes. This is not only from the viewpoint of the end state of change (where the organization wants to get to), but also the process of change (the transition). Attempting to change too much or too quickly can make the cost of change, including the risks to safety, outweigh the benefits. This is even more so for organizations that have major hazards to control, and where the consequences of loss of control can be disastrous.

HOW CAN ORGANIZATIONAL CHANGE IMPACT SAFETY?

Change can create safety issues directly or indirectly if it is poorly designed or delivered. Reducing maintenance staff numbers, for example, has a direct impact, whereas the appointment of a new supervisor who lacks plant experience will have indirect impacts over time. The indirect impacts may occur due to the loss of credibility, oversight, or the dilution of effective frontline safety leadership.

Other diffuse, but equally important, impacts can result from what may at first appear to be minor changes to key roles. Changing to an electronic control of work system from a paper system, for example, may appear benign ("like for like"), but more supervisor time spent on dealing with the electronic system and interface (and so less time spent out on the job) can slowly erode frontline leadership and oversight.

Where humans are relied upon as part of the barriers to prevent or mitigate incidents, such changes can be very serious, but sometimes harder to identify and assess.

Most adverse consequences of organizational change are unintended.

> ...installations are complex, and it is inherently demanding to foresee all the implications that a change may have on safety.
>
> **IAEA (2003)**

Whilst this quote from the IAEA refers to nuclear installations, the examples of organizational change issues given in the IAEA guidance are specifically noted as not being unique to this industry. Other sources (such as HSE, 2003) confirm this for other major hazard sectors such as the chemical and allied process industries. In short, change in more complex organizations can produce surprises. Poorly managed change contributes to poor management of major hazards (Figure 19.1).

There are many well-known major accidents where poor management of organizational change and the resulting impact have significantly contributed to the event:

- Bhopal (1984),
- Hickson and Welch (1992),
- the National Aeronautics and Space Administration (NASA) space shuttle disasters Challenger (1986) and Columbia (2003),
- BP Texas City (2005),
- the Royal Air Force (RAF) Nimrod XV230 loss (2006),
- Esso Longford (1998).

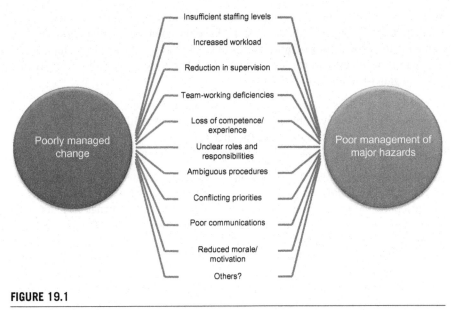

FIGURE 19.1

The impact of poorly managed change on the control of major hazards.

On an individual level, *"Change is frequently seen as a threat by individuals in an organization and can have a significant effect on their state of mind, their commitment to the organization and to their contribution to safety culture in particular."* (IAEA, 2003, p. 4).

In brief, some organizational changes can impact in unanticipated ways and occupy people's minds if not well managed. Extended periods of uncertainty, fears about job security, even just the appointment of a new manager, can create this effect. "Human" capital is a core asset of any business. It makes sense to consider the human needs, both in identifying appropriate change outcomes and transition periods.

COMMUNICATING THE RATIONALE FOR CHANGE AND ENGAGING STAFF

Whilst "not changing" may not be an option if the business is threatened, the reason for making changes, and the rationale for deciding on one particular change solution, needs to be carefully thought through. Some industries are contracting over time or subject to chronic governmental and regulatory uncertainty, and others are expanding and booming or racing to innovate to keep up with technology and the market.

Without a clear rationale for the change (and one that is clearly set out and explained), communication is immediately made more difficult. An organization needs good staff engagement to facilitate good two-way communication. Engagement includes involvement in key activities, but involvement does not necessarily mean that staff are fully engaged (that they really have the company's values embedded in all their activities). Communication needs to be effective with all staff including key contractors and other third parties. Having such arrangements also allows an organization to gain a more realistic picture of:

- where the business is (and so when, where, and if change is required);
- what solutions are available (a more involved and engaged workforce can contribute invaluable suggestions for change); and
- what will work best to make the transition to that change.

Even if the main driver is negative, the rationale for change should emerge from the communication and involvement process and be as balanced as it can be between threats and opportunities. In general, managing organizational change is better seen as an opportunity to improve health and safety performance rather than just aiming to maintain the status quo in that area.

The need to engage staff appropriately and actively is essential. This includes ensuring that any concerns are actively solicited, listened to, and feedback given on decisions made on these concerns. It also includes harvesting what is often a rich source of good ideas and information about what will work and what will help the planned changes. Effective engagement also means that staff are less likely to perceive the organization's values as drifting away from their own established ones. It is easier for staff to feel threatened and be fearful of change if they do not feel a part of it.

THE PROCESS OF MANAGING ORGANIZATIONAL CHANGE

Having decided on the need and rationale for change, the organization needs to make sure it:

- fully understands the existing situation (the baseline);
- selects and designs the appropriate solution(s); and
- designs, manages, and paces the process of the transition.

There are always both risks and opportunities in the choice of change and the transition to the change, so achieving the right balance is essential. There are, of course, risks in *not* making timely changes.

Typical issues that arise with *the end state or outcome* when the correct balance is not achieved include:

- staffing arrangements that are inadequate for foreseeable process upsets and sickness/absence;
- unrealistic expectations about operator reliably and behavior;
- losing technical competence and experience (or losing this too quickly);
- leaving maintenance inadequately resourced.

Typical issues with the transition include:

- lack of rigorous approach by sites;
- ad hoc risk assessments;
- failing to understand the baseline for change;
- not soliciting, or not listening to, legitimate concerns raised by operators;
- not resourcing the transition adequately, for example, for assessment, training; and
- a lack of objectivity (oversight) of the process of change.

While it seems obvious, a similarly thorough procedure to that usually in place for managing engineering change is required to set performance standards for managing organizational change and to map out the full process required (proportionately). It is useful to have both procedures together or closely linked so that anyone carrying out an engineering change process is encouraged to think about any allied organizational changes. The key contents may include:

- the current "organizational baseline";
- a definition and some examples of organizational change;
- lessons learned from previous changes;
- a simple categorization system for proposed changes;
- suitable assessment methods;
- key issues, such as competence, supervision, workload, staffing, roles, intelligent customer capability;
- clear accountabilities;
- recording of decisions;
- involvement and consultation requirements;
- in-built independent challenge at the key stages;
- prestart-up reviews;
- monitoring arrangements for implementation;
- a formal approval process; and
- postimplementation review/lessons learned.

In managing the change process itself, the key is to pace it and preferably phase changes appropriately. The plan should avoid overcomplexity and workload peaks for those involved. Additional cover may be necessary for the increase in work during the transition, for example, for training and documentation amendments. Staff with new roles will require support and the chance to consolidate new knowledge and skills in a structured way.

Uncertainty should be minimized, or at least clearly bounded (e.g., "Staff will be told about Y on X date"). The key is to maintaining good and consistent communication with all stakeholders, aiming for redundancy and diversity in communication channels and media, but avoid saturation or "empty" communication.

Key performance indicators need to be selected, set, and tracked (guidance can be found in the Canadian Society for Chemical Engineering, CSCE, 2004).

The transition risk assessments will require effective monitoring and review. Clear "go-live" criteria should be set for agreeing significant changes and ensuring that phases are ready to implement. This may include management approval and sign-off.

Finally, it is worth stepping back to ask two big questions:

"If we get this wrong (the change end state, the transition or both) how bad can it get?"

"How will we know that things are going wrong or we have made wrong choices?"

Being willing to admit that a change (or part of a change) is not working is critical. No one gets everything right all the time, but as is often said, the only error-free performance possible is that involved in doing nothing, and that in itself is often a bigger risk. Businesses do need to change, ideally proactively as well as reactively, and they need to make timely decisions about change. Following a good management of organizational change process will help deliver successful change by design and in implementation, and also identify, admit, and rectify the occasional failure.

RISK ASSESSMENT FOR CHANGE

There is no single risk assessment method for organizational change and it is better to think of having a toolkit of methods appropriate (and proportionate) for different categories and priorities of change. It will likely be necessary to assess both the desired change as an *end state* and the *transition* process to it.

The HSE guidance (HSE, 2003) uses a mapping and a scenario assessment approach to the *end state* (where the organization wants to get to). Key tasks and activities are mapped against specific roles and individuals to ensure that the existing ones are effectively maintained (or improved) and any significant new ones are covered. Assessments that do not include this level of detail will usually be inadequate and will make effective monitoring of change difficult. The scenario assessments build on the already identified major hazard or conventional accident scenarios identified through hazard and risk analysis. The process allows a structured assessment of the change against the control measures for these.

The CSCE provides an initial checklist screening process that also focuses on personnel and roles. For the transition, an initial "What If" checklist is provided to help determine the depth and rigor of assessment required. The guidance allows for further in-depth analysis similar to those advocated by the HSE. The US Center for Chemical Process Safety (CCPS) has also more recently produced guidance (CCPS, 2013) which takes a broadly similar approach. Their book provides more detail, including extensive case studies, and is arranged by the main change categories, such as organizational hierarchy, personnel, and task allocation.

In all cases, a team approach is required with sufficient competence and experience to maintain a sensible and informed oversight of the process. Without this kind of intelligent engagement, no risk assessment method will deliver effectively.

MANAGING PSYCHOLOGICAL ADJUSTMENT

Managing cognitive, behavioral, and emotional adjustment to planned organizational change is also essential. This essentially tackles, "How does handling this change proactively, thoughtfully, and sensitively benefit teams? How does it benefit individuals or the business as a whole?" Change is part of everyone's life and, arguably, is also an inevitable feature of the world of work. Accordingly, whilst the change process can be a period of energy, vigor, opportunity, and creativity, it can also carry perceived risks for those intimately involved. Its impact on employee morale, health and safety, and performance can all be influenced by how the process and outcomes for individuals and teams is being managed. There are a number of challenging psychological states that can come "into play" across planned organizational change processes. There are, however, simple but practical actions a Change Leader can apply to help everyone navigate their way through an often complex and labyrinthine journey.

UNDERSTANDING THE PSYCHOLOGY OF CHANGE

The literature offers many attempts at understanding what it is that change requires of individuals, or what internal processes people may go through when dealing with significant change in their lives. One of the most enduring and influential models, first described in the 1930s and later perfected by Professor William Bridges in the 1990s, is the three-stage model of change (Bridges, 2009). This suggests that change involves transitioning between three phases: from an initial phase of "unfreezing," letting go of what is currently known and/or habitually done, into a phase of "experimentation and adjustment", before reaching a final phase of "refreezing" into the new reality.

This has proven to be an insightful but accessible way to understand change from the perspective of those immediately affected by it. It is an appropriate model to use when change is introduced as an intentional and planned program, not just something that merely "happens" in the course of everyday life.

As with other models of change, it recognizes that successful progression through the stages (in this case, "Letting Go," "Neutral Zone," and "New Beginnings") will vary in pace between individuals/groups, and that the smoothness of the transition will be influenced by internal factors (e.g., personality, experience) and external factors (e.g., team culture, management of change). For example, if what has to be let go is not valued, then the transition into the change will be rapid and positive. People in leadership positions tend to have greater control over the change, more understanding of the benefits and greater confidence in their ability to cope with the change taking place; they are, therefore, potentially more able to transition more smoothly and quickly through the stages. Those who have less control, and a perception of more to lose, may move at a considerably slower pace.

A key learning point for Change Leaders is that even the most positive, well-intentioned, and advantageous change cannot simply be "pushed" through since that fails to recognize the inevitability that people will have to "let go" of something that is currently a source of comfort and predictability. The Change Leader cannot simply impose their own perspective on the value of the change and then expect others to adopt it as their own. Rather, they must seek to put themselves "into the shoes" of those impacted by the change agenda, and anticipate a wide range of reactions, some positive, others less so. The key to successfully managing transition, therefore, is to know what to look out for by way of presenting emotional, psychological, and behavioral signals, but to do so from the perspective of those actually exhibiting the signs (Tables 19.1–19.3).

> The concept of "resistance to change" implies that resistant people do not properly understand the change, hold views that are objectively wrong and generally unhelpful, and are obstinate/unjustified in their criticisms.

Table 19.1 The Bridges Transition Model—Letting Go

What Does the *Letting Go* Stage Look Like?	Suggested Management Actions
During this initial phase, some people experience emotional upheaval if they feel "compelled" to "let go" of something that currently provides comfort, for example, knowledge of the tasks, colleagues, working environment. This may present in different ways, but most noticeably as resistance to those introducing the change.	Guiding people *through* Stage 1 primarily involves understanding others' reactions and accepting their legitimacy. Allow time and space for "letting go" and provide opportunities for people to express their feelings.
If there is a clear attachment to what is being left behind, it is common for people to experience fear, denial, anger, sadness, frustration, uncertainty, and a sense of loss.	The key is two-way communication; that is, open communication to clarify what is happening, why, and how, and listening empathically to concerns. People often fear what they do not understand, so explain the rationale for the change, and help them to see how their knowledge and skills are an essential part of getting there. Provide reassurance about the support that will be provided (such as, training and resources) to work effectively in the new environment.
People have to accept that something is ending before they can begin to accept the change. Failure to acknowledge these emotions is likely to lead to further resistance throughout the process.	An acknowledgment of what people are losing, as well as reassurance about what they will be able to retain and even enhance, can help them to let go and move on. Complete solutions to concerns may not be known, so resist the temptation to offer them. If anything, "selling" the problems is more appropriate at this stage. Mark the endings by celebrating the successes of the past.

Table 19.2 The Bridges Transition Model—Neutral Zone

What Does the *Neutral Zone* Look Like?	Suggested Management Actions
During this stage, people affected by the change can be confused, uncertain, and impatient. They may also experience a higher workload as they acclimatize to new and imperfect systems and ways of working. This phase can be thought of as the bridge between the old and the new; people will still be attached to the old, whilst also trying to adapt to the upheaval and uncertainty. There can still be residual resentment and skepticism towards the change initiative. Therefore, morale and productivity may not be high, especially as there can be ongoing anxiety about one's role, status, or identity. Despite these adjustment issues, this stage can also be one of great creativity, innovation, and renewal. This is a great time to encourage people to try new ways of thinking or working.	Guiding people *through* Stage 2: This stage can feel very uncomfortable and unproductive, so people need a solid sense of direction and the opportunity to express their feelings and concerns. Meet frequently and remind people of team goals, to create more sense of ownership and unity. Establishing short-term goals can boost energy and create a sense of progress. Help with managing workload, for example, with extra resources to manage the change can make it feel more manageable. Offering positive feedback, encouragement to experiment with new ways of working and finding ways people can contribute, will help in moving things forward. A key facet that can help or hinder progress is encouraging managers to role model the behaviors they are keen to encourage in others.

Table 19.3 The Bridges Transition Model—New Beginnings

What Does the *New Beginnings* Stage Look Like?	Suggested Management Actions
The last transition stage is a time of acceptance and renewed energy. People are now better able to embrace the change initiative. They are building the skills they need to work successfully in the new environment, and are starting to see early wins from their efforts. At this stage, people are likely to experience a heightened sense of optimism with a corresponding rise in energy. There will be greater openness to learning and a renewed commitment to the group or their role.	Guiding people *towards* Stage 3 involves: Providing clarity of *purpose* by continually clarifying and reminding people of the compelling case for the change. Creating as clear a *picture* as possible of what the new world will look like—what will the improvements be (can they see a place that is already "there"?). Keep communicating the *plan* and the steps to get there. Involve people as much as possible to provide a sense of control and to improve the ability to predict how things will develop (these are essential components of well-being) Finding ways to involve people in developing new ways of working so that they have a *part to play* and feel able to contribute their skills and experience productively.

UNDERSTANDING INDIVIDUAL REACTIONS TO CHANGE

Change is often viewed by Change Leaders as an objectively positive transition from a less desired state to a more desired state. In these circumstances, when people do not react with enthusiasm for the change proposed, this can be seen as unexpected, unhelpful, and unjustified. However, most reactions to change, even those that might be perceived as "negative," are, in fact, understandable, predictable, and, often, logical or justifiable. This is not to suggest that people's views cannot and should not be challenged and changed, but that "resistance" to change needs to be understood from the people's perspective to truly be able to turn the situation around. Change for many can be an invigorating experience, and Change Leaders should anticipate there being significant variation in individual reactions, with much depending on the circumstances of the change itself, and the personal interpretation of what is happening. Change is, however, more likely to have a negative impact on adjustment and well-being if it is:

- associated with the threat to or loss of something valued: for example, job loss may threaten our incomes, but could also mean the loss of esteem, of colleagues and friends and, indeed, our sense of who we are;
- seen as something over which staff have little control or influence: for example, change in a team which we have some influence within may be easier to cope with than change at a higher organizational or structural level where we have little or no control; or
- perceived as something staff do not agree with: bearing in mind this can be part of our reaction to the change itself.

Therefore, it is not given that people will be resistant to change per se, but are more likely to do so when it is perceived as negative, associated with a personal loss, and/or outside of personal control. In their paper "Managing Change in the Nuclear Industry—The Effects on Safety," the Atomic Energy Authority note that "*Change is frequently seen as a threat by individuals in an organization and can have a significant effect on their state of mind, their commitment to the organization and to their contribution to safety culture in particular.*" (IAEA, 2003, p. 4).

One such condition that is more likely to result in resistance is when the individual feels personally threatened by the process itself or the perceived outcomes. The body's natural stress response is triggered by the perception of threat, whether that be real or otherwise. This stress response is not conducive to higher order thinking, information processing, or data recall, all of which can have health and safety implications; clarity of thought and good decision-making can be essential in emergency situations. When people feel threatened by change, both a negative thinking bias and a narrowing of attention field can develop; neither of which are helpful for those imaginative and creative thought processes that a change agenda can strongly benefit from. In order to understand people's reactions with a view to providing better support, this should be explored from their perspective, and with the avoidance of assumptions about how well people are adjusting. The following models illustrate the dynamic nature of psychological adjustment. It can help to interpret the states people are experiencing from the behaviors typically observed, and provide guidance on ways to act to promote more positive responses to change.

UNDERSTANDING GROUP REACTIONS TO CHANGE: CONSIDERING CULTURAL FACTORS

Although it is useful to understand and manage psychological adjustment to change for the individual, it is also useful to see resistance within the broader cultural context into which the change is being introduced. In this sense, resistance is not just individual, but has a clear "collective" dimension. Also, resistance is not necessarily always, or only, about the *current* change; it can and will be influenced by *previous* changes, successes, and failures and the way they have been interpreted and are being talked about.

Background Conversations
(Ford and Ford, 2009)
Complacency
"we're doing well now, no need to change";
"if it isn't broken, don't fix it";
"we're happy as we are".
Resignation
"this is too hard, too much, too soon";
"it won't work, so I won't try"
"in an ideal world then yes, but we just can't do it…"
Cynicism
"who are they kidding, this will never work!";
"I can't believe what they are saying/doing/expecting…!";
"in an ideal world yes, but the bottom line is…"

Using the term "background conversations," Ford and Ford (2009) suggest ways in which the specific, existing organizational context can influence how people interpret and react to a planned change initiative. Three "background conversations" are found to be the most common in organizations.

In the first of these "background conversations," named *Complacency*, the collective mind-set is underpinned by taking sanctuary in the perception of historical success. The conversation is of an "internal" protective nature.

The second "background conversation," *Resignation*, is similar to "Complacency" in that the focus is internal, but this time reflecting a perception of historical failure.

The third "background conversation" is categorized as "*Cynicism*." In this case, again the focus is on the perception of historical failure, only this time it is concentrated on factors external to the group or community.

The Change Leader must acknowledge that the prevailing cultural norm will also need to be considered when positioning communication and considering what support will be required. If these "background conversations" reinforce one position, and the change proposition emphasizes another, then efforts to secure motivation and engagement need to go beyond an appreciation of individual adaptability and be sensitive to wider organizational influences. By recognizing the strength and tone of the "background conversations," appropriate counterarguments can be presented that acknowledge, but challenge the concerns. The key is to acknowledge the historical perspective; the successes and failures of the past that reinforce the cultural status quo.

UNDERSTANDING BARRIERS TO CHANGE

The theme of collective resistance is also tackled by authors Kegan and Lahey in their book "Immunity to Change" (2009). Their central tenet is that desire and motivation may not, in themselves, be sufficient to cement in place the key behaviors that deliver sustained change. Even where there is an intellectual recognition of the value the change proposition brings, people can still be drawn towards maintaining the status quo. This, they argue, is a consequence of either unacknowledged worries or concerns about the change, or the merits of the competing motivation preserving the status quo are not fully surfaced, acknowledged, and challenged.

As a basic example, the benefits of following a good diet are commonly recognized, at least at an intellectual level. However, many efforts at dieting fail, or are short-lived. One possible competing motivation to following a good diet is the short-term comfort experienced by eating the wrong type of foods. It is only by surfacing and examining that competing motivation that we can understand and overcome ingrained obstacles, and then successfully make what is known as "adaptive change."

The key is to search for these self-protective cognitions, surface them, and then examine the counterproductive thoughts and behaviors that sustain the status quo. The liberating influence comes from seeing something that *was not previously* known—that all these behaviors have a perfectly good reason behind them, that they are highly productive on behalf of other competing but shared motivations. Kegan and Lahey offer a four-stage process that guides teams and individuals through a process of self-examination, thereby helping them identify and adjust assumptions that may be holding them back. This technique was used successfully with an Extended Corporate Management Team (ECMT) leading organizational restructure (Case Study—Box 19.3).

PREVENTATIVE MEASURES: PLANNING FOR CHANGE

If we can identify human factors risks early in the change process, then appropriate preventative measures can be introduced that ultimately avert problems from surfacing further down the change journey. One such proactive approach is a Future Focused Stress Risk Assessment. This assessment is designed to identify likely stressors in advance, and build stress prevention measures into the project or change plan. This assessment was used by a petrochemicals company when commissioning an extension to a chemical process technology demonstration plant and illustrated the advantages of establishing concerns upfront, estimating their potency to cause harm and then instituting control measures for those identified as the "big ticket" items (Box 19.4).

BOX 19.3 CASE STUDY: ECMT LEADING ORGANIZATIONAL RESTRUCTURE

Background

The pressure of having to deliver "more" with "less" required the ECMT of a large public sector service organization to fundamentally review its current governance and service provision. One significant change was the need to move from a management structure where each individual service head essentially focused on what was happening in their own directorate, to one that emphasized corporate responsibility and decision-making. This change in emphasis was deemed appropriate as efficiencies needed to be realized across the organization and, therefore, problem solving, decision-making, and communication needed to be coordinated across organizational boundaries to a much greater extent. Despite each manager intellectually subscribing to the overall change goal, the behaviors required to sustain the new approach did not consistently follow. Indeed, those behaviors that sustained the "silo" approach were even more in evidence, thus creating tensions and undermining collective effort.

Methodology

A Keil Centre Chartered Psychologist facilitated an exploration of the issue with the team, using Kegan and Lahey's "Immunity to Change" methodology, following a four-stage process.

Step One: Our Collective Improvement Goal

Following a brainstorm session, the group agreed their goal. This had to be something with a big payoff that they were collectively motivated to achieve. In this case, it was defined as the need to "*create a culture of mutual trust and support*." They then identified specific, concrete behaviors that would underpin the achievement of that goal. An example behavior was "consider who else would be impacted by a decision, and communicate accordingly."

Step Two: Telling on Ourselves

This element was about identifying those behaviors the ECMT engaged in, or failed to engage in and which operated against the achievement of the main goal. The *whole group* had to acknowledge its *collective* ineffectiveness, rather than it being about "those who don't get it." Examples included: "We let our individual agendas trump our collective agenda," "we assume bad intent," "we form cliques," and "we avoid difficult conversations with each other."

Step Three: Surfacing Hidden Commitments

This started by considering the "worries" the group would have if they tried to do the opposite of all the behaviors identified during Step Two. These are the things that "justified" the Step Two behaviors. For example, "*We worry that if we trust others to make decisions then our own self-interest is not protected*" and "*Those within our Directorate will think we have 'sold out'.*" By then reviewing these worries, they can be converted into something that explains why they offer "collective self-protection." So the above becomes, "*We are committed to having a say in every decision in order to protect our own self-interest and those of our own immediate team.*" In this way the group could see exactly why trying to succeed merely by eliminating the Step Two behaviors was not going to work, that is because these behaviors are serving an important "protective" purpose.

Step Four: Uncovering Collective Assumptions

The competing commitment in Step Three is typically the result of a "big assumption." This is an idea held to be true even though, until it is challenged, there is no way of knowing for sure if it is. One such "big assumption" for this team was unearthed as: "*If we are not personally involved in making a decision, then it cannot be a good one.*" What then followed was a facilitated discussion around the following questions:

- *Do we feel that this aspect of our mind-set is seriously impairing our effectiveness?*
- *Do we feel like it could make a big difference if we were able to release ourselves from this group belief?*
- *Do we feel that we owe it to ourselves to see if we can alter this?*

Having then secured affirmative commitment on these questions, it was then a relatively simple case of the team brainstorming "thought" experiments and "action tests" that would yield information or experiences that shed doubt on, or disprove, the big assumption.

BOX 19.4 CASE STUDY: IDENTIFYING AND MANAGING POTENTIAL STRESSORS IN ADVANCE OF ORGANIZATIONAL CHANGE

Planned Change

Commissioning an extension to a chemical process technology demonstration plant in a petrochemicals company.

Challenge

Plant commissioning is recognized as a demanding time. The size of the operator team was increased whilst staff reductions were made in other areas. The new staff were less experienced and the site staff reductions led to considerable job insecurity. The Applied Technology team had already taken steps to eliminate physical hazards and improve safety culture. They then wished to take action to prevent avoidable stress by identifying and mitigating any work-related sources of stress arising from the project.

Solution

A project team was formed, including the project manager and representatives of the two main project groups working with the aim of securing the cooperation of a cross-section of the workforce to prevent work-related stress. A Chartered Psychologist briefed them on stress. They then identified likely sources of stress that might arise from the commissioning project. The team members then prioritized their stressors.

Stressors were then sorted by:

(a) their relevance to the project; (b) whether they were currently well controlled; (c) whether they were likely to cause stress.

This yielded a set of "top five" stressors on which there was a consensus about their potential to cause harm. Examples of the top-five stressors included high workload, job insecurity, demands for unnecessary detail, and pressure from senior managers. The team identified how or why each stressor caused harm, and shared ideas about what organizational and individual actions would mitigate the effects of the top-five stressors, and identified relevant, practical control measures. This involved discussing some personal issues, such as the effects of job insecurity on other family members.

The most striking example of a stressor, which was effectively identified and controlled, was "unnecessary detail." This stressor, specific to the demands of plant commissioning, referred to the effects of other people not specifying the amount of technical detail they required, and the timescales involved. As a result, staff worked long and hard to promptly produce detailed technical information, which was often not required. Subsequently, a phrase was coined which became the watchword for dealing with unnecessary detail. This phrase "the minimum requirements" is now used by all team members to challenge others on the level of detail and deadlines attached to work they require. Adoption of the "minimum requirements" concept has had lasting benefits for managing workload amongst the Applied Technology team.

Results

The stress prevention project was evaluated by interviewing a sample of people from the Applied Technology team. These interviews focused on how the project had impacted upon their perceptions and personal experience of stress, and whether and how their behavior or the behavior of others had changed.

In summary, this relatively simple, low cost stress prevention project was conducted by a cross-section of employees, with minimal external input. The project's design and execution exceeds the requirements of UK legislation and regulatory guidance on preventing risks to health and safety arising from psychosocial hazards at work. Framing stressors as a hazard to be controlled during a change program, just like the more familiar process and chemical hazards, was a logical extension to existing risk assessment processes, and opened a mature debate about otherwise delicate topics, such as the effect of management style on others. The project normalized discussion of stress and stressors amongst the team, and facilitated team spirit and open communication.

Comments

This intervention succeeded in taking a preventative and holistic approach to stress at work at the beginning of a change project. It shows how at the design stage, future hazards can be identified and removed or reduced. The effects should be sustainable. It also illustrates how employees can be involved in the change risk assessment and management process.

Another organization, which used the Future Focused Risk Assessment, was relocating operations from central Europe to the United Kingdom, and some employees were required to uproot from one country to another. Work–life balance was one of the "big ticket" items identified by their assessment. The process provided a structure for the organization to examine the potential risk and consider which control measures were feasible to incorporate into the change plan (Table 19.4).

Table 19.4 Trying to Balance Work and Home Life

1. *How/Why is this stressor going to cause stress?*	3. *What specific actions will help to achieve this?*	4. *When and how will this be reviewed?*
During the transition: • Frequent travel; • Working evenings and weekends; • Less time and energy for home life, hobbies, family; • Not taking enough exercise; • Sense of there being "no light at the end of the tunnel"; • Vicious cycle of feeling one has to work longer and longer to keep up with decreasing reserves of energy and motivation	• Managers "peer review" activity and gap closure— hold accountable through one to one discussions; • Processes and policies that encourage people to take their holiday entitlement; • Sensible distribution of work, expertise, and capacity taking resourcing levels into consideration, including work–life balance as an appraisal item for all staff; • Including "stress management" as an appraisal item for all people managers (and making this known); • Assessing those for whom homeworking is an opportunity to maximize time at home (through avoiding air travel, commuting, or being trapped at the office). This was a project team priority; • Consequence management and coaching for people managers driving the wrong behaviors; • Flexible start/finish times where appropriate to help employees cope with external pressures, such as childcare, a sick family member.	• Monitoring changing patterns in presenteeism and absenteeism; • Monitoring the taking of holiday entitlement; • Monitoring stress levels with a "stress thermometer"; • Make "work–life balance" agreements a regular agenda item during the transition and act on failures.
2. *What can be done to prevent or manage this?*		
• Leadership by example with respect to working hours; • Proactive intervention by both colleagues and management to comment when an individual seems to not be getting the balance right; • Concrete intervention—"did that intervention change anything?"; • Reinforcement of a "Brother's keeper" culture—intervention that is everyone's responsibility; • Promoting awareness through keeping stress, and stressors on the agenda; • Improved use of technology to rationalize or mitigate the need for travel.		

THE ROLE OF COMMUNICATION

Communication is a common key element in all aspects of organizational change. Its purpose may be to communicate the rationale for the proposed change, engage staff in the change process, or generally support their transition, but the important characteristics are that the same; it must be consistent, supportive two-way communication.

Van Vuuren et al. (1991) examined how communication would be received, processed, and interpreted when the receiver perceived there to be a threat to their job security. They found that if individuals sense something personal to them is under threat, it is less likely that they will find communication about the goals and objectives of a change program meaningful. Where perceived uncertainty about job role and security exists, receivers will filter the messages to pick out any nuance, inconsistency, and so on that might provide any clues as to "what will happen to me." Consequently, there is a high risk that organizations will underestimate how much and how often it needs to repeat the message, and will consider less what the audience will be listening out for.

Communication should be regular and two-way, erring on the side of overcommunication. It also has to be sensitive to what will actually be "heard" and processed by the receiver. In other words, what people "need" and "want" to know.

- Firstly, Change Leaders must recognize employees want to *understand and predict outcomes*; "why is this happening, when, how…?"; "what process will be used…?" and most importantly, "what will the impact be on me, personally?";
- Secondly, people will also want to know what *remains under their control*; "how will things operate in my team?", "how will this affect my career and my development?";
- The third area relates to people's *sense of esteem and self-confidence*. The concerns might include: "how will I manage through the change?" and "how will I cope with the new systems and processes?";
- Finally, are issues more to do with a *sense of belonging*: "how will this impact on the team and my role within it?", "how can I retain my support network?", and "how can I relate to others as a survivor?".

Ford and Ford (2009) emphasize the importance of the flow of effective, targeted conversations between the Change Leader and their team. Effective change happens, they argue, because leaders are able to talk to their teams about the right things at the right time. Through these conversations, they achieve clarity of focus, a sense of commitment to the change, and effective action to support the change process. In fact, the Fords argue that the power of effective conversations underpins leaders' effectiveness in nearly every task that requires them to ask someone else to do something for them.

The Fords emphasize four distinct conversations, namely initiative, understanding, performance, and closure in leading change. Their argument is that all four demand attention in order that the initiative is driven through to a successful conclusion. However, cultural factors, leadership styles, and organizational pressures can result in some conversations being either underplayed or overplayed, both of which can undermine progress, confidence, and successful conclusions (Table 19.5).

Table 19.5 Conversation Types in Leading Change

Purpose	Main Focus	When Successful
Initiative conversations		
To create a future. Ideas generators. To propose something new or different, such as a goal, idea, strategy, policy, procedure	*What* are we aiming for? *When* do we want success? *Why* it matters...	• Creates an "active and intentional" approach to the future; • Generates momentum and energy.
Understanding conversations		
To include and engage. To create mutual understanding for what is proposed, and to be able to relate this to current jobs and ideas.	An opportunity to hear and learn; Emphasizes involvement and inclusion; Understanding does not mean acceptance, or result in action	• Relates the change/desired future outcome to people and their circumstances; • Generates understanding, if not always acceptance and commitment.
Performance conversations		
To ask and promise. Action generators. Result in agreements to take actions and/or produce results. Secure agreement for the future.	An exchange of promises; Commitment to the what, when and why; Clarity on specific deliverables and timescales.	• Clarity over roles, responsibilities, and accountabilities and commitment to act; • Action taken as a result of agreements on requests and promises.
Closure conversations		
To create endings. A summary of the project's status. The review stage.	Acknowledgment: *Say what's so* Apologies: *For mistakes and misunderstandings* Appreciation: *Recognize accomplishment and achievement* Amendments: *Recognize, report, repair, and recommit*	• Strengthens commitment and builds engagement; • Addresses performance issues and challenges and builds accountability.

SUMMARY OF CHANGE LEADER SUPPORT ACTIONS

- Engage in a continuous cycle of communication and listening: Understand change from the perspective(s) of those involved to anticipate and more effectively manage reactions;
- Realize that mixed views on any changes initiative are inevitable: Discussing these differences openly is more important than imposing the one "correct" way to view the change; Acknowledge that not everybody will react as expected/positively from the start; anger, anxiety, and other "negative" emotion may be part of a normal reaction to change and may need to be expressed before people can move on;
- Talk to individuals face-to-face regularly to see how their thinking and feeling is changing/progressing;
- Set reasonable goals: The Change Leader may not be able to help all people through this change positively; Be aware of your own tensions, and look after yourself too.

SUMMARY: MANAGING ORGANIZATIONAL CHANGE

Change is inevitable, but growth is optional. Even those organizations who rigorously analyze, plan, and execute structural change can still fail to achieve the desired outcomes through a failure to predict and manage the human factors aspects of the program.

The next chapter addresses some of the key issues that can arise through poor management of organizational change; namely insufficient staffing, excessive workload, and insufficient competence.

KEY POINTS

- Organizational change requires the same rigorous management as engineering change.
- The adverse impacts of organizational change can be difficult to identify. Traditionally, this area has received less attention and so has been a major contributory factor in many major accidents, as well as more conventional ones.
- Good two-way communication and staff engagement arrangements are essential to effective management of organizational change.
- Organizational change is best viewed as necessary and desirable for a successful business (and for such a business to remain so) and as an improvement opportunity rather than just reactive to events.
- There is good guidance available describing proportionate organizational change processes and risk assessment.

REFERENCES

Bridges, W., 2009. Managing Transitions: Making the Most of Change, third ed.. Da Capo Press.

Canadian Society for Chemical Engineering (CSCE), 2004. Managing the Health and Safety Impacts of Organizational Change. CSCE, Ottawa, ONT.

Center for Chemical Process Safety (CCPS), 2013. Guidelines for Managing Process Safety Risks During Organizational Change. Wiley, Hoboken, NJ.

Ford, J., Ford, L., 2009. The Four Conversations: Daily Communication That Gets Results. Berrett-Koehler Publishers, San Francisco, CA.

Holman, P., Cady, S., Devane, T. (Eds.), 2007. The Change Handbook: The Definitive Resource on Today's Best Methods for Engaging Whole Systems, second ed. Berrett-Koehler Publishers, San Francisco, CA.

HSE (2003). Organisational Change and Major Accident Hazards. Chemical Information Sheet No. CHIS7. Available from: HSE: http://www.hse.gov.uk/pubns/chis7.pdf (accessed 27.01.16.).

IAEA (2003). Managing Change in the Nuclear Industry: The Effects on Safety, INSAG-18, A Report By The International Nuclear Safety Advisory Group, Austria: IAEA. Available from: http://www-pub.iaea.org/MTCD/publications/PDF/Pub1173_web.pdf (accessed 27.01.16.).

Kegan, R., Lahey, L., 2009. Immunity to Change: How to Overcome it and Unlock Potential in Yourself and Your Organization. Harvard Business Press.

Kübler-Ross, E., 2007. On Grief and Grieving: Finding the Meaning of Grief Through the Five Stages of Loss. Simon & Schuster Ltd.

Quote Investigator, (2016). Entry for Gaius Petronius Arbiter/Charlton Ogburn Jr. Available from: Quote Investigator: http://quoteinvestigator.com/2013/11/12/reorganizing/ (accessed 27.01.16.).

Van Vuuren, T., Klandermans, P.G., Jacobson, D., Hartley, J.F., 1991. Job Insecurity: Coping with Jobs at Risk. Sage, London.

Wilkinson, J., & Rycraft, H. (2014). Improving Organisational Learning: Why Don't We Learn Effectively from Incidents and Other Sources? Proceedings of The Institute of Chemical Engineers' Hazards 24 Conference, Edinburgh.

Staffing the operation

20

J. Edmonds

LIST OF ABBREVIATIONS

°C	Degrees Centigrade
°F	Degrees Fahrenheit
App	Application
CC	Carbon Copying
CEO	Chief Executive Officer
CMS	Competence Management System
CRR	Contract Research Report
DEI	Duke Energy International
DIF	Difficulty, Importance, Frequency
EHS	Environment, Health, and Safety
EI	Energy Institute
HAZOP	Hazard and Operability
HR	Heart Rate
HR_{MAX}	Maximum Heart Rate
HRR	Heart Rate Recovery
HSE	Health and Safety Executive
HSG	Health and Safety Executive Guidance
ISA	Instantaneous Self-Assessment
KPI	Key Performance Indicator
MAH	Major Accident Hazard
NASA TLX	National Aeronautics and Space Administration Task Load Index
NOPSEMA	National Offshore Petroleum Safety and Environmental Management Authority
NOTECHS	Non-technical skills
OGP	International Association of Oil and Gas Producers
ORR	Office of Rail Regulation
OSD	Open Safety Dialogue
OTA	Operational Task Analysis
PPE	Personal Protective Equipment
PSF	Performance Shaping Factor not in alphabetical order with PPE
RPE	Rating of Perceived Exertion
SMT	Self-Managed Teams
SWAT	Subjective Workload Assessment Technique

Human Factors in the Chemical and Process Industries.

TGA	Training Gap Analysis
TLAP	Time-Line Analysis and Prediction
TNA	Training Needs Analysis
TOA	Training Options Analysis
VACP	Visual, Auditory, Cognitive, and Psychomotor
VO_2	Volume of Oxygen
W/INDEX	Workload Index

The structure of an organization has a profound influence on the behavior, performance, and ultimately, the safety of the people within it.

The past few decades have seen substantial changes to organizational structures for reasons of technical advances, commercial constraints, societal and culture changes, and many other factors. For example, technology has made it possible to automate certain activities that traditionally would have been undertaken by people, or perform activities that may not even have been possible 30 or 40 years ago. The nature of jobs has changed and there are relatively fewer people engaged in order to meet the needs of an organization. Indeed, some sites have few people present for most of the time, or even none, such is the case for a remotely operated site.

Many roles have been combined, necessitating the need for multiskilling the workforce. Some of these changes have emerged from structured job design and led to improvements. This includes the level of control and responsibility that people have over their work, an increase in job variety, and job enrichment. These are typical intrinsic factors that are known to be related to greater job satisfaction and can motivate people, and improve their performance. The converse is also true in that some less well-managed changes have left people with less control, degraded skills, and limited job variety. This can lead to reduced job satisfaction, lower motivation, and ultimately a reduction in performance.

There have also been changes in the levels of supervision and management within organizations. Many organizations are now less hierarchical with fewer layers and less rigidity in their structure. Some organizations have moved towards flatter structures with fewer supervisory and managerial levels and the use of Self-Managed Teams (SMT) has also gained momentum.

Organizational structures are by no means static in their nature. They continue to evolve and change over the life of the organization. This is often not well managed, and the organizational design and resulting job design can often be left to evolve itself through custom and practice, rather than being actively and objectively designed with human capabilities and limitations in mind. In coping within such a dynamic situation, it is necessary to have a well-designed, well-understood, and appropriate organizational structure for managing ongoing change if an organization is to perform well and perform safely.

This chapter does not cover all the aspects of organization and job design but it does focus on providing an insight to answer some critical questions, including:

- Staffing and Workload (20.1): How many and what type of people does the organization need? How does the organization ensure that the workload is well balanced, particularly in relation to safety critical work?
- Training and Competence (20.2): How does the organization ensure the competency of the workforce?
- Effective Supervision (20.3): How does the organization ensure that supervision (which is critical to safe performance) is effective?

There is a lot of cross-over of these issues with the topics discussed in the other chapters within the organizational performance section of this book, such as safety culture, the management of organizational change, fatigue and stress. These issues are also important considerations that influence safe performance (and human failure) and need to be effectively considered during the design of new or modified systems.

Staffing and workload

J. Edmonds and C. Munro

Staffing and workload are critical factors in achieving safe and effective performance. These factors can be overlooked during times of staff reductions, and may result in jobs not being reorganized to take account of having fewer people to perform the activities. This can lead to the remaining staff trying to work harder, quicker or longer, and possibly taking short cuts to compensate. Some tasks may also be delayed or not done, the competence of the workforce may be reduced, error rates may increase, and the ability to deal with a crisis may be compromised. This can negatively impact the organizational performance and lead to serious health or safety consequences. There are a number of major accidents where staffing has been a contributory factor. Two examples are presented in Box 20.1.

BOX 20.1 CASE STUDY EXAMPLES OF INADEQUATE CONSIDERATION OF STAFFING AND WORKLOAD

Excessive workload was identified as a key contributory factor in the accident that occurred at the Texas City Refinery in 2005. After a 25% reduction in staffing, only one board operator was available to control and monitor three complex processing units, as well as dealing with the start-up of the isomerization unit. The ensuing critical error, being unaware of the dangerous level of the raffinate in the isomerization tower, was made more likely when there was too much work for the board operator to deal with, high workload increased the likelihood that this error would occur.

The explosion at the Buncefield oil storage facility, also in 2005, had several human factors failings, one of which was related to chronic staff shortages. This was exacerbated by the continuous increasing of the site throughput and lack of facilities and agreed arrangements for storing incoming fuel. There had been a lack of organizational focus on whether the staff could cope with a changing profile of the throughput and the increased complexity for product movements caused by the site limitations.

Staffing is concerned with having the optimal number and type of people (staff and contractors) to consistently perform at the required standard in all operational scenarios. Workload must be considered to ensure that the staffing profile is suitable to enable activities to be safely conducted.

This chapter provides an overview of what is meant by physical and mental workload in the context of the human capabilities and limitations. Key tools and techniques are described for measuring and predicting workload, and assessing an organization's staffing arrangements. Finally, possible solutions to staffing and workload issues are discussed.

HUMAN CAPABILITIES AND LIMITATIONS

The term "workload" refers to the total demand (or load) placed on a person as they perform a task. In simple terms, this can be the quantity of work or quality (in other words, difficulty) of work. The demand can be physical or mental, but in reality any activity is comprised of both elements. Over the last century, there has been an increasing trend towards mental over physical task elements as a result of changes in technology, but both are still relevant. High physical workload is not generally an issue for tasks such as control room monitoring, which is a predominantly cognitive activity. Some plant side activities are physically demanding; and some can be both physically and mentally demanding, such as performing a complex line up with multiple manual valve changes and temporary connections. Other examples of mentally demanding tasks might include driving a crane under difficult circumstances, or repairing a complex instrument.

People have physical limitations and these are often determined by work rate and effort. It is only possible to maintain maximum effort for short durations. If a longer duration is required, the work rate will need to be reduced and the work–rest cycle carefully controlled to minimize potential consequences such as muscular strain/injury and exhaustion. This can be exacerbated in different work environments, and is discussed in more detail later in the chapter.

People also have mental limitations, and it is useful to understand the cognitive activities involved in a task. The model of information processing (based on Wickens, 1992) is shown in Fig. 20.1. This shows the stages of cognitive activity. People receive information through their senses (visual, auditory, tactual, smell, and taste), then process and act upon that information. The person's attentional resources limit their processing capacity and this influences the information that is attended to and processed. It seems logical that where someone has "spare capacity" (or sufficient attentional resource), then the task demands can be accomplished and performance is good. However, there is more to the capacity/resource argument which warrants further discussion.

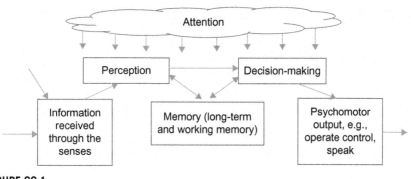

FIGURE 20.1

Human information processing model.

The multiple resource theory proposes that mental processing does not use one single resource but instead uses several mental processing resources (Wickens, 1992). Cognitive resources are limited and excess workload occurs when the individual performs two or more tasks that require the same resource. For example, people cannot look at two things at once if they are in completely different locations or say two different statements at once. These tasks must be sequential. However, people can hear an alarm whilst visually monitoring a display screen (or walk and whistle at the same time) because these tasks use different resources (Visual, Auditory, Cognitive, and Psychomotor, commonly referred to by the acronym VACP (or VCAP)).

This becomes more complex when the input modality of information (received through the senses) is differentiated from the output modality (e.g., pressing a control, saying something). The stages of the information process therefore have a bearing on the ability of the person to manage the task. The person attends to external stimuli (predominantly visual and auditory), cognitively processes the information and then performs physical actions (psychomotor). The processing bottleneck could be at the input end, the output end (e.g., visio-spatial or verbal-acoustic), or in the middle (cognitive processing). If the task demands different resources at different stages, then the person can achieve a considerable degree of simultaneous activity, but there will still be limitations to what can be achieved.

The multiple resource theory provides the opportunity to identify when it is appropriate for tasks to be performed concurrently or whether one may adversely affect the other(s). Some of the modeling techniques that are discussed in this chapter draw on the notion that the VACP demand of a task can be predicted.

It is important to recognize that workload does not have a simple linear relationship with performance, so performance does not necessarily decrease as workload increases. Firstly, performance can be adversely affected by sustained low workload, as well as high workload. Secondly, people develop compensatory behaviors for handling task demands, such as slowing the task, task shedding, or rapid task switching in high workload situations (Keller, 2002). In low workload situations people may also develop compensatory behaviors to relieve the boredom (such as thinking about something else, watching television, adding their own tasks to make it more stimulating). These behaviors may or may not lead to a performance decrement.

There is a nonlinear relationship of workload to performance and optimal performance requires a degree of pressure (demand). The "Yerkes Dodson Pressure-Performance Curve" illustrates that performance can be adversely affected by low pressure (arousal) as well as over pressure (arousal). A modified version of the "Yerkes Dodson Pressure-Performance Curve" is shown in Fig. 20.2, based on research by Teigen (1984).

Without enough pressure a person can become bored, frustrated, unmotivated, and lose confidence. In this state, a person can lose situational awareness and alertness. Conversely, if the pressure is too much, a person can become anxious, "tunnel-visioned," fatigued, or emotionally drained. Both of these states are "stress states" effectively, where the demands or pressure on someone mismatches their ability to cope. Stress effects are discussed in more detail in Chapter 23, Managing Performance under Pressure, and fatigue is discussed in Chapter 22, Managing Fatigue.

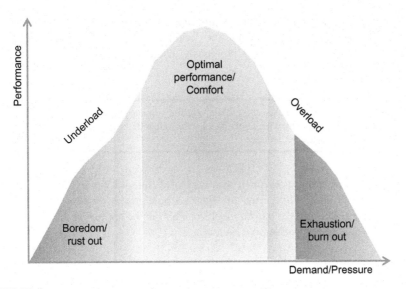

FIGURE 20.2

Performance under pressure: Yerkes-Dobson curve.

Stress is a Performance Shaping Factor (PSF) that makes error more likely. The optimal state is where someone is able to effectively balance the demands made of them with the internal and external resources that they have. The internal resources can include someone's skills, knowledge, and experience, and even personality. See Box 20.2 for an illustration of how individual capabilities can affect the ability to cope.

BOX 20.2 THE CASE OF THE LEARNER DRIVER

Take the case of a learner driver approaching a busy and complex junction. The situation could cause the person to be overwhelmed by the perceived high workload, due to lack of experience, skill, and familiarity with the road layout. The experienced driver, who is familiar with the road layout, may not find the task of driving at this point particularly demanding at all. In this situation, the task may not pose high workload, but the *inability to cope* with the task may cause the experience of high workload.

As the learner driver becomes more experienced, many aspects of the task require less conscious thought, so certain elements, such as changing gear, operating the steering wheel, using the indicators, and so on demand less attentional resource. The attention can be freed up and allocated to other less predictable aspects, such as attending to what other drivers are doing.

The external resources that people can draw on might include the support from others, ability to delegate tasks to others, adequate time to do the job, the way the task is organized, and having the tools and materials required.

In essence, there are individual differences both in the reaction to pressure and the ability to cope, so the "optimal" zone on the Yerkes-Dobson curve can differ from one person to another and is affected by the task, work equipment, and work context. People often cope with being outside the "optimal" zone for short durations given the opportunity for recovery, or as previously discussed, by developing compensatory behaviors.

> The aim is not to force the person to adapt to unacceptable workload, but to design the task, work context, and organization to meet the capabilities and limitations of the person. In this manner it is possible to achieve optimal performance.

The next sub section presents a summary of different tools that can be used to assess workload and staffing, and the final section presents a discussion on how workload and staffing issues can be resolved. This can be applied at the design stage (predictive) and during operation (reactive).

TOOLS FOR ASSESSING WORKLOAD AND STAFFING ARRANGEMENTS

A multitude of different techniques are used to measure or predict the human response to work scenarios and enable assessment of different aspects of these scenarios. It is vital to understand the question that needs to be answered to ensure the selection of the right tool(s). Undertaking complex workload modeling might be interesting, but is likely to be a long-winded method to use when others will suffice and provide a more direct answer. It is also sometimes necessary to use more than one method.

This section presents an overview of the different types of technique available. They have been grouped as follows:

- time-based assessment methods;
- tools for workload measurement and prediction;
- assessing staffing arrangements;
- tools for physical workload measurement.

The objective of these tools is to provide clarity regarding whether there is a workload or staffing issue, and if so, what aspects of the work or system are causing the issue.

TIME-BASED ASSESSMENT METHODS

Average task times

The most basic form of assessment is to identify the temporal nature of task performance. A simplistic approach to this might be to calculate staff numbers on the basis of how many tasks they can accomplish in the given time frame. The data to calculate this includes average task times and task frequencies, and staff availability (minutes per hour taking account of breaks and so on). This approach may be sufficient for a work scenario, such as a customer call center, where the task is highly repetitive and the demand can be easily predicted. However, the task demand within the chemical and process industries is not easy to predict as the tasks performed are complex and vary considerably from day to day, minute by minute (sometimes), and by scenario. This simplistic form of assessment is not typically sufficiently comprehensive for this type of industry, although it may be useful to provide an initial gross indication.

Timeline analysis

Timeline analysis can be used to model existing scenarios or for predicting future scenarios (albeit that the predictive use generally benefits from further "live" or "simulated" testing). This means that it can be used during the system design phases as well as during operations. The technique is also useful for assessing individual tasks, task allocations, different operational scenarios, and team performance.

Regardless of whether the assessment is for a current operation or a new system, the first stage is to collect observational data of a current/similar work scenario. In the absence of some or all of this data, subject-matter experts (e.g., operators) and designers may provide an estimate of considerations such as task structure, task duration (and task time variation), frequency, and task nuances.

The work activity is then broken down into its constituent parts in the order the tasks are undertaken and graphically plotted against a timeline (see Fig. 20.3). This allows the timing and functional requirements of the system to be determined (Kirwan and Ainsworth, 1992). In other words, what does an operator need to do, when, and how quickly is it undertaken? This is more difficult to determine if aspects of the task performance are highly variable and so the variances of task durations will need to be defined. Consideration of the impact of the task time variation on the performance of subsequent tasks is also required.

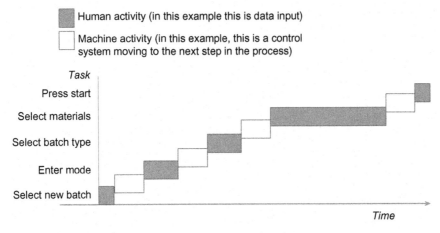

FIGURE 20.3

Example of a timeline.

TOOLS FOR WORKLOAD MEASUREMENT AND PREDICTION

Workload modeling techniques

The timeline technique can be extended to include a subjective assessment of an operator's workload, defining the degree to which a person's capacity is used. These assessments are generally undertaken by operational experts using rating scales, ranging from 0% to 100% and the judgments are then used to plot the workload (capacity utilization) over time (see Fig. 20.4). When the subjective workload is presented directly with the timeline breakdown, the demanding task elements can be determined. As a general rule, sustained workloads of 50–80% are regarded as acceptable (Kirwan and Ainsworth, 1992).

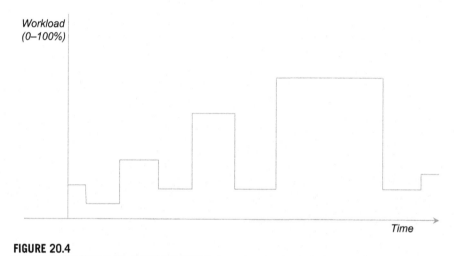

FIGURE 20.4

Illustration of a workload profile.

Another extension of timeline analysis is to identify the task components involved in a task: Visual, Auditory, Cognitive, and Psychomotor (i.e., VACP or VCAP). So each task in the timeline is related to a V, A, C, and P component. The Time-Line Analysis and Prediction (TLAP) tool uses visual perception, auditory perception, spatial cognition, verbal cognition, manual response, and voice response. The Workload Index (W/INDEX) is a similar tool (Megaw, 2005) (Fig. 20.5).

This can be embellished with a demand rating of the task resource components. Specific rating scales are used to determine the VACP demand (such as Keller, 2002, who uses a 7-point rating scale for each component (originally based on Bierbaum et al., 1987). The 7 points on each scale are accompanied by a description to help anchor and define each numerical value; the higher the value (maximum 7) the greater the demand on the resource component. The lower the value (minimum 0) the lower the demand. Using this scale the task of pressing the start button on a mixing vessel might have a rating as shown in Table 20.1.

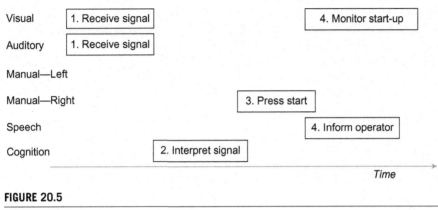

FIGURE 20.5

Illustration of a task component breakdown.

Table 20.1 Example VACP Rating for Pressing the Start Button on a Mixing Vessel

Component	Rating	Definition of the Rating for this Task
V	5.0	Visually locate/align (selective orientation)
A	4.3	Verify auditory feedback (detect occurrence of anticipated sound)
C	1.2	Alternative selection
P	2.2	Discrete actuation (button, toggle, trigger)

If this is the only task being conducted at a specific point in time, then there is no major demand on any resource component. Performing another simultaneous (or overlapping) task, which requires similar resources, presents the potential for excessive workload.

Workload modeling is normally scenario driven and detailed profiles of the scenarios need to be developed on a timeline or flowchart before they can be analyzed. Several different scenarios can be reviewed to analyze the effect on workload. Different team members can also be modeled to show simultaneous activity.

Sequential tasks can be analyzed without the need of complex modeling tools. In this case, a workload profile can be derived for each resource component over time. The total workload over time can be determined by adding the individual scores together for each time slot.

Where tasks are simultaneous and/or highly variable, then a discrete-event simulation tool, such as the MicroSaint software package (now owned by Alfasoft) is often used. These tools use algorithms to aggregate the workload to enable understanding of the overall demands on the person, but they do not predict the demand of the individual information processing resources.

Workload modeling is typically undertaken by human factors experts due to the complexities of the techniques. Expert judgment is required to determine what signals excessive workload. This may be based on the task demand and also the task conflict, and how this might lead to compensatory behavior or poor performance. Workload modeling techniques are normally time-consuming and most of the work to date has been in the military and air traffic control applications.

Task performance

Task performance can be used as a measure of workload. Typically this is measured by reviewing the performance of the task and how well it is being executed using the following types of parameters:

- reaction times for specific aspects of the task;
- the amount of errors made by the individual or the accuracy of task performance;
- the individual's ability to perform additional or secondary tasks (multitasking).

Wilson and Corlett (2005) discuss a method for quantifying mental workload called the Secondary Task Technique, which estimates the spare capacity in mental resource in terms of the attention and effort allocated to performing the primary task. The technique involves measuring task performance when a secondary task is added to the performance of a primary task. Fig. 20.6 illustrates the impact on an individual's primary task, and their spare mental capacity resource when a secondary task is introduced. In addition, there are likely to be PSFs further affecting the situation. The individual's performance is represented by the glass and their spare capacity has been

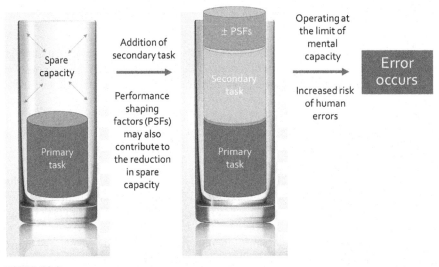

FIGURE 20.6

The effect of secondary task performance on spare capacity.

Reproduced with permission of Rondale ©123RF.com.

utilized by the secondary task, which, with the possible addition of PSFs, results in an increased potential for errors. This is due to the reduced availability of mental spare capacity, with the individual effectively at or beyond their mental processing limits. Clearly this is an issue if the primary and/or the secondary tasks are safety critical in nature.

Subjective workload assessment

A number of subjective workload tools are available to enable individuals to rate or rank their current workload. People are generally good at perceiving their own level of workload, albeit there will be individual differences in that perception. Subjective workload tools are typically administered prior to, during, and after completing a task. They are useful and generally easy methods to administer. There is a word of caution, in that just by the fact of asking someone to rank their workload can impose a distraction and an increase in the workload during task execution.

The following methods are examples used to measure subject workload:

- Instantaneous Self-Assessment (ISA) is a 5-point scale to assess mental workload and was originally designed for use with Air Traffic Controllers. The scale ranges from "1—Underutilized workload, with high spare capacity" through to "5—Excessive workload, with no spare capacity." Feedback from the user can be gathered via a keypad, verbally to an assessor or more recently an application (app) designed for use with smartphones. Data are normally gathered every 2 minutes throughout the entirety of the task. Data analysis enables identification of the points of the task which are more demanding;
- Subjective Workload Assessment Technique (SWAT) was originally designed to assess pilot's mental workload. SWAT is used to assess and rank an individual's perception of time, mental effort, and psychological stress in relation to mental workload. These three factors are rated by participants on a scale from 1 to 3, and an overall workload score between 1 and 100 is calculated. For example, the factor of time includes "often have spare time," "occasionally have spare time," and "almost never have spare time" (adapted from Stanton et al., 2005). Similar scales are used for the factors of mental effort and stress on the 3-point rating scale. The scale factors are generic and have been widely used in a range of industries, including the petrochemical industry. Stanton et al. (2005) states that it is not sufficiently sensitive enough for assessing mental workloads;
- National Aeronautics and Space Administration—Task Load Index (NASA-TLX) is the most widely used mental workload assessment tool and is used in a variety of industries. It measures mental workload during a task. It is a multidimensional rating tool used to derive an overall workload rating based upon a weighted average of six workload subscale ratings. The participant is asked to rate the frustration, effort, performance, temporal demand, physical, and mental aspects of a task by placing a marker on a line that has a scale of 1 to 20; low to high. NASA-TLX has been the subject of multiple validation studies (Stanton et al., 2005);

- Psychophysical rating scales measure perceived effort. The Borg Rating of Perceived Exertion (RPE) enables individuals to rate their level of exertion after a physical activity using a scale from "0—nothing at all" through to "10—very, very heavy."

ASSESSING STAFFING ARRANGEMENTS

A structured technique was developed on behalf of the Health and Safety Executive (HSE) to assess an organization's staffing arrangements to manage high workload process operational situations. The tool is presented in a Contract Research Report (CRR) (HSE, 2001) and is referred to in this chapter as CRR348/2001. Further guidance on the tool is presented by the Energy Institute (EI, 2004).

During the 1980s and 1990s, the HSE recognized that the impact of changes in the process industry on staffing (such as downsizing, staff reductions, and changes to the organizational structures) were not being adequately assessed.

The tool is used to assess scenarios with the potential for major accidents (i.e., highly demanding events, such as the loss of containment). It assesses the ability of teams to cope, particularly where staff changes are made. It asks the following questions:

- Does the organization have enough people, with the right skills, who can work together to successfully deal with high demand situations?
- does the organization have the systems to maintain or improve the situation?

The tool has two parts and is conducted in a workshop forum with relevant representatives from within the organization. The two parts are:

- physical assessment which is used to assess the ability of staff to detect, diagnose, and recover hazardous scenarios. This is undertaken for a number of site specific "high workload" scenarios; and
- ladder assessment which is used to benchmark the site against industry best practice in the human factors of staffing arrangements.

Different operational high workload scenarios are defined at the start of the physical assessment. Safety reports and past events can be used to aid the definition of the scenarios. A scenario definition includes a description of the pertinent aspects, such as the timing and timeline of the scenario, location-specific details, resource details, and any other ongoing parallel activities. Typically communications between team members are also considered.

The physical assessment includes a set of eight decision trees which are talked-through step by step to identify where the site passes or fails the set criteria. The criteria are designed to test the site's physical arrangements. Effectively, they are designed to identify whether staff will be in the required location to respond to the event, whether there are enough staff to deal with the issue, and whether the response is feasible with in the time available. It also assesses whether the communications will be adequately supported. The key questions are presented in Table 20.2, with some examples of common weaknesses.

Table 20.2 Topics within the Physical Assessment

Key Questions	Examples of Weaknesses
1. Is the control room continuously manned?	Not having someone available to detect and respond to an event at all times
2. Does the control operator undertake tasks away from the console?	Prevention of critical information being received caused by being away from the control console, for example, issuing permits, or dealing with critical information presented on different consoles
3. What else do control operators have to do?	Simultaneous demands, such as telephone calls, visitors, alarms, and other distractions
4. Does the control operator need additional information for diagnosis and recovery?	Not having sufficient data to perform the diagnosis and recovery, such as process data, drawings, documents, procedures, or job aids
5. Does the control operator need to call for assistance?	Having insufficient support with diagnosis and recovery or difficulties in communication for diagnosis and recovery
6. Who executes the recovery?	Inability to execute the recovery, due to where critical staff are and what they are doing at the critical points
7. Does the control operator need to communicate with field operators?	Insufficient means to communicate with critical team members, for example, poor reliability of communications equipment, lack of communication backup in the event of failure
8. Other activities of the control operator	Inability to concentrate due to distractions caused by, for example, alarm floods, or other distractions

The site's failure points are then used to identify improvements in its ability to manage that particular scenario. This can be presented as a baseline (as the site is now) in comparison to planned changes. This will then be repeated for several scenarios.

The ladder assessment is a generic assessment of the site's arrangements and is carried out only once. A set of criteria are presented for different benchmarking topics, including:

- situational awareness;
- team-working;
- alertness and fatigue (workload and health impact);
- training and development;
- roles and responsibilities;
- willingness to initiate major hazard recovery;
- management of operating procedures;
- management of change;
- continuous improvement of safety;

- management of safety;
- automated plant/equipment.

Each of the topics is presented as a "ladder" with a set of rungs. The rungs represent a description of the work system attributes from poor to good and a minimum acceptable level is defined for each ladder. The aim is for the site to strive for continuous improvement rather than defining a minimum standard. This enables the site to identify the next rung (and relevant) criteria to aim for by knowing the relevant level of attainment.

TOOLS FOR PHYSICAL WORKLOAD MEASUREMENT

Tasks that are physically demanding, both in short intense time periods or cumulatively over an extended period of time, can have adverse physiological and psychological effects. They can cause physical stress and fatigue, musculoskeletal injury, and affect mental capability. For example, personnel who are tired are more likely to make errors, which can have safety consequences.

Physical workload can be measured objectively (e.g., using physiological measurements as described in this section) and subjectively (e.g., Borg's RPE discussed previously). The results can be used to ascertain the acceptability of what people are being asked to do.

The context in which the task is being conducted is an important factor to consider as it affects performance and the ability to cope with the physical task. These factors include the working environment (indoors and/or outdoors), such as temperature, humidity, and seasonal variations which can also be exacerbated by the use of Personal Protective Equipment (PPE). The ability to cope will also be affected by the time of day, gender, age, and the existing health and fitness of the operator.

Physiological assessments can be made using the following:

METABOLIC RATE AND ENERGY CONSUMPTION

Metabolic rate is measured in kilocalories over time and represents the amount of energy used by an individual to complete a task. Energy consumption is measured in kilojoules which rises as physical work increases. Increased energy expenditure on a particular task is derived by subtracting the resting energy expenditure from the overall energy consumption. Typical values for tasks are as follows (Kroemer and Grandjean, 1997):

- light work, sitting (e.g., an office worker)—9600 kJ per day (male) and 8400 kJ per day (female);
- heavy manual work (e.g., a tractor driver)—12,500 kJ per day (male) and 9800 kJ per day (female);
- heavy bodily work (such as a Shunter)—16,500 kJ per day (male) and 13,500 kJ per day (female).

Kroemer and Grandjean (1997) advise against workers saving their "break" times to leave early at the end of the shift as it "leads to overstress, particularly in older workers." Rest pauses should be structured and distributed evenly across the day's work schedule to avoid exceeding the total energy expenditure of 20,000 kJ for a day, particularly for heavy work. In addition to lunch breaks (usually 30 minutes or more), rest pauses of 10–15 minutes are recommended in the morning and the afternoon for moderately heavy work (Kroemer and Grandjean, 1997).

OXYGEN UPTAKE

The Volume of Oxygen consumption (VO_2) is the amount of oxygen consumed by an individual, and can be used in conjunction with other measurements to assess workload. Tasks that are more physically demanding require a higher VO_2. There are also formulae which enable estimates of VO_2 to be determined and there are differences between men and women. The following figures from Wilson and Corlett (2005) provide an indication of typical VO_2 rates (liters of oxygen per minute):

- light energetic work (such as laboratory work, computer work) $= -0.7$;
- moderate heavy energetic work (such as brisk walking with speed >4 km/h) $= 0.7-1.1$;
- heavy energetic work (such as loading of aircraft) $= 1.2-2.1$;
- very heavy energetic work (such as intensive shoveling of snow) $= 2.2$;
- maximal energetic work (such as pedaling with load of 400 W) $= 5.7$.

When considering physical tasks, personnel should never be required to work at their maximum VO_2, continuously. By measuring the VO_2 tasks can be redesigned to reduce the more demanding aspects by including more rest breaks, job rotation, and work pacing.

THERMAL MEASUREMENT

Physical work causes an increase in body temperature. Thermal measurements, typically skin and/or core body temperature, measured in degrees Fahrenheit (°F) or degrees Centigrade (°C), can be used to assess the acceptability of the workload given the work context. The assessor can compare measurements taken from individuals as they perform physical work with normal physiological ranges. Core body temperature should be maintained around 37°C and skin temperature should never exceed 40°C or fall below 15°C. Body temperature is affected by the amount and type of physical work, the environmental conditions, the clothing worn, and the fitness of an individual.

HEART RATE

Heart rate (HR) and Heart Rate Recovery (HRR) are useful nonspecific measures of physiological strain. (Strain can be caused by several factors, including heat strain, physical strain, psychological strain, and other factors.)

Heart rate is measured in beats per minute and measures how hard the heart has to work. It is compared with the Maximum Heart Rate (HR_{Max}) which is calculated as 220 beats per minute minus a person's age (so for a 40 year old, this would be 180 beats per minute). However, there is a rule of thumb that 110 beats per minute is an acceptable limit for an 8-hour work period for a 20–30 year old (but this should be reduced for an older person). More than 110 beats per minute can lead to exhaustion, reduced coordination, and error.

HRR is correlated with cardiorespiratory strain. A person's pulse is taken for the first 3 minutes after a physical activity using the number of beats in the second 30 seconds for each minute (the values for the 30 seconds are doubled to represent the full minute):

- HR1= Heart rate in the 1st minute after the physical activity;
- HR2= Heart rate in the 2nd minute after the physical activity;
- HR3= Heart rate in the 3rd minute after the physical activity.

The values are then compared to criteria as follows (Wilson and Corlett, 2005):

- "Normal Recovery" is where HR1 minus HR3 is more than or equal to 10 beats per minute or if HR1, 2, or 3 are below 90 beats per minute;
- "Workload is not Excessive" is where HR1 is less than 110 beats per minute and where HR1 minus HR3 is more than or equal to 10 beats per minute;
- "Inadequate Recovery" is where HR1 minus HR3 is less than or equal to 10 beats per minute and HR3 is more than 90 beats per minute.

POSSIBLE SOLUTIONS TO WORKLOAD AND STAFFING ISSUES

The tools that have been described in this chapter provide the opportunity to assess workload issues, and more often than not, the answer becomes obvious from the results that are presented. However, a few general considerations are presented in this section.

The focus of attention for avoiding workload and staffing issues is to design the work system well. It is about good human factors and the most effective time to achieve this is during design, albeit there are many opportunities to improve workload during the operational stages.

ALLOCATION OF FUNCTIONS

The allocation of functions refers to how functions are allocated between the "machine" and the "human" and has been discussed in previous chapters (Chapter 2, What Part Does Human Factors Play within Chemical and Process Industries? and Chapter 9, Overview of Human Factors Engineering). In the event of a workload issue, there may be an opportunity to redress it through a different allocation of function so that the "human" does more or does less. The allocation may be fixed, that is, the human does "x," the machine does "y," or it may be dynamic such that in certain scenarios, the hardware/software may perform a specific function which is reverted to the human in other situations. It is possible for this to either be human controlled (such as auto-pilot) or machine controlled (possibly triggered by the detection of errors). Care is required in the allocation of functions as the workload issue can be shifted to somewhere else within the system. For example, the operator's task might be simplified, but the maintainer's could be made much more demanding.

WORK EQUIPMENT

There are many opportunities to improve human efficiency through effective design of the human interface. In fact, many issues are caused by people having to adapt to inadequate design (poor work tools and equipment) and spurious workload.

Poor alarm design demonstrated this issue prominently in the case of the Three Mile Island accident in 1989 where the control room operators were flooded with over 500 alarms in the first minute and more than 800 in the first 2 minutes of the unfolding scenario (Wickens, 1992). Software navigation that requires convoluted routes to perform a task, when a single mouse click should suffice is another example of where there can be a lack of human-centered design. Likewise, other examples of how unnecessary workload can be designed in to the control system interface include:

- poor information presentation, including lack of clarity, ambiguity, inaccuracies;
- inconsistencies with the operator's mental model of system operation;
- incompatibility between the controls and displays;
- information that requires complex mental manipulation;
- high demands on memory;
- system response delays;
- poor error tolerance and recovery.

There may be opportunities to provide job aids to aid mental processing, or even provide the operator with prewarnings. An example is the use of trend displays which show the operator how the system is starting to deviate from normal parameters. This provides the opportunity to act to avoid a process upset (and avoid what is often a high workload scenario). This principle may be appropriate for other aspects of the control system interface.

WORK AREA

Poor work area layout means that traffic flow and coworking is poorly supported, requiring extra effort and activity to undertake work tasks. For mentally dominant tasks, such as control room operations, this is often exacerbated during high workload scenarios and the stress response of those working in such an environment can become palpable.

For physically dominant tasks, the operator has further to move, additional movements or possibly more strenuous movements to cope with.

WORK ENVIRONMENT

A poor working environment can exacerbate adverse aspects of the work scenario, for example:

- noise distraction or inability to hear auditory signals;
- poor lighting making visual signals difficult to detect or interpret;

- vibration causing visual difficulties and fatigue; or
- thermal discomfort which can cause both physical challenges and effects on mental processing capabilities.

Interventions need to consider moving the work to a less stressful work environment, improving specific aspects of the work environment, reducing exposure to the environmental stressors or protecting people from them. With regard to PPE, there may be an opportunity to select or design more suitable PPE for the task. Two standards provide guidance on the ergonomic issues for PPE (British Standards Institution, 2003, 2007).

JOB/TASK DESIGN

The intricacies of the task may be the root of the issue, such as performing simultaneous tasks requiring the same resources, thereby creating a conflict or excessive demand. In this case, it might be possible to redesign the task so that it uses different resources (e.g., an auditory signal rather than a visual one, a speech response rather than a manual one).

It may be possible to redesign a task so that it does not need to be simultaneous. It may be possible for it to be sequential or for the scheduling to be changed to avoid time pressures or clashes with other tasks. It may be possible to allow more time to do the task.

It may be appropriate to reschedule tasks to build in adequate work–rest cycles. Work–rest cycles and taking breaks are important for any type of job as it provides an opportunity for mental and physical recovery. The general principle is to have short frequent breaks rather than long work periods followed by a long rest break. The time of day affects the body's ability to cope due to circadian rhythms. Avoiding complex and critical tasks in the early hours of the morning is preferable. It is also important to consider shift duration due to fatigue (see Chapter 22: Managing Fatigue).

If the issue is underload, then it may be possible to add tasks or reconfigure them. There is a need to ensure that jobs are interesting and satisfying and not mentally under-stimulating. Repetitive physical tasks can lead to musculoskeletal injury. In both cases, these are candidates for interventions, such as:

- Job enrichment—redesigning task elements of the job to make them more challenging;
- Job rotation—moving between different jobs during a shift so the job is time-shared;
- Job enlargement—increasing the scope of the job by extending the range of duties. Typically, this is done by combining activities.

Monotony can also be relieved through providing social opportunities, that is, reducing social isolation and providing opportunities for people to interact with each other. It can also be relieved by task variation and physical activity, both of which could be designed in to the job.

RESOURCE ALLOCATION

It is important to maintain an even allocation of tasks to different job roles across the team. Discrepancies can lead to a reduction in overall team performance and cause conflict within the team. Workload across the team should also be monitored so that it remains balanced and equal.

Another issue can be interrole dependency, where one role is dependent on the task performance of another team member or team. This can create bottlenecks and reduced time to deal with a task. If this is an issue, opportunities to decouple tasks should be reviewed.

SKILL, KNOWLEDGE, AND EXPERIENCE

Training and competence development can aid the ability of people to cope with task demands. An expert can assimilate and react quicker to situations and maintain a steady state workload. Training can be used to help operators to develop strategies for normal and abnormal scenarios and "overtraining" provides the opportunity for responses to become more "automatic." This is one of the key points behind emergency rehearsals and drills so that the response becomes more rapid.

RECRUITMENT AND SELECTION

Alongside the need for good training, effective recruitment and selection of the right personnel can make a great difference to the ability of the individual and team to cope with the required workload. The number of people available to perform tasks is clearly a critical aspect to get right.

SUMMARY: STAFFING AND WORKLOAD

This chapter provided an overview of a complex area of human factors, to define what it is, how it can be assessed, and different strategies that can be used to attain optimal performance. Workload and staffing are not the only aspects of organizational performance, but they are critical to safety. The related factor of training and competence is discussed in the next sub chapter.

KEY POINTS

- Workload and staffing can have a profound effect on an organization's efficiency and safety performance, and there are key issues that need to be addressed.
- Ensuring the optimal number of the right type of people directly affects the workforce's ability to cope and perform effectively in different operational scenarios.

- Human capabilities and limitations need to be understood to identify how this can reduce performance. The relationship between workload and performance is not a simple linear relationship.
- There are several types of techniques available to measure and assess workload and staffing, many of which are appropriate for the chemical and processing industries.
- Resolutions to workload and staffing issues can be resolved using a multitude of different interventions, but these can be summarized as applying good human factors. Solutions may require a number of these to be applied.

REFERENCES

Bierbaum, C.R., Szabo, S.M., Aldrich, T.B., 1987. A Comprehensive Task Analysis of the UH-60 Mission with Crew Workload Estimates and Preliminary Decision Rules for Developing a UH-60 Workload Prediction Model. Technical Report ASI690-302-87[B], Vols I, II, III, IV. Fort Rucker, AL: Anacapa Sciences, Inc.

British Standards Institution, 2003. Protective Clothing. General Requirements. BS EN 340:2003.

British Standards Institution, 2007. Personal Protective Equipment. Ergonomic Principles. BS EN 13921:2007.

Energy Institute, 2004. Safe Staffing Arrangements—User Guide for CRR348/2001 Methodology: Practical Application of Entec/HSE Process Operations Staffing Assessment Methodology and its Extension to Automated Plant and/or Equipment. Available from: <www.energyinst.org/technical/human-and-organisational-factors/human-factors-staffing-arrangements-toolbox> (accessed 22.01.16.).

Health and Safety Executive, 2001. Assessing the Safety of Staffing Arrangements for Process Operations in the Chemical and Allied Industries. Contract Research Report CRR348/2001. Available from: <www.hse.gov.uk/research/crr_pdf/2001/crr01348.pdf> (accessed 22.01.16.).

Keller, J., 2002. Human performance modelling for discrete event simulation: workload. In: Yücesan, E., Chen, C.-H., Snowdon, J.L., Charnes, J.M. (Eds.), Proceedings of the 2002 Winter Simulation Conference.

Kirwan, B., Ainsworth, L.K., 1992. A Guide to Task Analysis. Taylor and Francis, London.

Kroemer, K.H.E., Grandjean, E., 1997. Fitting the Task to the Human: A Textbook of Occupational Ergonomics, fifth ed. Taylor & Francis, Philadelphia, PA, (Reprinted 2000) (Textbook).

Megaw, T., 2005. The definition and measurement of mental workload. In: Wilson, J., Corlett, N. (Eds.), Evaluation of Human Work, third ed. CRC Press, Boca Raton, FL.

Stanton, N., Hedge, A., Brookhuis, K., Salas, E., Hendrick, H., 2005. Handbook of Human Factors and Ergonomics Methods. CRC Press, Boca Raton, FL, (Edited book, Handbook).

Teigen, K., 1984. Yerkes-Dodson: a law for all seasons. Theory Psychol. 4, 525–547.

Wickens, C., 1992. Engineering Psychology and Human Performance, second ed.. HarperCollins.

Wilson, J., Corlett, N., 2005. Evaluation of Human Work, third ed CRC Press, Boca Raton, FL.

Training and competence

20.2

J. Edmonds

There is often an assumption that people will consistently meet a set performance standard on all occasions. One of the key roles of the human factors discipline is to challenge this assumption and highlight the potential risks associated with not meeting a performance standard, particularly when credit is being taken for 'competence' as a risk control measure. People are typically the last line of defense. Therefore, effective training and competence development are fundamental for achieving safe and effective performance. The aim is to develop and maintain competence to increase the reliability of staff performance.

Several case studies cite inadequate training and competence as contributory factors to the accident event. Some examples are presented in Box 20.3. Aside from major accidents, incompetence can also contribute to single fatalities, personal injuries, and ill health.

BOX 20.3 CASE STUDY EXAMPLES OF INADEQUATE CONSIDERATION OF TRAINING AND COMPETENCE

Piper Alpha (1988)—staff lacked knowledge of what they should do during the event of a fire or explosion because of inadequate emergency training.

Flixborough (1974)—an unsafe plant modification was made by staff with insufficient engineering competency.

Kegworth (1989)—an inadequate conversion course was provided and pilots had no simulation training or any opportunity to practice the learning or develop the necessary skills to deal with the abnormal event. The only training provided was delivered by slide show with a voice-over.

Buncefield explosion (2005)—training was unstructured with little instruction on upsets and emergencies, and the competence management system (CMS) was not well developed.

Esso Longford Gas Plant explosion (2008)—occurred as a result of cold metal embrittlement in a heat exchanger which ruptured on restart due to the cooling effect of the loss of oil circulation. Operators did not understand the danger of cold metal embrittlement despite training and assessment of this issue.

BP Texas City (2005)—training for board operators was inadequate. Simulators were unavailable for operators to practice handling abnormal situations, including infrequent and high hazard operations such as start-ups and unit upsets. There was also an issue of training being moved to online, rather than face to face.

Competence is defined as, "*the ability to undertake responsibilities and to perform activities to a recognised performance standard on a regular basis. It involves a combination of practical and thinking skills, experience and knowledge, and may include a willingness to undertake work activities in accordance with agreed standards, rules and procedures*" (Office of Rail Regulation, ORR, 2007).

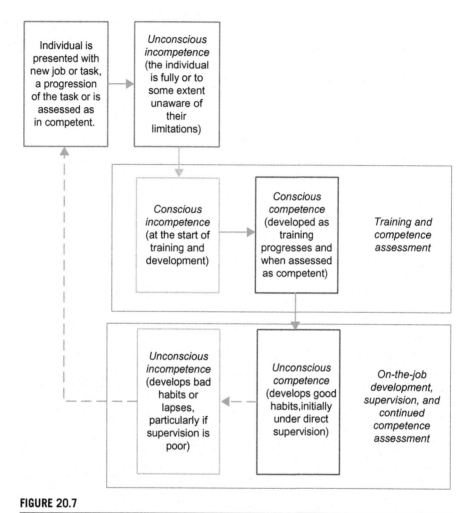

FIGURE 20.7

Stages of competence development.

Competence is a continuum ranging from novice, through competent and proficient, to expert. This is progressive but it can also be regressive, as shown in Fig. 20.7. The dashed line represents the regression. This diagram is based on ORR (2007), but it makes the distinction where training, development, and competence assessment fit within this continuum. It also outlines where effective supervision (see Chapter 20.3: Effective Supervision) becomes important in the maintenance of staff competence. An example of the stages of competence is provided in Box 20.4 in relation to the task of driving.

There is a need to keep staff at the point of being unconsciously competent and avoid the person becoming unconsciously incompetent again. In reality, there may

---

BOX 20.4 EXAMPLE OF THE STAGES OF COMPETENCY DEVELOPMENT FOR THE TASK OF DRIVING

A learner driver starts by being unconsciously incompetent. At this stage, the learner does not appreciate the complexities of driving a car; the psychomotor coordination of using the vehicle controls, the road navigation, negotiation of other road users, and the rules of the road. Once they get out on to the road for the first time, it is fairly obvious that the task is not as easy as they first thought; they become consciously incompetent. As the person learns the task of driving, becomes aware of the rules of the road and gains in experience, he or she will become consciously competent, actively thinking about pressing the lever down for indication, or changing gears. On passing the driving test (including knowledge and skills) and with further practice, the person becomes unconsciously competent, indicating and changing gears with no conscious effort. The stage of unconscious competence is effectively when tasks require less conscious effort to reach the required performance. A link was made in Chapter 20.1, Staffing and Workload, relating to workload where the example of the learner driver was also discussed. As people become more practiced at driving, the functions such as controlling the vehicle become less dependent on conscious thought, freeing up the person's mental capacity to attend to other tasks. This is the point in the cycle where the driver needs to be and avoid the next stage where the person becomes unconsciously incompetent again. This next stage happens if the person develops bad habits, such as using only one hand on the steering wheel, or when their knowledge lapses, and they cannot recall the meaning of road signs. It can also occur if the driver does not stay current with changes to the task of driving, for example, road signs or even new technology in vehicles, such as cruise control or other similar functions which require competence to use safely.

---

be a mix of competence levels across the team, and there could be changes to the competency standard, such as through the introduction of new tasks or systems.

A common oversight is to focus only (or primarily) on the technical competencies of a role to the detriment of the equally important non-technical competencies. Even if non-technical competencies are included, they are sometimes poorly defined and/or not integrated into the training, development, and assessment. Competency involves three aspects:

- The *underpinning knowledge* of the job might include the production processes and chemical reactions, the engineering systems, work processes, and operating procedures;
- *Technical* competence, is task related and enables the job to be functionally completed, such as a welder's skill and ability to make an effective joint. Technical competence is typically job or occupation specific and the objective is to meet minimum standards of performance;
- *Non-technical* competencies, are behaviors that contribute to superior performance and require knowledge and skill during the conduct of the role; examples include the ability to schedule tasks effectively, work well in a team, coach or motivate others, and communicate effectively.

All three aspects, knowledge, technical competence, and non-technical competencies, are necessary for effective and safe performance.

Training is the means by which a basic level of competence is enabled. This may not always be classroom-based training, but may also include on-the-job mentoring and direct supervision, which becomes less direct over time. There is a need to consider both individual and team competence, focusing on the required knowledge, skills, and attitudes for all team roles. Training Needs Analysis (TNA) is a structured approach to assessing training requirements and identifying the optimal means to meet them. Training alone should not be regarded as sufficient for developing competence; on-the-job development, refresher training, coaching, and mentoring are also significant elements for developing and continuing to develop competence.

This chapter provides an overview of the structure and content of an effective CMS. It includes a summary of techniques to define technical and non-technical competencies, and outlines a structured approach to TNA. A section is also presented related to different types of learning.

COMPETENCE MANAGEMENT SYSTEMS

Competence management is defined as "*the arrangements to control, in a logical and integrated manner, a cycle of activities within the organisation that will assure, and develop, competent performance. The aim is to ensure that individuals are clear about the performance that is expected of them, that they have received appropriate training, development and assessment, and that they maintain, or develop, their competence over time*" (ORR, 2007).

The ORR developed a comprehensive CMS for railway safety critical work (ORR, 2007) that is equally applicable to the chemical and process industries. It is presented as a cycle with five phases, as summarized in Fig. 20.8.

Within the five phases, there are 15 principles, which are summarized in Table 20.3.

PHASE 1—ESTABLISH THE REQUIREMENT

Establishing the requirement for competency involves identifying activities that might affect safety. Safety critical tasks are those tasks which, if not performed or if performed incorrectly, could contribute to a major accident or impair a safety critical element. A failure of one of those tasks is where a person initiates the incident, fails to mitigate it, fails to recover it, or causes escalation of it. Safety critical tasks are discussed in Chapter 6, Human Factors in Risk Management. Safety critical task screening is used to identify which tasks performed on a site are safety critical. In the chemical and process industries, this is applied to different categories of task, as follows:

- Operations, which includes:
  - Activities for maintaining a safe operating envelope (i.e., operations in normal, abnormal, and upset conditions);
  - Activities requiring breaching of the normal pressure-containment envelope (e.g., receiving a pig); and

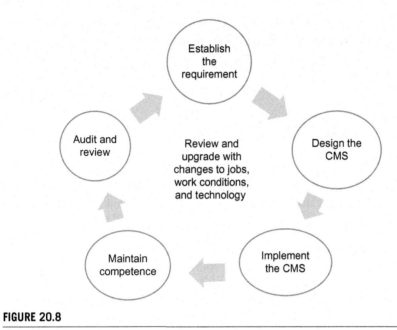

FIGURE 20.8

Five phases of the CMS.

Table 20.3 Five Phases of a CMS

Phase	Description	Principles
1	Establish the requirement	• Identify safety critical activities and assess the risks • Select standards of competence
2	Design the CMS	• Develop procedures/methods • Define how to meet standards • Establish training, development, and assessment • Maintain managers' competencies
3	Implement	• Select/recruit staff • Train, develop, and assess • Control activities undertaken
4	Maintain competence	• Monitor/reassess competence • Update individuals' competence • Manage substandard performance • Keep records
5	Audit and review	• Verify/audit the CMS • Review and feedback

- Activities for managing risk across interfaces and/or safety critical communications (e.g., shift handover; combined operations).
- Maintenance, which includes:
  - inspection;
  - test; and
  - repair.
- Emergencies, which includes:
  - Emergency preparedness and
  - Emergency response.

The process includes further task and human reliability analysis, particularly for priority safety critical tasks, to identify their vulnerabilities and reduce those risks. Risk assessments, such as the Hazard and Operability (HAZOP) studies and reviews of Major Accident Hazard (MAH) scenarios, should also be reviewed as part of this process to identify stated dependence on competence.

It is then necessary to either select the competence standards (where they already exist) or develop new standards of competence to define the performance and knowledge requirements for the safe performance of tasks; these need to include management, technical, and non-technical competencies. A summary of techniques for defining competencies is discussed later in the chapter and practical examples are provided in Chapter 20.3, Effective Supervision.

Although national vocational qualification standards may be part of the competence standard, there may be a need to customize the standards of competence to include all modes of operation and scenarios that are relevant to the site.

A competence standard must reflect the type of competence required, including the correct way to perform the task and the underpinning skills, knowledge, and attitudes/behaviors. This is explained in Fig. 20.9.

Knowledge	**Skills**	**Attitudes and behaviors**
Underpinning principles and theoretical understanding of a process or procedure	Ability to perform the task—can be cognitive (e.g., calculation), psychomotor, and include soft skills	The mentality that one brings to a task and which influences the behavior of the person (e.g., in relation to safety)
For an electrical maintainer this may include knowing the class of equipment and the tests to be performed.	*For an electrical maintainer this might include the ability to calibrate test results, and recognize/diagnose and identify the modifications required.*	*For an electrical maintainer this might be deciding whether it is safe to do a test and ensuring others do too.*

FIGURE 20.9

Competencies required to meet a standard.

Competence standards should be prepared so that assessments can be performed against them. They need to:

- provide a testable description of competence. For example, "the ability to use a specific tool in a particular circumstance or combination of circumstances without supervision";
- define measureable criteria for judging performance. For example, by defining the frequency, accuracy, volume, amount, and so on;
- specify the level of performance evidence, taking account of the individual's status; for instance, whether they will be a supervised or an unsupervised technician (where the standard will need to be higher).

PHASE 2—DESIGN THE CMS

There are four principles that need to be met in designing the CMS. First, there is a need to develop the procedures and work instructions for operating the CMS itself; this is to ensure the CMS is consistently applied and achieves the intended results.

The second step is to establish how each competence standard is to be met and assessed. A comprehensive guide to the development and assessment of competence standards is provided for the hazardous industries (HSE, 2003). Table 20.4 provides examples of the assessment methods from the guidance that are deemed relevant for different types of competency.

Table 20.4 Example Assessment Methods by Competency Type

Type of Competency	Example Methods of Assessment
Physical/sensory-motor competences	Practical demonstration assessments (e.g., moving a load from A to B using crane controls)
Ability to carry out a prescribed procedure of work	Practical demonstration assessment where the candidate completes the task using specific technical performance markers, such as performing an isolation procedure
Cognitive skills	Verbal talk/walk-through assessment of displayed information (e.g., decisions and action taken in the event of a specific abnormal scenario, such as loss of cooling for a reactor vessel)
Knowledge of equipment, plant, and processes	Verbal or written assessment (often a formal test or qualifications) (e.g., accurate recall of safe operating envelopes, pressure limits, temperature limits
Interpersonal and team management skills	Observation of actual behavior in the work setting using behavioral performance markers (e.g., communication of relevant instructions)
Safety behaviors and attitudes	Observation of actual behavior in the work setting using behavioral performance markers (e.g., complying with site safety rules)

The third step is to establish the methods of recruitment, training, and assessment for both staff and contractors that will be used to enable the competence requirement to be met. The recruitment methods need to include reference to the aptitudes and specific baseline abilities for each role to ensure that competencies are accurately and consistently assessed.

The requirement for training, development, and assessment needs to be established, noting that the training required for new recruits may differ from that for existing staff and contractors. Training needs analysis is discussed in more detail later in the chapter.

The fourth step is to establish the competencies for managing and operating the CMS.

PHASE 3—IMPLEMENT THE CMS

This is the phase where the planning and preparation work from Phase 2 is put into practice. Selection and recruitment using the competence standards for the relevant job roles is undertaken as vacancies arise. The selection and recruitment process needs to include the use of role-specific written job descriptions, where the safety-related aspects are specified. The selection and recruitment methods for identifying candidates' existing capabilities should then be implemented.

Each successful candidate then needs a training and development plan, with a training delivery timetable, that enables assessment against the relevant competence standards. This training and development can include training assessment, on the job assessment, and mentoring/coaching.

It is important to ensure staff and contractors only perform the work if they have adequately demonstrated they are competent to do so (supervised or unsupervised as relevant). There may be a need for additional training and development, or if this is unsuccessful, then redeployment or an other performance management strategy may need to be implemented.

PHASE 4—MAINTAIN COMPETENCE

As indicated in Fig. 20.7, there is potential for staff to lose competence by picking up or developing bad habits or by forgetting aspects of their training. There is therefore a need to monitor and reassess staff through planned and unplanned observations. It is important to ensure that assessors are suitably competent to assess staff on all aspects of the relevant competence requirement.

One of the potentially adverse effects of organizational change (see Chapter 19: Managing Organizational Change) is an increase or alteration to the tasks carried out by individuals or teams without first assessing that staff are sufficiently competent to take on the new roles. A change such as downsizing can mean that there is a large loss of competence purely because a number of competent people leave the organization; this can leave a major gap in maintaining the safe performance of the organization's activities. A change such as the introduction of new technology or a change to the work context needs to be accompanied by an update in staff competence in relation to that change if safe performance is to be maintained.

There is also a need to respond to substandard performance. This requires a method for improving competence and a procedure for removing those who consistently fail to meet the competence standard. However, it is also important to understand the reasons for any perceived lack of competence as it could be due to factors outside of the person's control and not reflective of inability. These might include aspects of the job, for example, a change of equipment; attitudes within the organization, for instance behavior driven by the work culture; or the individual themselves, for example, their attitude to work, whether they have an injury, whether age is a factor, or whether there are personal issues affecting their behavior.

Finally, record keeping in relation to competence is essential and this can be used to plan further training and development in an organized and systematic way.

PHASE 5—AUDIT AND REVIEW THE CMS

The final stage in the cycle is to audit and review the CMS to ensure it maintains integrity. This includes checking the quality of competence assessments (including those operating and managing the CMS) and ensuring the assessment process remains effective.

Audits should be performed across the whole system to assess compliance with the procedures established in Phase 2, and to ensure the competence requirements remain current. Any deficiencies or deterioration in the CMS should be rectified and reassessed.

Putting the whole process in place can seem daunting; however, it is better to make a start and see it as a process of continuous improvement, than to not tackle the issue at all.

TECHNIQUES FOR DEFINING COMPETENCIES

There are a number of techniques that can be used for defining technical and non-technical competencies. Further information about this topic is provided in Chapter 20.3 on Effective Supervision, which also presents some practical examples of how these work.

Technical competency is generally job or occupation specific and relates to the physical capability to perform a given task requiring skill-based knowledge and competence. The job analysis techniques used to define the required competencies include: task observation or walk-through; functional (task) analysis; document review (such as procedures); and critical incident review (which is described in Box 20.5).

> **BOX 20.5 CRITICAL INCIDENT REVIEW**
>
> Critical incidents are reviewed, in this case, to identify where specific capabilities or behaviors have either positively or negatively impacted on safe task performance. The details of the incident are reviewed to identify the contribution of the behavior or performance on the safe or unsafe execution of the task. This is undertaken for a number of critical incidents to identify a list of capabilities or behaviors.

Non-technical competency is typically generic and applicable across jobs or occupations; it relates to tasks where behavior-based knowledge and people skills are required. The job analysis techniques used to define the required competencies include: structured job analysis; questionnaire; observation; critical incident review (Box 20.5); and the repertory grid technique (described in Box 20.6).

> **BOX 20.6 REPERTORY GRID TECHNIQUE**
>
> The Repertory Grid Technique (Kelly, 1955) is used to identify how a group of people perceive a particular issue. In this case it would be used to define a list of role attributes. Examples of incumbent or previous role-holders are identified, including those who are both more and less effective at performing the function. By comparing and contrasting those identified in both categories, attributes and behaviors that define effective from ineffective conduct of the role can be identified.

STRUCTURED APPROACH TO TRAINING NEEDS ANALYSIS

Training and development are the means by which a basic level of individual and/ or team competence is enabled. Safety critical work should include appropriate formal training in addition to on-the-job training and development. This may be in different forms, such as classroom-based training or simulation. The current thinking in learning and development uses the concept of the "70:20:10" model: 70% of development comes from performing job tasks and encountering problems that allow experimentation, practice, and honing of skills; 20% comes through social learning and coaching; and 10% of development is delivered through classroom and online learning. This model emphasizes the need for active, experiential learning rather than receipt of passive instruction. It is important to reflect on how training is increasingly moving to online solutions because it is sustainable, cheap, and easy to roll out. A large proportion of safety training is delivered online, and the remainder is generally presented in classroom settings. This suggests the need for the industry to refocus attention on delivering experience-based learning. This topic is readdressed later in this chapter and in Chapter 20.3 on Effective Supervision.

TNA is a structured approach, initially developed in the military setting (see Ministry of Defence, 2015), to assess training requirements and assess appropriate training methods to meet them. It is typically used to identify and support training needs created by the introduction of new or modified systems and equipment; it is particularly useful in the design stages prior to the introduction of the new system and fits within the CMS.

TNA is an iterative process and provides an audit trail for training related decisions. TNA is typically undertaken in five stages:

- Scoping document;
- Operational Task Analysis;
- Training Gap Analysis;
- Training Options Analysis;
- Training Plan.

SCOPING DOCUMENT

Dependent on the scope of the TNA (e.g., a new role, a new system, a new equipment item) there may be a need to prepare a scoping document to clearly identify what the TNA aims to achieve. This may include the following type of content:

- overview of the new system or equipment being introduced, including the proposed capability or technology and associated equipment;
- description of the operational roles that will be affected by the new/changed equipment, including the impact on skill requirements for both individual job roles and teams;
- previous or similar training undertaken by the roles affected, including relevant competency frameworks;
- description of the scope of the TNA;
- constraints affecting the resources for and scheduling of training;
- any assumptions about how the training might need to be delivered. For example, when the new equipment uses emerging technology, the analysis may be more subjective than objective and may need to be stated as an assumption. This particularly applies to software projects when iterative design actions could influence the training need and vice versa;
- areas of risks, such as technical, financial, contractual, and other impacts/ considerations; or
- relevant policies, including health and safety requirements, minimum qualification levels for prospective job holders, and any accrediting or legislation authority issue.

OPERATIONAL TASK ANALYSIS

Tasks undertaken within a specific role are analyzed; initially this is presented as a task inventory for each job role to identify the tasks associated with the new system or equipment being introduced. This is referred to as the Operational Task Analysis (OTA) and it is necessary to include tasks that interface with parent or related systems.

It is possible that task analysis data may be available from a previous safety critical task analysis, from another human factors analysis at an earlier stage in design, or from a similar or predecessor system that can be modified. This is useful as it can help to outline the change in the tasks that need to be undertaken. Collective tasks, as well as individual tasks, should also be recorded within the task inventory (i.e., team, subteam, and/or other group tasks).

The next stage is to establish the operational performance, conditions, and standards required of the new or changed individual and collective activities. This is the competency requirement. These are defined in Table 20.5.

Table 20.5 Requirement Definition

Requirement	Description
Operational performance	Define what the trainee must be able to do. State the specific activities the trainee must demonstrate after completing the training using action verbs, such as "list," "explain," and "define."
Conditions	Explains how the trainee must perform. This may include the ability to use a specific tool, equipment or resource, particular circumstances or combinations of circumstances, and/or how to do the task, such as using a specific job aid or referring to a procedure without supervision.
Standards	Determine how well the trainee must perform to achieve competency. Standards may be expressed in terms of frequency, accuracy, volume, amount, situation, or a combination. Only include time limits if they are required on the job.

If a safety critical task screening assessment has not already been performed, the training requirement can be assessed using techniques such as DIF analysis (Difficulty, Importance, Frequency). DIF analysis enables priorities for training to be identified. The premise of DIF analysis is that if tasks are difficult and important, and only moderately or infrequently performed, training will be necessary. Consideration of recertification and annual checks will also be required. However, if the task is not difficult, not important, and only moderately frequently performed or infrequently performed, then training is not necessarily required.

The TNA as used by the military assumes that every task is analyzed, regardless of its safety criticality. However, if only safety critical tasks are within the scope of the TNA, the DIF analysis may not be necessary.

TRAINING GAP ANALYSIS

The Training Gap Analysis (TGA) is used as a measure of the gap between existing skills, knowledge, and attitudes and those required of the new system or change (see

Fig. 20.9). The skills, knowledge, and attitudes required may vary between different roles performing the same "new" task. For example, one job role may need to attain an expert level of task performance, whereas another role may only need to attain a basic understanding of the principles involved. It may be that certain job roles already have a level of competence for the new task whilst others do not.

There is also a need to consider skill fade, that is, the degree to which the learning decays over time. There are some useful tendencies to note: complex cognitive skills, such as performing a calculation, tend to be more prone to skill fade than psychomotor skills such as learning to ride a bicycle; tasks performed infrequently are more prone to skill fade particularly if they are important and difficult or complex. High "skill fade" activities should be selected for more intensive training, practice, and refresher training.

At this stage of the analysis, there should be clarity regarding the activities which are new or changed for each job role and the competencies required for each new or changed role. The training objectives can therefore be clearly stated.

TRAINING OPTIONS ANALYSIS

Prior to undertaking a Training Options Analysis (TOA), it should be established whether the training requirement can be met using the existing training resources.

If not, the TOA is undertaken by reviewing different training methods and media for each task and considering the advantages and disadvantages of each delivery method. This can be undertaken for each training method and media in combination, or independently. The criteria on which to base the judgment include:

- training effectiveness;
- cost-effectiveness;
- risk;
- on-the-job training/workplace requirements.

Table 20.6 presents examples of some generically described training methods and whether they are suitable for training knowledge, skills, and attitudes. The methods are linked by possible training media that can be considered for their delivery. It does not consider all training methods and media. There is a useful guide to the advantages and disadvantages of different training methods in Railway Safety and Standards Board (2008).

TRAINING PLAN

The final stage is to prepare the training plan, which is used to assist training package designers and support the input to the competence assessment scheme. As well as presenting the detail of the analysis, the plan is used to define the following:

- *Implementation Plan*—defining the resources required by both instructors and trainees for the training, including classroom, facilities, equipment, and materials;

Table 20.6 Example Training Methods and Media

KSA and Media Selection	Training Method Selection							
	Discussion	Lecture	Simulation	Tutorial	Embedded training	Theory lesson	Practical lesson	Self-study
Knowledge (K)		Y	Y	Y		Y		Y
Skills (S)			Y	Y	Y	Y	Y	
Attitudes (A)	Y		Y		Y			Y
Media selection								
Face to face	Y	Y		Y		Y	Y	
Computer-based training			Y	Y	Y	Y	Y	Y
Forum	Y			Y	Y	Y		Y
Interactive electronic technical manual					Y	Y	Y	Y
Immersive learning environment (3D/gaming)			Y	Y	Y	Y		Y
Simulator			Y		Y			Y
Teleconference	Y	Y		Y				Y
Virtual learning environment	Y			Y		Y		Y
Vodcast (video broadcast)	Y	Y		Y		Y		Y
Webinar	Y	Y		Y		Y		Y
Web documents	Y							Y

- *Delivery Schedules*—presenting the schedule for implementing the training strategy, indicating responsible parties. It includes the key tasks to be completed, such as:
  - when to set up training facilities and schedule participants;
  - development of the competence assessment scheme;
  - defining how the training will be evaluated;
  - other activities essential to training;
  - dates on which tasks and activities must be finished; and
  - a milestone plan.

The evaluation and review of the training should be included within Phase 5 of the CMS to ensure that the right competence is being developed and that the training effectively delivers the standard of competence expected.

TYPES OF LEARNING

There are different levels of learning: individual, team, and organizational. There are also particular changes that can result from learning, such as a different ways of thinking, a gain in knowledge, development of routines, or a change in performance (Argote and Todorova, 2007).

Learning is most effective when the learning method is *active* (i.e., one that gets people involved) and when the type of experience is *direct* (*i.e., the person is directly involved in the experience*). However, indirect methods can be effective if participants are engaged with interpreting and understanding cause and effect relationships. The least effective method is passive and indirect, for instance a PowerPoint presentation or a prejob brief with no interaction (Argote and Todorova, 2007). This is summarized in Table 20.7. When developing a training program it is useful to remember that training using active participation and involving trainees in a direct experience,

Table 20.7 The Effect of Types of Experience and Learning Method on Learning

		Type of Learning Method	
		Passive	**Active**
		Less mindful	**More mindful**
Type of experience	*Direct* Own experience	More effective	Most effective
	Indirect Others experience	Least effective	More effective

From Ronny Lardner, Keil Centre and Ian Robertson, BP Shipping Ltd – TOWARDS A DEEPER LEVEL OF LEARNING FROM INCIDENTS: USE OF SCENARIOS in Hazards XXII, 2011, SYMPOSIUM SERIES NO. 156.

such as completing an exercise in simulator, is much more likely to be effective in its purpose (Lardner and Robertson, 2011).

Burke et al. (2011) undertook a meta-analysis of how safety training and workplace hazards have impacted the development of safety knowledge and safety performance. They identified that more engaging and interactive styles of learning were associated with better knowledge acquisition and safety behavior performance. Specifically, they found that highly engaging training and less engaging training is comparable in terms of its effectiveness when the hazardous event/exposure severity is low. However, when the severity of the hazardous event is high, the training is considerably more effective in promoting safety knowledge acquisition and performance when it is highly engaging. The implication of this is that it is worth designing engaging training methods particularly for high severity/hazardous topics.

Training will also be more effective if it allows for different learning preferences, for example, using a mixture of visual and auditory materials and opportunities for practice.

SUMMARY: TRAINING AND COMPETENCE

Having established the links between training, competence, and safety deficiencies, the purpose of this chapter was to outline how it is possible to develop and maintain competence, utilizing effective training needs analysis methods.

Some of the issues raised in this chapter regarding competency are reviewed in detail in the next chapter, in relation to the role of supervisors.

KEY POINTS

- Developing competence is a part of organizational design and is critical to safety.
- Providing the right training and developing the required competencies has a direct influence on the reliability of human performance.
- There is a need to consider both technical and non-technical competencies.
- Competency should be managed as a CMS.
- A structured approach to training can enable effective training solutions to be derived which are justifiable and auditable.
- The effectiveness of learning can be influenced by the type of learning methods used and the type of experience created for the trainee.

REFERENCES

Argote, L., Todorova, G., 2007. Organisational learning. Int. Rev. Ind. Organisational Psychol. 22, 193–234. (Chapter 5, Chichester: Wiley.

Burke, M., Salvador, R., Smith-Crowe, K., Chan-Serafin, S., Smith, A., Sonesh, S., 2011. How hazards and safety training influence learning and performance. J. Appl. Psychol. 96 (1), 46–70. American Psychological Association.

Health and Safety Executive, 2003. Competence Assessment for the Hazardous Industries. Prepared by Greenstreet Berman Ltd. Research Report RR086.

Kelly, G., 1955. The Psychology of Personal Constructs. Norton, New York, NY.

Lardner, R., Robertson, I., 2011. Towards a Deeper Level of Learning from Incidents: Use of Scenarios. In Hazards XXII, Symposium Series No. 156. Institution of Chemical Engineers (IChemE).

Ministry of Defence, 2015. Defence Systems Approach to Training—Direction and Guidance for Individual and Collective Training, Joint Services Publication (JSP 822). Part 2: Guidance. Available from: <https://www.gov.uk/government/uploads/system/uploads/attachment_data/file/491471/20151210-JSP_822_Part_2-DRU_VersionFinal-O.pdf> (accessed 27.01.16.).

Office of Rail Regulation (ORR), 2007., second ed Developing and Maintaining Staff Competence, Vol. 1. Railway Safety Publication., HSE, ISBN 0717617327.

Railway Safety and Standards Board, 2008. Understanding Human Factors—A Guide for the Railway Industry.

Effective supervision

E. Novatsis and J. Mitchell

Providing an effective level and quality of supervision is critical for safe and reliable operations in the process industries. There have been many major accident events, including BP Texas City and Esso Longford, where inadequate supervision was identified as a contributory factor. Supervision can be considered a management function; a control in the organization to manage health and safety risks.

The level of supervision required for different work environments should be based on the experience of the workforce and the risk profile of the work being conducted. Organizations also need to be clear about what supervisory responsibilities are required to operate safely, and then be clear about how these are allocated.

People in supervisory roles need appropriate technical competence, as well as non-technical competencies, to perform effectively. Yet they are often insufficiently prepared for such an important role, and in particular, for the non-technical features. Organizations must have a valid non-technical framework of skills and behaviors that describes what effective supervision means in the organization, and then rigorously select, develop, and measure performance against these criteria. Ensuring these elements are in place will assist organizations to deliver supervision effectively.

THE SUPERVISORY ROLE
THE IMPORTANCE OF SUPERVISION

Supervision involves controlling, influencing and leading a team to ensure that activities are performed correctly (Health and Safety Executive, 2004). Task management and team management are two overarching responsibilities of supervisory roles. Specific activities include planning work, allocating tasks and priorities, ensuring compliance to standards and procedures, and motivating and involving the team. A supervisor can be anyone who has a significant level of supervisory responsibilities, and may go by various job titles such as team leaders, leading hand, lead technicians, and superintendents. Although supervision is often seen as an individual's job, in practice supervision can be delivered by one or more individuals both within and external to a team (Health and Safety Executive, 2004). Organizations can deliver supervision in different ways; the important requirement is to clearly define all supervisory responsibilities and then ensure these are allocated to specific individuals.

The level and quality of supervision provided by an organization influences its safety performance. Health and safety regulators (e.g., NOPSEMA, Australia; and Health and Safety Executive, United Kingdom) identify supervision as a key influencing factor that must be proactively managed in organizations. Incident analysis in the process industries routinely shows problems with supervision or work direction as a causal factor in serious incidents (e.g., OGP, 2013; NOPSEMA, 2013). Moreover, leadership style and team management skills of supervisors have been shown to influence safety outcomes (see Health and Safety Executive, 2012, for a summary). Supervisors have such a pivotal role because they have a direct impact on the behaviors and performance of their teams. Providing effective supervision is essential to setting teams up for safe and reliable performance. This includes ensuring the amount or level of supervision is suitable for the work environment. The Health and Safety Executive (2007) indicate that the required level of supervision should be based on the experience of the workforce and the risk profile of the work being conducted.

TECHNICAL COMPETENCE AND NON-TECHNICAL COMPETENCIES

When defining supervision requirements, an important distinction to make is between technical competence and non-technical competencies. Competence concerns specific job-relevant knowledge and skills that are required to achieve a set performance standard (SHL, 2013). This is different from competencies, which are sets of behaviors that underpin successful performance; they concern how the person achieves the required outcomes in their role (SHL, 2013). Given the similarity of terms, these are differentiated as Technical Competence and Non-technical or Behavioral Competencies in this chapter. Table 20.8 summarizes the key differences between Technical Competence and Non-technical Competencies, which are subsequently elaborated.

Table 20.8 Key Differences between Technical Competence and Non-technical Competencies

	Technical Competence	**Non-technical Competencies**
Focus	Jobs or tasks that people do	People who do the job
Level of performance	Minimum standard	Superior performance
Outputs	Key roles and tasks. Minimum knowledge and skills required	Behaviors that contribute to superior performance
Application	Job or occupation specific	Generic, apply across jobs, or occupations
Appropriate job analysis methods	• Task observation • Studying documentation • Functional job analysis	• Studying documentation • Incident analysis • Focus group and interviews using critical incident technique and repertory grids • Task observation • Structured job analysis questionnaire

TECHNICAL COMPETENCE

Technical competence is job or occupation specific and hence, for supervisors, differs depending on the discipline they are supervising: maintenance, operations, engineering, and so on. Technical competence is focused on the jobs or tasks that people do and describes the minimum standard of performance for the job. It describes the knowledge and skills required to perform the supervisory job effectively and safely. Technical competence is typically determined using job analysis techniques including task observation, studying documentation associated with the role, and functional job analysis; a task-oriented technique that focuses on the activities performed in a job.

Technical competence is not addressed further in this chapter given the diversity of subject areas. However, there are some company specific and regulatory knowledge areas required across supervisory roles in the chemical industries. These areas include:

- Legislative requirements in the country of operation, including due diligence expectations of supervisors;
- The organization's Environment, Health, and Safety (EHS) Policy and EHS management system requirements and processes (e.g., permit to work; emergency response; hazard management; incident investigation);
- The organization's Human Resource management system requirements and processes (e.g., performance management and development, fitness for work).

Technical knowledge in their chosen discipline is an important requirement for supervisors, since a core feature of their role is helping to resolve technical issues. Performance issues faced by supervisors, however, do not usually concern the technical aspects of their role, but rather the non-technical areas. Supervisors in a particular discipline can have similar technical capability, yet their performance varies due to differences in their non-technical competencies.

NON-TECHNICAL COMPETENCIES

Non-technical or behavioral competencies relate to how knowledge and skills are applied in the role; *how* a supervisor performs the role. They describe the critical positive behaviors required to achieve superior performance, and in some cases, the negative behaviors that should be avoided. The techniques appropriate for defining behavioral competencies include studying documents, incident analysis, focus groups and interviews, task observation and structured job analysis questionnaires (Hayes et al., 2007; Flin et al., 2008). These methods are described further in Table 20.9 and are appropriate for identifying the non-technical or behavioral competencies in any role. Different methods or techniques draw out different information. Therefore, a combination of these approaches is recommended to develop a comprehensive and relevant set of behavioral competencies for supervisors in the organization. The process involves generating a list of behaviors using the methods described in Table 20.9. The behaviors are then refined and grouped into logical themes.

Table 20.9 Key Techniques for Identifying Non-technical or Behavioral Competencies

Techniques	Description
Document review	Reviewing existing literature on the supervisory role to identify critical behaviors that influence performance. • External documents—includes research on supervisory behaviors and safety outcomes, and industry guidance on effective supervision in safety. • Internal documents—includes company information such as safety management system documents, supervisory training, and leadership frameworks.
Incident analysis	Reviewing industry and company incidents and near misses to identify critical supervisory behaviors that contributed to those events.
Focus groups/interviews	Conducting structured interviews and focus groups with those in an organization intimately familiar with the target job function and its key responsibilities and accountabilities to identify examples of exemplary supervisory behaviors from their personal observations and experiences. This method utilizes two job analysis techniques: • Repertory Grid Technique (Kelly, 1955)—whereby a number of examples of incumbent or previous supervisors are identified, including those who are both more and less effective at performing the function. By comparing and contrasting those identified in both categories, behaviors that define effective from ineffective supervisors can subsequently be identified; • Critical Incident Interviewing (Flanagan, 1954)—a sample of individuals familiar with the role and the organization identify events with either positive or negative outcomes, and then identify the behaviors that supervisors displayed during those events that were contributory factors to the outcome.
Job analysis questionnaires	Asking in questionnaire format a series of structured questions about the duties and responsibilities in a role, including supervisory requirements.
Task observation	Direct observation of supervisors performing their daily tasks. It can be helpful to "shadow" different supervisors for a shift.

Typically, organizations develop one set of non-technical competencies for all supervisory positions. This is because a high degree of commonality exists, unlike technical competence. An example of a non-technical supervisory competency framework that was developed in the oil and gas sector (Hayes et al., 2007) and adapted in several other companies and industries (e.g., Hunter and Lardner, 2008; Novatsis et al., 2012) is summarized in Table 20.10. Lardner et al. (2011) tested the validity of a version of this competency model in an Australian oil and gas company and established that these supervisory competencies were associated with personal and process safety performance. Facilities where supervisory behaviors were rated as more positive had better personal and process safety performance, as measured by self-report data and actual incident data.

Table 20.10 Example of a Non-technical Supervisory Competency Framework

Supervisor Competency	Example Behaviors
Ensure compliance	visiting the worksite regularly to monitor performance and check compliance; emphasizing safety over production and schedule
Encourage the team	seeking the team's ideas for safety improvements; acting on safety concerns; managing and developing the team
Promote risk awareness	planning and prioritizing work, helping the team identify and manage hazards and risks; helping the team solve problems and make decisions
Involve the team	supporting and encouraging safety activities, helping the team to learn from events; initiating discussions about performance improvement

Other studies linking supervisory behaviors to safety outcomes support the behaviors comprising the supervisory competencies defined in Table 20.10. For example, Zohar (2002) found that increased supervisor safety reward and monitoring behavior was linked to positive safety outcomes; Simard and Marchand (1994) reported that a participative style of involvement in safety activities was linked to better safety performance; and HSE (2001) identified valuing subordinates, visiting the worksite frequently, a participative style of management, and effective safety communication as critical supervisory safety behaviors.

A full example of one of these competencies, "Ensure Compliance" is shown in Table 20.11. It can be helpful to include a description of both what effective supervisors should aspire to (positive behaviors), and also what will undermine their performance and what to avoid doing (negative behaviors). It is recommended that organizations develop their own model to ensure that it contains the right language and emphasis to suit the industry and organization (Box 20.7).

Table 20.11 Example Behavioral Indicators for the Non-technical Competency Ensure Compliance

Ensure Compliance	
Positive behaviors	*Negative behaviors*
Explain and reinforce to the team that compliance with rules and procedures is expected at all times	Set a poor example by breaking rules or procedures
Spend time with the team to check compliance with procedures and rules	Get too involved in the team's tasks and be unable to supervise properly
Help the team ensure that production, cost, or schedule do not override safety	Encourage ways of working that prioritize production or schedule above safety
Ensure that team members have adequate training, skills, and experience for the task	

BOX 20.7 PRACTICAL TIPS FOR DEVELOPING A NON-TECHNICAL COMPETENCY MODEL FOR SUPERVISORS

- Use a range of methods or techniques for collecting data
- Involve a range of people in the focus groups, including managers, supervisors, and people reporting to supervisors
- Include company examples and terminology
- Focus on behaviors that differentiate excellent from poor performers
- Keep the competency structure or framework simple
- Review other competency models in the organization to ensure alignment
- Use simple language and avoid jargon
- Ensure the behavioral indicators are concise
- Include a numbering system for easy reference
- Check and validate the framework with supervisors

LEADERSHIP STYLE

Another factor that characterizes supervisory effectiveness is an ability to demonstrate different leadership styles. Research shows that the most effective leaders are able to vary their leadership style to match the situation (Goldman, 2000). There are many theories and models of leadership and these are not reviewed here. However, one leadership model is described as an example of the repertoire of leadership styles that will enable supervisors to effectively direct, encourage, and develop their teams. Spreier et al. (2006) describe the HayGroup's leadership model, which encompasses six leadership styles.

Directive—involves demand and control behavior, where the leader tells people what to do, when to do it, and what will happen if they fail. Supervisors might use this style in an emergency situation or when they have to manage a poor performer. If used too much over time it can hamper creativity and initiative.

Visionary—involves the leader expressing their challenges and responsibilities with regard to the organization's strategy and goals, which helps gain support and commitment from people. A supervisor might use this style to motivate people during a busy or difficult time such as a shut down or start up.

Affiliative—involves emphasizing the emotional needs of the person over the job. A supervisor may use this style to support a team member through a personal crisis or during high stress circumstances in the workplace.

Participative—involves being collaborative and democratic, where the leader includes people in the decision-making and problem-solving processes. A supervisor might use this style to encourage a disengaged team member by asking for their ideas and experience.

Pace Setting—involves leading by example, setting high standards and ensuring they are met, even if the leader has to do some work themselves. This style can be effective in the short-term, yet will demoralize people if used consistently. Supervisors may find it useful to pace set when the team has to achieve a short-term deadline.

Coaching—involves mentoring and developing team members, giving them opportunities to try new skills and develop their capabilities. Supervisors would draw on this style when needed to challenge their direct reports to develop their capabilities and skills.

Organizations need to educate supervisors about different leadership styles, how to determine the ones most appropriate for different situations, and the impact that leadership style has on culture and team performance. Such education can be provided in either general leadership training or specific EHS leadership training programs.

This section has outlined key elements of the supervisory role. In preparing supervisors for their pivotal role, organizations need to ensure that they are trained and competent in both the technical and non-technical aspects defined for the role.

CHALLENGES FACED BY SUPERVISORS

Setting supervisors up for effective performance also means preventing, detecting, and addressing challenges that hinder them. For example, Conchie et al. (2013) identified role overload, production pressure, and characteristics of contracting workforces as key factors that hinder supervisors. Some examples of these types of issues based on the authors' experiences are described in Table 20.12, along with what organizations can do to help prevent this challenge. Identification of these issues comes from conducting safety culture assessments and delivering safety leadership programs for supervisors in various organizations in the process industries.

Table 20.12 Common Challenges Faced by Supervisors and How to Address this Challenge

Supervisory Challenge	Addressing this Challenge
Administration—High amount of administrative tasks, which prevents supervisors from spending enough time with their teams in the worksite or work area.	• Formally assess changes to the workload and responsibilities of supervisors; • Periodically revisit the workload and responsibilities of supervisory roles to ensure the scope has not expanded; • Ensure external resources required to manage the job demands are sufficient; • Email only essential communication and avoid Carbon Copying (CC) where it is not necessary
Onshore/head office initiatives—Overload of programs and initiatives originating from onshore/head office personnel increasing workload.	• Require that onshore/head office teams consult with the sites to discuss requirements, suitability, and timing of improvement projects. This should be done early and planned into annual activities.
Safety behavior conflicts with back-to-back—Working with a back-to-back supervisor on an alternating shift (e.g., supervisors share a role, where one works 2 weeks, and is replaced by another who works the next 2 weeks), who demonstrates safety behaviors inconsistent to their own, making it difficult to lead the team.	• Ensure a non-technical competency model is in place and used to develop and manage the performance of all supervisors.
Contractor management—Needing to influence and supervise different contracting groups that may have different EHS standards and culture.	• Ensure contract selection includes assessment of EHS expectations and performance to ensure good company fit; • For project work, shut downs or award of long-term contracting agreements, implement supervisory "lead-in" programs, where supervisors of both companies can discuss and agree their expectations and approach to EHS management and any differences that emerge.
No formal supervision title—Performing supervisory responsibilities in practice, but not being recognized in title. This can lead to confusion about accountabilities (for the supervisor and the team) and resentment about the extra duties performed.	• Provide clear roles and responsibilities
Role in the middle—Having to convey management views to their team and their team's views to management can leave supervisors feeling pressured and stuck in the middle, especially on contentious issues.	• Provide supervisors with the skills to manage potential conflicts between management directives and team member responses; • Ensure non-technical competency models are in place for managers and individuals.

SELECTING EFFECTIVE SUPERVISORS
SELECTION METHODS

Selection processes should be aligned to, and assess, the defined technical and non-technical competencies required for supervisory roles. Organizations need to design a rigorous selection process for supervisory positions. When assessment of non-technical competencies is weak, it can compromise the organization's ability to identify, appoint, and retain high performing supervisors. Some of the common mistakes made by organizations in selection are: promoting the person with the best technical competence, using only one method to assess non-technical competencies, and asking questions about what the person would do, rather than what they have done; past behavior is the best predictor of future behavior.

No selection method is perfect. Therefore, using a combination of methods can significantly improve the probability of making good recruitment and promotion decisions. Two important factors in determining what methods to use are the degree to which they are reliable and valid. Reliability of a selection method or tool refers to its consistency. In others words, the degree to which the assessment measures the same way each time it is used under similar conditions by the same participants. Validity is about understanding what the assessment or method measures and how well it does so. In a selection context, some key types of validity are:

Face validity—refers to whether the method/tool appears valid and acceptable to supervisors being administered the tool. For example, a supervisor being asked an interview question about how he/she has promoted risk awareness with a team in the past is likely to appear reasonable to that supervisor.

Content validity—refers to whether the content of a method/tool covers a representative sample of the domain to be measured. For example, a coaching role play for supervisors that does not assess the candidate's ability to provide constructive feedback to a direct report does not fully represent the non-technical competency of coaching and developing others.

Predictive validity—refers to whether there is a relationship between results or scores on a selection measure and scores on relevant performance criteria, in this context, supervisory job performance in the future.

Selection methods differ in terms of their validity. Pilbeam and Corbridge (2006) provide a summary of the predictive validity of selection methods based on the findings of various studies, and this is adapted in Table 20.13. There are large differences between methods, which helps focus attention on the ones that are likely to better predict future performance.

Some of the more effective methods to select supervisors are described as follows. It is recommended to use a combination of methods to increase the likelihood of selecting a supervisor that will perform effectively.

Behavior-based interviews—sometimes also called criterion-based interviewing. This method is a structured interview where candidates are asked to provide

Table 20.13 Predictive Validity of Different Selection Methods

Correlation Coefficient	Selection Method
(Certain prediction) 1.0	
0.9	
0.8	
0.7	Assessment centers for development
0.6	Skillful and structured interview (behavior-based); Ability tests
0.5	Work sampling
0.4	Assessment centers for job performance; Biodata; Personality assessment
0.3	Unstructured interviews
0.2	
0.1	References; interests; years of experience
(No prediction) 0.0	Graphology; Astrology; Age

Adapted from Pilbeam and Corbridge (2006). People Resourcing: Contemporary HRM in Practice, third ed. Financial Times Management, Prentice Hall, London.

specific examples from their recent work experience. People are required to explain the situation, task, actions they took, and the outcome. This method aims to identify if a person has demonstrated the non-technical or behavioral competencies required in the role in the past, which is an indicator of their ability to demonstrate these behaviors in the future. For example, a key non-technical competency for supervisors is to "Ensure Compliance" as discussed earlier. A typical behavioral question would be "Give me an example of how you have ensured that a team you have supervised complied with standards, procedures, and rules." The question is typically followed by a series of prompting questions that help elicit specific behavioral indicators from the competency. For instance, "How did you ensure that your team understood the standards, procedures, and rules?," "What did you do to help your team follow the rules and comply with procedures?," and "What challenges did you face and how did you address them?" Behavior-based questions are typically included within a broader structured interview that also includes questions to elicit the candidate's motivation to work in a supervisory role.

Psychometric tests—there are many types of psychometric test available, including ability tests (e.g., verbal, numerical, and abstract reasoning), personality questionnaires, and motivational questionnaires. Well-designed test batteries can provide valuable information, however, organizations need to ensure they do not weight the results too heavily, but always treat them as one part of the overall selection process.

Role plays—aim to put the candidate in a realistic work situation, and provide opportunity for them to demonstrate specific behavioral competencies.

For example, a supervisor may be asked to explore a performance issue with a direct report, to identify their effectiveness on the behavioral competency of "Coaching and Developing Others." Background information of the direct report's situation is provided to the candidate and they are given time to prepare. A role player in the role of a direct report plays out the scenario with them. The role plays are usually audio and video taped and are scored afterwards to identify the effectiveness of specific behaviors against a scoring criteria linked to the non-technical competencies in question.

Written scenarios or situational judgment tests—provide the candidate with a realistic written work scenario that a supervisor is likely to encounter in their role, and ask them to document how they would respond to the situation. Like role plays, the scenarios are designed to elicit behavioral competencies and are scored using specific criteria.

Group exercises—aim to put a group of candidates in a realistic team-based work situation, and provide the opportunity for them to demonstrate specific behavioral competencies.

Assessment centers—use of a combination of the above selection methods. Assessment centers need to be carefully designed to ensure that the combination of methods are selected to assess the range of non-technical competencies required in the role. Whilst assessment centers have a reasonable predictive validity, they tend to be costly and time consuming (Box 20.8).

BOX 20.8 PRACTICAL TIPS FOR SELECTING EFFECTIVE SUPERVISORS

- Ensure both technical competence and non-technical competencies are assessed
- Use the organization's non-technical competency framework for supervisors in the selection process
- Use appropriate methods to assess the non-technical competencies
- Use a range of selection methods to increase predictive validity
- Ask for the validity and reliability information from providers of psychometric and behavioral assessments
- Use the same rigor for internal and external candidates
- Ensure that appropriately trained and experienced personnel perform the assessments, including structured interviews
- Ensure that cost, time, and practicality are accounted for when designing selection processes

---

DEVELOPING EFFECTIVE SUPERVISORS
DESIGNING DEVELOPMENT PROGRAMS

Before thought is given to *how* effective supervision can be developed, it is firstly essential to have a clear understanding of (a) what competencies, skills, and knowledge are needed to be effective in the role and (b) what the key supervisory competency, skill, and knowledge gaps are in the organization. As discussed, establishing the development needs can be supported through an activity such as a job analysis, which helps clarify the key components of the role and how to discriminate between good and poor performance. This involves conducting workshops with individuals in that role to ascertain the perceptions of role, tasks, context, and indicators of effective and poor performance. For many organizations the key competencies, skills, and knowledge of the supervisor role will already be documented, however the current context and issues may not always be apparent. It is essential to understand this information beforehand so it can be considered and managed during the planning of any interaction. Organization-based research such as safety culture assessments, incident investigations, audits, and 360-degree feedback should be used to highlight the current skill/knowledge gaps that need to be addressed. Collating this information alongside stakeholder interviews should help establish a set of key aims and objectives that are targeted and identify any potential blockers.

Once the aims and objectives are clear, it is necessary to select a method of delivery. Currently, a popular model in learning and development, led by Charles Jennings, is 70:20:10. As introduced in Chapter 20.2, Training and Competence, the idea behind this model is that 70% of development comes through on the job tasks and problems that allow experimentation, practice, and honing of skills, 20% come through social learning and coaching, and 10% through classroom and online learning. The model is meant as a reference, and emphasizes the need for experiential learning rather than passive, dependent instruction. Many successful classroom/online-based learning approaches acknowledge this concept and incorporate various experiential methods into their training so that the approach becomes a *blended* approach. The options highlighted and expanded on in the following paragraphs target *specific skills training*, *on-the-job programs*, and *supervisor development centers*. Although each of these approaches has a slightly different focus by nature, they can all be developed so they incorporate a blended approach with experience at the core.

SPECIFIC SKILLS TRAINING

The "development needs" stage may identify a specific skill or knowledge area as a priority focus. As an example of an intervention to develop individual-based skills, Lovby and Dahl (2004) describe how they introduced the Open Safety Dialogue (OSD) method in Statoil, to improve worksite interactions. These interactions and dialogues are essential activities associated with the non-technical communication and coaching skills discussed earlier in this chapter. The OSD method involves two supervisors engaging in an open and honest conversation with an employee, leading to the sharing of the employee's opinions on the potential risks and consequences of their ongoing work. Two key features of their approach were that there was no-blame, and the conversation ends with a formal agreement of behavioral change, which is subsequently followed up. As part of this program, they conducted one-day courses at site with a small group of supervisors per instructor. The training involved some theory on why the current method will be effective, followed by the structured step-by-step introduction of the OSD method. This was followed by a practical element where participants engaged in real-life dialogues, under instructor supervision, based on what they had learnt during the course. After the course, participants established their own personal target for how many OSD's they will engage in during the year, which was written in the form of a commitment with the Chief Executive Officer (CEO).

The reflections from Lovby and Dahl (2004) shed an interesting, and honest, light upon some of the common issues that crop up during such supervisor development interventions. They note:

- Some supervisors developed their own standards and method of delivery, different from the taught OSD method. Many, especially those with strong technical knowledge, reverted to a "fault finding" approach rather than guiding employees in a positive manner to assess their own behavior. This suggests a need for coaching, verification, and observation on the approaches being applied;
- The time needed to train supervisors sufficiently should not be underestimated. They reflect that a 1-day course was often insufficient to train supervisors in the simple 4-question OSD method;
- Reduced time for supervisors to visit the workplace and engage themselves, due to an increased administration burden and supervisor role changes. This suggests the need for a strong prioritization message from the leadership team along with the lifting of any constraints that make the time required difficult;
- Extraordinary tasks will not get done unless reporting is required and controlled. In order to counteract negative feelings about this extra burden, reporting should be simple and quick. In this case, the supervisors were only required to report the confirmation of an OSD being conducted (with the option of more extensive reporting if they wished);
- The need for anonymity for supervisors and workers. No workforce participant names were recorded and OSD results were reported by organizational unit, rather than by individual. Additionally, no negative performance quotas were applied.

Despite these difficulties, Lovby and Dahl (2004) noted that both supervisors and the workforce appreciated that the OSD method offered them an opportunity to have more contact and understanding of operational risks. They also observed that, when used correctly, the OSD method revealed significant "at risk" behaviors with potential to cause severe injuries. They established a standardized rating factor for the number of OSDs completed and early indications were that there is only stable and positive results (in terms of injury frequency) when a minimum critical mass of OSDs were completed.

Hunter and Lardner (2008) describe how an experiential "team"-based approach was used to develop supervisory "team management" safety behaviors on a large pharmaceutical site in the United Kingdom. A research-based competency model of safety behaviors was developed, which reflected the behaviors that were required at the site to ensure a positive safety performance. This model was introduced to the site in a series of "reinduction" workshops that emphasized the need for behavior change, starting with managers and supervisors. The workshops were held initially with managers and supervisors and their personal responsibilities with respect to safety culture and performance were highlighted. In particular, they were asked to ensure their team engaged in the workshops, develop and support team action plans, and personally complete 360-degree feedback on the positive safety behaviors in the model. Following these initial workshops, the supervisors (and managers) attended the same workshops and led team-based activities to promote safety culture, culminating in individual commitments to safety behaviors, team safety action plans, and a range of follow-up activities. This approach had a positive impact on site safety Key Performance Indicators (KPIs) and illustrates how the supervisor's safety role can be developed through organizations providing supervisors with a structured and planned opportunity to lead their team in safety-related programs. Hunter and Lardner (2008) noted that the sessions worked extremely well because the supervisors were "primed to facilitate."

PRACTICAL "ON-THE-JOB" PROGRAMS

As referenced earlier in this chapter, supervisors who primarily adopt a participatory and coaching approach to safety can successfully develop employee motivation and ability. Studies have shown that supervisors who display behaviors such as offering support, initiating safety discussions, and offering positive feedback have a positive effect on safe worker habits (Niskanen, 1994) and accident levels (Simard and Marchand, 1997). One way to develop these supervisory behaviors is through a focused program with an emphasis on practical "on-the-job" activities. Mitchell et al. (2014) describe how a Latin American power generation company, Duke Energy International (DEI), developed supervisor behaviors through their program "My shift, My responsibility." DEI, through their safety culture survey, had identified supervisory behavior gaps in four areas: Setting and recognizing safety responsibilities; Planning and risk identification; Positive safety conversations; and Encouragement of near-miss reporting. DEI provided training to supervisors on all of these topics, supplementing this training with practical activities that supervisors were instructed to complete during their shift. Supervisors were provided with an activity workbook that encouraged them to reflect and record what activity they had completed, and could be used to discuss progress with their line manager. As an example, supervisors were asked to set a safety goal with each of their team members, track progress, and provide coaching.

Another "on-the-job" program, "Not on My Shift," used a similar approach to engage supervisors in safety activity. Birch (2010) described how Wood Group Engineering (North Sea) used a carefully designed workbook to encourage and prompt supervisors to consider and discuss their teams' and their own safety behaviors during their shift offshore. Supervisors were asked to record and feedback examples of positive Health, Safety, and Environmental performance and any incidents that occurred while on shift, paying particular attention to behaviors. One of the keys to this approach was supervisor involvement and buy-in from the initial development of the workbook to implementation of the approach.

The benefits of both of these programs are that they encourage supervisors to take ownership and responsibility for their own and their team's safety behavior. They are prompted to behave in a certain way and reflect on their actions. In behavioral terms, the workbook acts as an antecedent to positive supervisor behaviors and the consequence is provided through the immediate benefit of that particular activity (e.g., employees motivated) and also through support from managers and EHS personnel (e.g., positive feedback on completion of shift book). The considerations when designing these programs are similar to those outlined by Lovby and Dahl (2004). In other words, provide sufficient time, training, structure, coaching, monitoring, support, enthusiasm, and a quick and anonymous way to report.

SUPERVISOR DEVELOPMENT CENTER

Another way to develop the effectiveness of supervisors is through supervisor development centers. In their best practice guidelines on assessment and development centers, The Psychological Testing Center describes how in development centers "a group of participants takes part in a variety of exercises, observed by a team of trained observers, who evaluate performance against a number of predetermined job-related behaviors." With development centers (as opposed to assessment centers) the emphasis is not on passing or failing, but on helping participants to identify development priorities against a set of predefined job criteria. As with all development initiatives, job analysis and other scoping activities (e.g., stakeholder interviews) should be utilized to establish the key aims and objectives of the development center.

Leach et al. (2011) outlined how they implemented a development center with National Grid in order to develop the effectiveness of supervisors in reducing procedural noncompliance. The development center adopted three activities: (1) job analysis workshops; (2) a review of existing training; and (3) stakeholder discussions. Through these activities they were able to gain considerable insight into existing training methods and topics, issues relating to the supervisors role context and priorities, and the competencies (observation, challenging, and coaching) considered key for improving noncompliance. In reflecting on this stage they identified that job analysis workshops were "essential for ensuring the intervention has maximum effect."

Development centers compose a combination of methods that enable participants to reinforce, practice, and build upon methods and skills they have learnt. Common methods used within development centers include presentations, group discussion, knowledge tests, role play, situational judgment tests, and psychometric assessment. Some of these methods can be applied in advance (e.g., reading; tests; psychometrics) and then discussed either before or during the centers. Ideally, the competencies and skills being developed should not only be developed using one method, but repeated and built upon during the program. As an example, Leach et al. (2011) explain how they used picture-based scenarios to setup role plays, allowing supervisors to practice safety conversation techniques. In these scenarios, one participant would play the role of the supervisor who was required to intervene in a work-based scenario, with two other participants playing the role of noncompliant employees, and a fourth person observing the scenario and offering feedback. They emphasize that the scenarios increased with difficulty during the development center, therefore allowing supervisors to practice more sophisticated behaviors as the center progresses. The more realistic these scenarios become, the more seriously the participants take them and the greater the learning potential. This can be achieved through the use of actors who can respond more accurately and sensitively to the situation and so build a more challenging, emotionally charged environment that requires focus and the use of complex techniques to manage efficiently. Leach et al. (2011) describe how they used live mock-ups to enhance the realism of safety interventions, setting them in realistic job locations. They noted that the realistic settings provided an increased challenge, particularly when it came to observation skills (in the picture-based scenarios they were told what behaviors to challenge).

A key component of a development center, which sets it apart from a regular training course, is observation. Assessors independently observe, record, classify, and evaluate participant performance throughout the various activities. In a development center, the purpose of these observations is purely developmental. Following the development center, the assessors would summaries and evaluate the independent behavioral evidence obtained in a wash-up session. Following this wash-up, participants are offered feedback to support their development. If there is a lack of feedback and support for continuous learning subsequent to the center, this can result in a failure to produce long-term benefits (Doyle, 2003). Leach et al. (2011) describe how following their supervisors development center, they produced a development needs report for each supervisor that covered five sections: demonstration of key competencies; leadership style; attitude on course; overall comments; and mentoring activities (offering advice for line managers on how they can support their supervisors).

Development centers have potential to provide many aspects of good practice in learning and development. They should offer active (not passive) training opportunities, experiential learning, specific behavioral feedback, and ongoing long-term benefits. However, there are significant time, availability, and cost restraints associated with enabling all of the targeted personnel to attend lengthy development centers. Additionally, development centers take time to develop and require highly skilled facilitators to effectively lead, observe, and feedback during the course.

ASSESSING SUPERVISOR SAFETY PERFORMANCE

Assessment of supervisor's safety performance includes both technical competence and non-technical competencies. In order to properly assess the performance of supervisors it is vital to have a clear understanding of the behaviors that are required for effective supervision, as discussed earlier. Established domain-specific behaviors for assessment purposes already exist for job roles such as pilots (Flin et al., 2003) and surgeons (Yule et al., 2006) and provide a good starting point for developing similar approaches for supervisors in other industries. There are a number of methods that highlight how well supervisors are performing. They include:

Observation—This can range from observations made by assessors during development activities to on-the-job observations made by managers (e.g., observing prejob briefs; safety conversations).

360-degree feedback—Formal 360-degree feedback on safety behaviors allows supervisors to understand how their behaviors are perceived by others, including their subordinates. This method requires careful managing, but can provide valuable understanding and does not rely solely on an individual's self-awareness.

Informal feedback—Managers can collect feedback through informal discussions with a supervisor's team on the messages that are being given and behaviors that are being demonstrated.

Personal reflection—Supervisors assess their own safety performance and highlight their own development needs. This requires self-awareness and personal motivation, however, can be enhanced by managers adopting a coaching approach to encourage learning and understanding.

Coaching—Safety coaches can observe, discuss, and advise supervisors on safety performance and development needs.

Testing—This might include testing a supervisor's knowledge in a quiz or survey, or testing their responses to a number of safety scenarios. The main caveat with scenario testing is that even if a supervisor selects the correct response it does not necessarily follow that they would act in the correct way. For this reason, scenario testing is best applied as a development activity rather than an assessment.

Leading indicators (goals)—This involves keeping record of proactive safety behaviors (such as the number of near misses reported, safety conversations held or the number of safety awards given) or the achievement of set goals.

When assessing safety performance, it is important to be realistic about what can be expected. If supervisors are promoted due to their technical competence they may not possess natural strengths in non-technical competencies such as coaching and developing others. There may be a requirement to provide comprehensive training, support, and coaching in order to develop these competencies. Safety programs that rely heavily on these competencies may fail or may at least be inconsistent in their delivery (Lovby and Dahl, 2004) if the required support has not been given.

THE ROLE OF MANAGERS IN SUPPORTING SUPERVISORS

As discussed, supervisors often have conflicting roles and priorities that need to be managed carefully. An increasing administrative burden often means they have less time to actually supervise. In order to provide support, managers need to have a clear understanding and empathy with the conflicting priorities placed upon supervisors. Further to this, managers need to help resolve some of these conflicts and help prioritize activities where appropriate. Introducing new activities to improve safety or quality or production is all very well, but careful consideration needs to be given to what this means in practical terms.

The Health and Safety Executive guidance "HSG65: Managing Safety at Work" outlines some of the key ways that managers and leaders can support supervisors. Using this as a guide, the key messages are summarized below:

- Define the supervisors' roles and responsibilities, ensuring they are trained, competent, and understand the job;
- Ensure there are sufficient resources (including sufficient number of supervisors for the given risk) to get the job done safely;
- Reduce the administrative burden where possible;
- Audit and confirm work is being planned and completed safely;
- Coach and encourage the right behaviors, offering positive reinforcement and constructive feedback;
- Set clear priorities and goals;
- Set a positive example and encourage the same in supervisors;
- Have an open door policy and encourage feedback;
- Involve supervisors in key safety activities such as assessing risks and managing change.

SUMMARY: EFFECTIVE SUPERVISION

Effective supervision is essential for preventing major accident events and maintaining an engaged and productive workforce. The level of supervision required in different work environments and the supervisory responsibilities needed to operate safely must be specified. Organizations must also have a definition of effective supervision that includes technical competence and non-technical competencies. It then needs to rigorously select, develop, and measure performance against these criteria to deliver effective supervision.

Supervisors play a key role in ensuring effective safety critical communications, a topic addressed in the next chapter.

KEY POINTS

- Providing an effective level and quality of supervision is critical for safe and reliable operations;
- Ensure that all supervisory responsibilities are defined for the organization and allocated to individuals;
- The technical and non-technical competencies must be described for supervisory roles;
- A rigorous selection process must be in place for supervisors, which includes thorough assessment of non-technical competencies;

- Engaging and practical development programs should be in place to support supervisors in their skill and competency development;
- Organizations need to help manage the challenges in supervisory roles and provide strong management support to help supervisors be effective.

REFERENCES

Bartram, D., 2013. The Universal Competency Framework. SHL.

Birch, P., 2010. Not on My Shift. HSE Behaviour Standard User Conference, 2010.

Conchie, S.M., Moon, S., Duncan, M., 2013. Supervisor's engagement in safety leadership: factors that help and hinder. Safety Sci. 51, 109–117.

Doyle, C.E., 2003. Work and Organizational Psychology. An Introduction with Attitude. Psychology Press, East Sussex.

Flanagan, J.C., 1954. The critical incident technique. Psychol. Bull. 51, 327–358.

Flin, R., Martin, L., Goeters, K., Hoerman, H., Amalberti, R., Valot, C., et al., 2003. Development of the NOTECHS (non-technical skills) system for assessing pilots' CRM skills. Hum. Factors Aerosp. Saf. 3, 95–117.

Flin, R., O'Connor, P., Crichton, M., 2008. Safety at the Sharp End. Ashgate Publishing Limited, Hampshire.

Goldman, D., 2000. Leadership That Gets Results. Harvard Business Review.

Hayes, A., Lardner, R., Medina, Z., Smith, J., 2007. Personalising Safety Culture: What Does It Mean For Me? Paper presented at Loss Prevention 2007, 22–24, May 2007, Edinburgh, UK.

Health and Safety Executive (1997). Successful Health and Safety Management. Available from: <http://www.fbusurrey.org.uk/Successful_Health_and_Safety_management.pdf> (accessed 20.05.15.).

Health and Safety Executive, 2001. Effective Supervisory Safety Leadership Behaviours in the Offshore Oil and Gas Industry. HSE Books.

Health and Safety Executive, 2004. Different Types of Supervision and the Impact on Safety in The Chemical and Allied Industries. Research Report RR292. HSE Books.

Health and Safety Executive, 2007. Managing competence for safety-related systems <http://www.hse.gov.uk/humanfactors/topics/mancomppt1.pdf>.

Health and Safety Executive, 2012. A Review of the Literature on Effective Leadership Behaviours for Safety. Research Report RR952. HSE Books.

Health and Safety Executive, 2013. Managing for Health and Safety <http://www.hse.gov.uk/pubns/priced/hsg65.pdf>.

Hunter, J., Lardner, R., 2008. Unlocking Safety Culture Excellence: Our Behaviour is the Key. Hazards XX.

Kelly, G., 1955. The Psychology of Personal Constructs. Norton, New York, NY.

Leach, P., Berman, J., Goodall, D., 2011. Achieving Compliance Through People: Training Supervisors to Tackle Procedural Non-Compliance. Symposium Series No. 156. Hazards XXII.

Lardner, R., McCormick, P., Novatsis, E., 2011. Testing the Validity and Reliability of a Safety Culture Model Using Process and Occupational Safety Performance Data. Paper presented at Hazards XXII—Process Safety and Environmental Protection Conference. 11–14 April 2011, Liverpool, UK.

Lovby, T., Dahl, T.A., 2004. Open Safety Dialogue (OSD), Dialogue, Delivery & Commitment. SPE 86625, Society of Petroleum Engineers International Conference.

Mitchell, J., Bernard, M., Villagran, J.C., 2014. Developing Safety Leadership Behaviours in a Latin American Power Generation Company. Hazards 24, Symposium Series No. 159.

Niskanen, T., 1994. Safety climate in the road administration. Safety Sci. 17, 237–255.

NOPSEMA, 2013. Annual Offshore Performance Report Regulatory Information About The Australian. Offshore Petroleum Industry. NOPSEMA Communications. Available from: <http://www.nopsema.gov.au> (accessed 15.05.15.).

Novatsis, E.K., McCormick, P., Lardner, R., 2012. Developing Internal Human Factors Capability in an Australian Oil and Gas Company. Society of Petroleum Engineers, International Conference on Health, Safety and Environment in Oil and Gas Exploration and Production, SPE156684.

Pilbeam, S., Corbridge, M., 2006. People Resourcing: Contemporary HRM in Practice, third ed Financial Times Management, Prentice Hall, London.

SHL, 2013. The SHL Universal Competency Framework <http://www.assessmentanalytics.com/wp-content/uploads/2013/08/White-Paper-SHL-Universal-Competency-Framework.pdf>.

Simard, M., Marchand, A., 1994. The behaviour of first-line supervisors in accident prevention and effectiveness in occupational safety. Safety Sci. 17, 169–185.

Simard, M., Marchand, A., 1997. Workgroup's propensity to comply with safety rules: the influence of micro-macro organisational factors. Ergonomics 40 (2), 172–188.

Spreier, S., Fontaine, M., Malloy, R., 2006. Leadership Run Amok. Harvard Business Review, www.hbr.org.

The British Psychological Society, 2012. Design, Implementation and Evaluation of Assessment and Development Centres: Best Practice Guidelines. www.psychtesting.org. uk. PTC12/09.12.

The International Association of Oil and Gas Producers, 2013. OGP Safety Performance Indicators Data Report No. 2012. Available from: <http://www.ogp.org.uk> (accessed 15.05.15.).

Yule, S., Flin, R., Paterson-Brown, S., Maran, N., Rowley, D., 2006. Development of a Rating System For Surgeons' Non-Technical Skills. Med. Educ. 40, 1098–1104.

Zohar, 2002. Modifying supervisory practices to improve subunit safety: a leadership-based intervention model. J. Appl. Psychol. 87 (1), 156–163.

FURTHER READING

Conchie, S.M., Moon, S., Duncan, M., 2013. Supervisor's engagement in safety leadership: factors that help and hinder. Safety Sci. 51, 109–117.

Health and Safety Executive, 2001. Effective Supervisory Safety Leadership Behaviours in the Offshore Oil and Gas Industry. HSE Books.

Health and Safety Executive. 2004. Different Types of Supervision and the Impact on Safety in The Chemical and Allied Industries. Research Report RR292. HSE Books.

Lovby, T., Dahl, T.A., 2004. Open Safety Dialogue (OSD), Dialogue, Delivery & Commitment. SPE 86625, Society of Petroleum Engineers International conference, 2004.

The British Psychological Society, 2012. Design, Implementation and Evaluation of Assessment and Development Centres: Best Practice Guidelines. www.psychtesting.org. uk. PTC12/09.12.

Safety critical communication

21

J. Mitchell

LIST OF ABBREVIATIONS

ATC   Air Traffic Control
BOS   Behavioral Observation Scale
CoW   Control of Work
PTW   Permit-To-Work
RORO   Roll-On Roll-Off

This chapter describes the range of safety critical communications that occur in high hazard industries, what can go wrong, and how to reduce the likelihood of communication error. Safety critical communication can occur in any part of the operating cycle from normal operation to emergency response. The high likelihood of errors and misunderstandings during these communications is illustrated using a person-to-person communication model and the potential consequences of these communication errors are highlighted using industry examples. The key learnings from incidents involving a breakdown in safety critical communications are extrapolated and discussed. A number of communication methods and techniques that can be used to reduce errors in safety critical communication are discussed and their relative strengths and weaknesses highlighted.

WHAT ARE SAFETY CRITICAL COMMUNICATIONS?

Safety critical communication has a broad range of definitions. For example, it may include any communication:

- during safety critical tasks and activities;
- between individuals;
- within teams (e.g., between control room and field operators on shift);
- between teams such as operations and maintenance;
- between shifts (and between night shift and day staff);
- during a range of operational modes such as normal, abnormal, emergency, maintenance, start-up, and shutdown.

Critical tasks are those identified as having major hazard potential from safety cases or safety analyses and also those identified when preparing or conducting a task such as confined space entry or lifting and slinging work.

A key area that is probably the most familiar example of critical communication is shift handover. This is a long-established critical activity and major incidents such as Piper Alpha have highlighted the impact of getting it wrong. Another familiar area is the communication within Control of Work (CoW) activities, including Permit-To-Work (PTW) and isolations. In essence, a CoW system is a structured communication mechanism, to reliably communicate information about hazards, control measures, and so on. During these activities, good communication between management, supervisors, operators, and maintenance staff and contractors is essential.

More generally, the topic involves the two-way communication of key safety information within a company and between the company and others such as contractors and the public and with the regulator. It is a key element of a good safety culture that involves sharing lessons from incidents, engaging staff in safety, demonstrating leadership and trustworthiness, and the effective sharing of key safety information.

Good communication requires more than just procedural and other arrangements. It is an essential non-technical safety skill. Although everyone learns to communicate as they grow up and develop, the communication skills necessary for critical work are not necessarily reliably developed. In addition, there are specific methods and standards that can only be learnt. Therefore, organizations have to ensure these skills are taught and applied sufficiently.

A MODEL OF COMMUNICATION

A model of communication is presented in Fig. 21.1 (adapted from HSE, 2010), which illustrates the various steps taken by both the sender and receiver during any "person-to-person" communication. Whilst this is an information processing approach that does not necessarily capture all the nuances of real communication, it helps to outline the basic structure of the communication process. The model illustrates that in some cases, there is an opportunity for the receiver to initiate a communication back to the sender (illustrated by the dotted arrow in the model). When this opportunity is missing, the communication is described as *one-way*. In this case, the sender transmits their message to one or more receivers who in turn decode and establish the meaning of the message. Examples of "one-way" communication include e-mails, voicemails, posters, written instructions, and public address system announcements. The advantage of one-way communication is that the sender can easily and quickly reach a lot of people and feels more control over the communication of the message. However, anyone who has ever been upset by a poorly worded e-mail will attest to the fact that one-way communication is easy to get wrong. Receivers interpret the sent message in various and inconsistent ways and not always as intended. The senders can also select the wrong content and style of communication, especially if they are in a hurry. In simple terms, there are no opportunities to clarify that the message has been received, perceived, decoded or understood in the way intended by the sender.

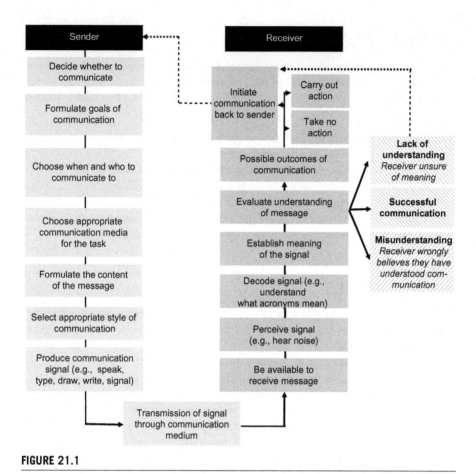

FIGURE 21.1

Simplified person-to-person communication model.

When the opportunity for a response does exist and there is a flow of information back and forth between senders and receivers, this is known as *two-way* communication. Examples include conversations, telephone calls, radio transmissions, and some e-mail communications (e.g., e-mails sent to a global list). In general, two-way communication provides the following benefits:

- it is less prone to errors due to the opportunity to correct any misunderstandings;
- it requires less preparation time as the message can be clarified and altered;
- there is mutual sender and receiver responsibility for the correct communication and understanding of the message, which acts as a second check;
- depending on the method, there is opportunity to convey meaning and understanding through the tone and body language of the sender.

The disadvantage of two-way communication is that it takes longer and creates the opportunity for the conversation to be side-tracked by other topics. This can be costly, especially when safety critical decisions need to be made.

BARRIERS TO COMMUNICATION

One of the things that stands out about Fig. 21.1 is just how complicated it is and how many steps there are. Each step has the potential for miscommunication. However, this potential can start before this communication model is even initiated. Reason (1997) identified three categories of communication problems:

- *System failures* where appropriate channels of communication do not exist, are not functioning as they should or are not regularly used;
- *Message failures* that are represented by the sender side of the model (Fig. 21.1) which culminate in the correct message not being communicated;
- *Reception failures* that are represented by the receiver side of the model, which culminate in the message not being perceived in time or misinterpreted in some way. This would include a failure to clarify the message adequately.

SYSTEM FAILURES

System failures can occur when an appropriate channel of communication does not exist. If these channels do not exist when the information to be communicated is safety-critical, the results can be devastating (see Box 21.1).

Sometimes there are communication systems (e.g., a shift handover procedure and logbook) in place, but they are not functioning in the way they were intended, or are not regularly used by personnel. There are many high-profile incidents in industry where this has occurred. For instance, one of the contributing factors to the BP Texas City refinery explosion was the lack of a comprehensive shift handover. In fact, the CSB (2007) investigation report concluded that at the time of the incident BP Texas City had no policy for effective shift communication, a lack of enforcement of a formal shift handover, and no requirement for logbook records. The shift logbook consisted of a one line entry—"ISOM: bought in some raff to pack raff with." The dayshift supervisor arrived 1 hour late for his shift and therefore did not receive a verbal briefing and had no opportunity for two-way communication. The handover

BOX 21.1 HERALD OF FREE ENTERPRISE DISASTER

The MS Herald of Free Enterprise was a Roll-On Roll-Off (RORO) ferry which capsized moments after leaving the Belgian port of Zeebrugge on the night of March 6, 1987, killing 193 passengers and crew. One of the root causes of this incident was the failure of the assistant boatswain to close the bow door (he had slept through his alarm) before dropping moorings. The bow door remained open as the ferry set sail and the boat filled with water and capsized minutes later. The captain (master) of the ship had no view of the bow door and no indicator light or other means for him to confirm that the doors were closed. The absence of a communication channel with deck crew meant that the captain had to make assumptions about the status of the rear door. This had previously been raised as an issue by a captain of a similar vessel, which had also gone to sea with bow doors open but ignored by the shore-based managers of the operating company, Townsend Thoresen.

between the two key operators occurred at the refinery gates and was inadequate in time and quality. This led to an unclear understanding of the plant status and the procedural steps that had been taken and the ones that remained. This is not an isolated case. A culture where there is an inadequate emphasis on safety critical communication and reliance on informal practices has been implicated in many other incidents, notably the Piper Alpha disaster in 1988. The characteristics of an effective shift-handover system are discussed later in the chapter.

MESSAGE FAILURES—SENDER ERRORS
DECIDING TO COMMUNICATE

The first sender error that can occur is a failure to communicate. This can result in critical information not being passed between team members, leaving them with an incomplete "mental model" (or situational awareness) of the situation they are facing. One of the key contributing factors linked with The Air France Flight 447 crash on the June 1, 2009, with the loss of 228 lives, was the absence of communication between two co-pilots. The plane they were flying was a dual control plane where it was possible for pilots to cancel out each other's actions by ascending (pulling back on the stick) and descending (pushing forward on the stick) at the same time. At the time both copilots were taking a different course of action (one ascending and one descending) in order to control an emergency, an engine "stall", but neither of them communicated to each other what they were doing. It was not until it was too late that the copilot who had been wrongly pulling back on the stick revealed this to the pilot and other copilot (Wise, 2011).

There are many reasons for the absence (or delay) of communication. For instance, there may be assumptions made by both parties that the other party understands what they are doing. Alternatively, there may be embarrassment about an individual's personal understanding of the situation that they might wish to keep hidden. One study that examined communication during a training simulation in the shipping industry found that teams stopped communicating relevant information with each other when they came under stress (Barnett et al., 2006). Whatever the reasons may be, the failure to exchange information and coordinate actions is one of the key factors that differentiates between good or bad performance (Driskell and Salas, 1992) and plays a critical role in safe operations.

THE FORMULATION OF GOALS

The second sender error is the lack of formulation of clear goals. During a simulated maritime activity, Barnett et al. (2006) observed that team leaders had difficulties in communicating their understanding of the situation to the team. If the situation is not understood by the sender it is likely that the goals of the communication will not be clear and the communication will subsequently be confusing. As an example, when preparing to give a presentation it is good practice to spend time formulating the message to convey on each slide before designing the content of the slide. If a presenter does not prepare and fully understand the end goal of the communication, the message can become confused. This is why people generally find it more difficult to present slides that other people have prepared than ones they have produced themselves.

CHOOSE WHEN AND WHO TO COMMUNICATE TO

The timing of communication is particularly important and requires an understanding of the receiver's needs and situation. The sender should put themselves in the shoes of others and ask:

- Is this the most relevant time to provide this information (in other words, is it too soon or too late?)
- Do they need this information to complete this task?
- Is this information critical to the task they are doing?
- Am I going to disturb the receiver's concentration during a critical task?

Lardner and Maitland (2009) describe how one of the root causes of an isolation error that occurred offshore was a supervisor being repeatedly called away (over the public address system) to complete a noncritical task while he was completing a complex isolation on a turbine generator. This noncritical communication disturbed his concentration, causing him frustration, which contributed to him returning to his interrupted work and isolating the wrong gas turbine. Orasanu (1993) found that the captains of high performing airline crews actually talked less during periods of high workload than during normal conditions. This active filtering of the communication by the sender is a beneficial skill that allows the receiver to attend to what is important and not become overloaded, albeit there is also potential to miss vital communication.

Communicating to the wrong person or people can also be a critical error. For instance, if a critical message is given to someone who does not have the accountability, qualifications, or capacity (e.g., if they are working on another task at the time) to give the required response, it could lead to situations going undetected or escalating. Knowing who to communicate to can be difficult when accountabilities are not clear. Box 21.2 describes how an unclear organizational structure and lack of a documented communication policy and accountabilities led to communication issues before and during the BP Deepwater Horizon disaster.

CHOOSE THE APPROPRIATE COMMUNICATION MEDIUM FOR THE TASK

The appropriate communication medium will in some cases be obvious. For instance, if a person is working side by side with someone on a task they are more likely to communicate verbally with them rather than through written means. However, there

---

BOX 21.2 THE BP DEEPWATER HORIZON DISASTER

On April 20, 2010, the BP Deepwater Horizon oil rig situated in the Gulf of Mexico had a major blowout and explosion resulting in 11 fatalities, a massive oil spill and considerable environmental damage. The immediate cause of the incident was the failure of the cement pumped to the bottom of the well to seal hydrocarbons. The chief counsel's report on the disaster revealed that the incident was not just due to technical and mechanical failures but highlighted the role of leadership decisions taken prior to the event. Due to management reorganization prior to the incident, the BP team were unclear about who was accountable for important practices associated with safety. In addition, there was a lack of clear guidance regarding relaying safety critical information between well leaders, experts, and shore leaders. This resulted in inadequate sharing of safety critical information between onshore engineers and offshore staff. As a result, the specific potential risks (cement failure) associated with the cement task were not relayed or discussed with offshore staff. In addition, the report highlights a failure to fully engage with BP's technical experts on the safety of the cement task. Finally, the well site leaders failed to discuss a safety critical negative pressure test with senior onshore engineers, both of whom were offshore at the time. All of these communication failures left BP vulnerable to disaster.

---

are times when face-to-face communication may not work. For instance in a noisy environment where spoken communication cannot be heard or when the sender or receiver are not in a close proximity. The main communication media and their various strengths and weaknesses are outlined in Table 21.1.

The richest form of communication is through face-to-face conversation. This method allows communication through words and other nonverbal cues such as tone, posture, gesture, and facial expression. The sender can also obtain immediate feedback on whether communication is being successful or not. A summary of the research on nonverbal communication identified that approximately 60–65% of the social meaning in communication is derived from nonverbal cues (Burgoon, 1994).

Nonverbal communication allows people to:

- reinforce what has been said in words. For instance, an individual might hold their hands out wide to emphasize how big an object is;
- intentionally or unintentionally convey information about their emotional state. Simple facial expressions such as a furrowed brow tell others a great deal about how the individual is feeling;
- provide feedback to the sender to confirm what has been said. This can confirm understanding or misunderstanding of the message. As an example, an individual might nod their head and say "I think I've got it" to confirm understanding of instructions. However, if they look confused and shrug their shoulders while saying "I think I've got it" it would indicate they were still unsure;
- regulate the flow of communication, for example, by signaling to others that they have finished speaking or wish to say something.

Table 21.1 Evaluation of Different Communication Mediums

Medium	Feedback from Receiver Directly Available through Medium	Location Dependence for Receiver	Speed of Transmission	Auditable at a Later Time	Number of Potential Receivers	Performance in Noise	Visual Cues Available from Sender to Receiver
Face-to-face	Immediate	High	Immediate	No	Variable	Poor	Yes (probably in line of sight)
Telephone	Immediate	High	Immediate	No	Single	Hardware related	No
Pager	Slow	Low	Immediate	No	Variable	Hardware related	No
Radio	Immediate	Low	Immediate	No	Variable	Hardware related	No
PA	Slow	Low	Immediate	Possible (due to multiple audience)	Multiple	Hardware related	No
Memos	Slow	High	Dependent on method, from potentially immediate via computer to slow via internal mail systems	Yes	Variable	Good	No
Log	Slow	High	Slow	Yes	Variable	Good	No
Hand signals	Immediate	High	Immediate	No	Variable	Good	Yes

Sometimes, nonverbal communication can suffice as the sole means of communication. For instance, in a noisy environment hand gestures can be utilized to communicate certain messages (e.g., thumbs up to indicate the job has been done correctly). Caution is required however, as these gestures can be culturally specific. For instance, giving the thumbs up is offensive in countries such as Greece, Russia, and Iran whereas it is a sign of approval in English-speaking countries.

Face-to-face communication has been found to be crucial for the development of trust, which has a significant impact on safety compliance and safety proactivity (Mitchell, 2008). Relationships can be difficult to develop where there are no opportunities for face-to-face interaction. For instance, teams that work remotely have fewer opportunities for face-to-face interaction and therefore the trust needed to work effectively as a team can be difficult to develop (Grabowski and Roberts, 1999).

The drawback of face-to-face communications is that there is often no record of the communication. For situations where a lot of communication is expressed and if some of it is to be used at a later date, it can be difficult for the receiver to remember everything that has been said and memory errors can occur. Shift handover communication is an example where the receiver may have to recall information from the handover later in the shift. It is therefore necessary to have a written record of the handover in addition to the face-to-face conversation providing both redundancy and diversity of the message for assurance purposes. In addition, a written copy of the handover provides a record of proof of what was communicated in the event of incident.

Methods such as e-mails are perfectly adequate for sharing information and communicating news. However, they should generally be avoided when communicating issues, concerns, grievances, and conflicts due to the lack of richness and the lack of instant feedback. They are also not suitable for anything that is time-critical as the sender cannot be sure that the receiver will either receive or read the message.

FORMULATE THE CONTENT OF THE MESSAGE

The general rule when formulating the content of the message is to make the message as unambiguous as possible. Any information that can be interpreted in multiple ways may cause confusion for the receiver. For instance, if an instruction is given to an employee to "cut all material in Area B" this is so vague that it could easily be misinterpreted. Similarly, if an employee is asked to "turn the big valve on Equipment B" this could cause confusion if there were more than one valve of a similar size or type. A high profile example of ambiguous wording causing a major incident is the Sellafield beach incident that occurred in 1983 (see Box 21.3).

SELECT APPROPRIATE STYLE OF COMMUNICATION

The style of communication refers to the tone and emotion of the message. In safety critical situations it is often necessary for the sender to be assertive (in other words, achieve the right balance between being passive and being aggressive). If the message is too passive the receiver may disregard the message or interpret it as not

---

BOX 21.3 SELLAFIELD BEACH INCIDENT

Sellafield is a nuclear reprocessing and decommissioning site located in Cumbria on the Irish Sea. The incident involved a tank full of highly radioactive material being accidentally discharged into the sea. One of the key causes of the incident was a written description of the tank contents that was changed from one shift to the next. The original contents description was "ejections from HASW." In the hand over notes, this was changed to "HASW washout." What had originally been interpreted as highly radioactive material was interpreted as being a low level effluent, suitable for discharge to the sea. In this case, the contents of the tank had originally been described ambiguously in terms of its origin rather than its nature. A clearer definition would have been to specify that the content was "highly radioactive liquid waste."

BOX 21.4 AVIANCA FLIGHT 52

On January 25, 1990, Avianca flight 52 crashed in New York State with 73 casualties. Due to congestion and bad weather Avianca flight 52 (en route from Columbia) had been placed in several holding patterns on their journey to New York and were starting to run low on fuel by the time they reached the New York air space. They were kept in a further holding pattern over New York. It was not until they were contacted by ATC to inform them that they would be in the holding pattern indefinitely, that the first officer radioed the controller with the message "ah well I think we need priority we're passing [unintelligible]." By this point, fuel levels were dangerously low. It was not until the ATC enquired 2 minutes later about how long they could hold for that they learned flight 52 could only hold for 5 minutes more. The plane subsequently made an initial attempt to land, which failed due to the impact of bad weather on their descent. They were then directed by the ATC away from the airport to make a second attempt. During this time the first officer failed to declare a state of emergency and identify to ATC the exact amount of fuel that was left in the plane. This meant that ATC did not fully understand the state of emergency and prioritize them accordingly. The flight subsequently ran out of fuel and crashed into a hill.

---

important. If the message is too aggressive the receiver may become offended and disengage from the conversation. An assertive style is achieved when there is a calm, objective approach that does not disregard others' opinions. The emotional style of the message can help to convey the urgency of the situation. For instance, if a message about an urgent situation is conveyed in a laid back, monotone style it could easily be misconstrued that the message is not urgent. The failure of the co-pilot on Avianca flight 52 to properly communicate the urgency of the situation to Air Traffic Control (ATC) led to the flight not receiving the necessary prioritization for landing. This is described in Box 21.4.

PRODUCE COMMUNICATION SIGNAL

The final step on the sender side is to produce the communication signal. All being well, the signal will be produced as planned. However, the clarity and quality of the signal depends on personal attributes such as the projection skills, confidence, and status of the sender and situational attributes such as the stress of the situation and

BOX 21.5 TENERIFE AIR CRASH

The Tenerife air crash involved two jumbo jets colliding on a runway at Los Rodeos airport in 1977. Five hundred and eighty-three people lost their lives, making it the world's worst civil aviation accident in history.

As a result of a diversion, due to a security alert at another location, the Los Rodeos airport was unusually saturated. In addition, the layout of the airport meant that planes were required to taxi down the runway before taking off in the other direction. Other contributing factors to what happened next were the weather (visibility was poor) and the fact that due to the delay and strict quotas on flying hours, the KLM plane was in a hurry to take off before their quota expired.

KLM 4805 had taxied all the way down the runway and had turned around ready for take-off. PanAm 1736 was taxiing down the runway and had been given instructions to leave on the third taxi-way to the left. There had been some confusion about which turn off the PanAm plane should take, partly due to the third turn off requiring a turn that would be too steep for the plane to make. The following radio communications between the two planes and tower control describe what occurred next.

17:05:45 (First officer to the control tower)—Uh, the KLM ... four eight zero five is now ready for take-off ... uh and we're waiting for our ATC clearance.

17:05:53 (Tenerife control tower)—KLM eight seven zero five uh you are cleared to the Papa Beacon climb to and maintain flight level nine zero right turn after take-off proceed with heading zero four zero until intercepting the three two five radial from Las Palmas VOR.

17:06:09 (KLM first officer)—Ah roger, sir, we're cleared to the Papa Beacon flight level nine zero, right turn out zero four zero until intercepting the three two five and we're now (eh taking off)

The KLM 4805 obtained clearance, but was not allowed to take off as yet. However, its captain, in a hurry, started advancing the throttle and said to his colleagues "we gaan" (we are going), presumably having forgotten that another aircraft was still taxiing down the runway. The control tower tried to clarify the instruction to stand by.

17:06:19 (Tenerife control tower)—OK

For the next communication the Pan Am first officer and the control tower were talking simultaneously, which meant the communication was heard in the KLM plane as a high pitched screech and message which was audible but difficult to detect the message clearly.

17:06:19 (PanAm first officer)—No ... eh...

17:06:20 (Tenerife control tower)—Stand by for take-off, I will call you.

17:06:20 (PanAm first officer)—And we're still taxiing down the runway, the clipper one seven three six

Following this communication, the control tower failed to confirm that the KLM plane had heard and understood its instruction to standby. Instead, they requested that the PanAm plane communicate when clear of the runway. Crucially, the control tower referred to the PanAm plane for the first time as papa alpha. Up to then this plane had been referred to as Clipper which may have led to the KLM crew filtering out this communication.

17:06:25 (Tenerife control tower to the PanAm 747)—Roger papa alpha one seven three six report when runway clear

17:06:29 (PanAm first officer)—OK, we'll report when we're clear

17:06:30 (Tenerife control tower)—Thank you

The KLM captain did not react. He may have filtered out the information or misunderstood and thought that the PanAm 747 had just cleared the runway.

17:06:32 (KLM first officer)—Is hij er niet af dan? {Is he not clear then?}

17:06:34 (KLM captain)—Wat zeg je? {What do you say?}

17:06:35 (KLM first officer)—EstIs hij er niet af, die Pan American? {Is he not clear that Pan American?}

17:06:36 (Angry KLM captain)—Jawel. {Oh yes.—Emphatic}

time restraints. As an example, described in Box 21.5, there are sections of the dialogue (from the Tenerife airport disaster) that took place between the KLM plane, PanAm plane, and tower controller. There were a number of examples in the lead up to this incident where the signal produced (particularly by the tower controller) was hesitant, incorrect, and inconsistent in their terminology for naming the planes. The human factors report produced by the Airline Pilots Association concluded that the inconsistent communication likely led to the crew aboard the two planes filtering out communications which were intended for them.

RECEPTION FAILURES—RECEIVER ERRORS
AVAILABILITY TO RECEIVE MESSAGE AND PERCEIVING THE SIGNAL

Communications breakdown if the receiver does not actually receive the message or fully perceive the signal (e.g., failing to hear a noise or read illegible writing). This can be due to environmental issues such as a noisy environment or a problem with technology, such as a phone being out of signal or a radio control being broken. The Tenerife aviation disaster outlined in Box 21.5 shows how a crucial instruction from the control tower to KLM 4805 advising them to "standby for take-off. I will call you" was missed due to a simultaneous communication from another plane (PanAm 1736). The simultaneous communication caused the communication to be transmitted as a high pitched squeal which made it harder to identify the message.

DECODE SIGNAL, ESTABLISH MEANING OF THE SIGNAL, AND EVALUATE UNDERSTANDING OF THE MESSAGE

If the message has been received and perceived correctly, the next few steps involve the receiver decoding the message (e.g., understanding acronyms) and establishing the meaning of the signal (e.g., they require me to do xyz). The most crucial step in the role of receiver is determining whether they have understood the message. Some people may naturally err on the side of caution and clarify what they have just heard in their own words. When it comes to safety critical communications this is an absolute necessity. Given the range of ways errors can be made in communication, as outlined in this chapter, it is always best to err on the side of caution. The transcript from the Tenerife disaster in Box 21.5 highlights that there were a number of opportunities for clarification of the situation. The KLM pilot, after hearing a high-pitched screeched message failed to clarify what the message was and whether they were actually cleared for take-off. The control tower failed to confirm with the KLM plane that they had understood the critical instruction to standby. The recommendations in the following section outline how both sender and receiver errors can be reduced and managed.

POTENTIAL SOLUTIONS FOR MANAGING SAFETY CRITICAL COMMUNICATIONS

The case studies outlined in this chapter highlight the types of communication failures that can occur and the criticality of communication skills for safe operations. These failures can occur during normal operations but also during specific communication activity such as during shift handover. This section covers recommendations to improve general communication skills but also how specific communication activities can be improved by developing processes and methods.

DEVELOPING SAFETY COMMUNICATION SKILLS

It has been identified in this chapter that certain communication skills are good practice. This includes, but is not limited to, providing unambiguous messages, choosing the correct method of communication, selecting the right tone and nonverbal cues, being assertive where necessary, clarifying understanding, checking the message has been understood, and so on. These skills can be taught and given their importance, they should be. In certain industries, training includes teaching set procedural communications. For instance, in the aviation industry, pilots conduct a set three-way communication whenever they perform a task (e.g., ATC: "Cleared flight level 120"; Aircraft: "Roger cleared flight level 120"; ATC "That is correct"). In addition, some communication is technical and involves standard phraseology. The importance of consistently using standard phraseology should be emphasized during training. The Tenerife disaster outlined in Box 21.5 demonstrates how communication failures can occur when standard phraseology is inconsistently applied (e.g., "we are at take-off" was used instead of "Roger, cleared take-off Runway 30").

Techniques such as three-way communication need not just be the preserve of high reliability organizations such as ATC and nuclear power stations. It is an excellent method for improving communication during a shift. Wachter and Yorio (2013) outline several human performance tools that have been applied successfully in the nuclear industry and can be applied within other organizations. Of the tools outlined in their review, the *three-way communication* and *concurrent verification* tools are particularly useful for reducing communication failures during normal operations.

- *Three-way communication*—an applied tool that can help ensure all team members have a shared and accurate "mental model" of the task or any changes that have occurred. The sender of the message states the message and the receiver acknowledges the message and repeats it back in a paraphrased form. The sender then acknowledges the senders reply.
- *Concurrent verification*—involves two workers working together to individually confirm the safety of an action or component before, during, and after an action. This tool is particularly important when the wrong action (e.g., pressing the wrong button) would cause an immediate harmful event. At each stage, the two

workers independently verify and agree on understanding and next steps and the verifier observes the performer during execution of the action.

Another important consideration for enhanced safety communication is the management of *social threats*. David Rock (2008) identified five social threats that, if not managed carefully, create a reaction in the brain that is similar to a physical threat. In other words, individuals will either retreat from the conversation (flight) or become defensive (fight). Either way, the physical reaction is such that the frontal part of the brain needed for engaging in complex conversations does not get the resources it requires. Rock identifies the following social threats that have this effect:

- *Status*—People feel threatened if they think their competence is being questioned. So if someone says "That doesn't seem very good" or "I don't think you are doing it right" conversations will break down quickly;
- *Fairness*—People are attuned to things that are unfair. If people suspect there is a lack of fairness (e.g., you are accusing me of not following the procedure when the procedure is incorrect!) they are likely to react defensively and are likely to feel annoyed;
- *Certainty*—People start to feel uneasy if they are unsure of what will happen next. This will distract them from any conversation as they will be thinking about what happens next;
- *Relatedness*—This is similar to building trust, which is best done face-to-face. If people trust you they are more likely to be honest. If they do not they are likely to be cautious; and
- *Autonomy*—People feel threatened if they feel they have a lack of autonomy or control. They might go along with things but they will not fully engage as they have lost control. People need to feel they are coming to conclusions, highlighting issues, and not being told.

The skills needed to manage these social threats can be taught and fit within good practice. For instance, if a worker is stopped by their manager and is told "you are doing it wrong," we know intuitively that this will not make them feel good. There is a danger that safety conversations can become too prescriptive and disingenuous if workers are given set scripts and questions to ask during conversations. A better approach is for workers to approach safety conversations in a nonjudgmental, caring, polite, and open way with a genuine interest in sharing information.

Of course, good communication skills are not limited to sending information. Listening is an active process and it is vital for workers to accept challenges and listen to each other. It is particularly important to build a culture where senior employees also listen openly to more junior employees. Part of this is being aware of internal personal reactions and emotions to communications and managing them. For instance, if a colleague tells me that "I am doing something unsafely" my initial defensive reaction might be to say "no I'm not." This is because my *status* has been challenged. However, if receivers can challenge their own initial reaction and replace it with a more positive response, a positive outcome can be achieved. Below are some *do's* and *don'ts* that support active listening:

DO: ask questions; paraphrase; offer encouragement; show interest using nonverbal (e.g., open body language, nod, and make eye contact) and verbal signs.

DON'T: interrupt; try and multitask; work ahead in the conversation; change the subject mid-conversation; get defensive; argue and focus on trivial facts; show disinterest using nonverbal signs (e.g., looking away, folding arms, yawning).

As well as providing training in communication skills, they can also be observed and assessed by the organization. Some companies will have devised a set of behavioral expectations that can be used to assess observed communication skills and identify strengths and weaknesses (see Chapter 18: Safety Culture and Behavior). Alternatively, a checklist (e.g., Beaubien et al., 2004) Behavioral Observations Scale (BOS) for communication can be used.

PREJOB AND POSTJOB BRIEFS

Wachter and Yorio (2013) reviewed the conditions when nuclear organizations found that *pre- and post-job briefs were effective*. This included when they encouraged workers to take ownership and to consider performance shaping factors, critical steps, stop-work criteria, and additional tools that could be applied during the shift to manage errors.

Pre-job briefs provide an opportunity to think about the task ahead and encourage workers to utilize their own experience in order to anticipate errors, worst-case scenarios, and evaluate controls and contingencies to catch and recover from errors. This conversation requires supervisors to facilitate the discussion but not give the answers. Workers disengage if they are simply told what to look out for. Instead, the supervisor should encourage and prompt team members to think through the task ahead. This can be aided by the use of a structured communication board or prompts.

The temptation could be to skip *Post-task briefs* once the task is complete. However, these briefs are essential learning opportunities to identify improvements and learning from complications that have occurred. These briefings can help uncover latent organizational weaknesses such as inadequate tools and resources, poor procedures, and skills shortages. Tannebaum et al. (1998) highlighted that during these posttask briefings team leaders should provide a self-critique early in the review. This is so that others also feel comfortable admitting their mistakes, feedback improvements in performance, and encourage active team participation. In addition, the whole team should be open to accepting feedback from others, keep feedback task focused, provide specific constructive suggestions, and discuss technical and nontechnical aspects of performance.

SHIFT HANDOVER

Many work processes are continuous and may involve a delay between making alterations to the process and the effects of these changes. Where work (e.g., maintenance) continues over more than one shift it can be hazardous and there have many high-profile incidents where this has occurred (HSE, 1996). Incoming shift workers are required to understand and control the process, as well as identify and analyze the effects of any actions taken in a previous shift. To do this, the incoming shift workers require an accurate mental model that relies on excellent shift communication. In a review of the literature on effective shift handovers published by the HSE (1996) a number of recommendations were made on how handovers should be conducted:

- shift handovers should be face-to-face and be repeated via more than one medium (e.g., verbal and written);
- shift handover should involve a high quality two-way communication dialogue with regular feedback and clarification;
- ambiguity should be reduced by carefully specifying the information to be communicated;
- unnecessary information should be eliminated to avoid overload;
- "cutting and pasting" in carrying-forward written information should be avoided as this can result in information inaccuracy.

In addition to recommendations made with respect to communications during handover organizations should also seek to improve the shift handover process. Recommendations include:

- prioritize shift handover communications and emphasize the potential for miscommunication as well as the consequences;
- where possible, design work so that maintenance work and other complex processes are completed during one shift to avoid the risk of miscommunication;
- allow additional time and priority for hazardous conditions, such as maintenance work that continues over more than one shift or deviations from normal working and/or where personnel require extra time such as personnel returning from an absence or staff in-experience;
- develop workforce communication skills and include these skills as a job requirement for new personnel;
- design logbooks so that they encourage the capture of key information (based on feedback from operators' needs) reliably and unambiguously;
- provide clear procedures which specify how to conduct an effective shift handover;
- involve operators when making improvements to methods of communication during handover. Their involvement should facilitate buy-in and use of the methods. One way to do this is through conducting an audit. The HSE provides a shift handover audit (HSE, 2006).

SUMMARY: SAFETY CRITICAL COMMUNICATIONS

Safety critical communications are a vital part of any business. It is important to understand the stages of the communication process and the errors that can occur. There are key lessons for both the transmitter of information and the receiver and it is possible to develop the necessary skills for maintaining safe communications.

The next chapter discusses fatigue which can pose a particular threat to safety critical communications. When people are fatigued communication suffers; people communicate less, the information is less detailed and errors can occur at any of the stages of the communication process. This is primarily due to the impact of fatigue on cognitive processing.

KEY POINTS

- Communication involves a message being sent from the sender to the receiver;
- Communication involves many stages and failures can occur at any stage;
- Investigations into major accidents have revealed that simple failures in communication can have a major impact on events;
- The key task of the sender is to formulate clear and unambiguous messages and communicate the message using the appropriate methods and using the correct tone and nonverbal cues;
- A key role of the receiver is to be cautious and check they have understood the message;
- It is the responsibility of both the sender and receiver to make sure both parties have the same understanding of the message and the same *mental model* of the task or procedure;
- Communication skills can be developed and tools can be introduced to improve the communication of safety critical information;
- People leading briefings should model good communication and involve others in open discussions;
- Shift handovers should be given sufficient time and priority and organizations should engage the workforce in ensuring that the process is working and the right information is being passed between shifts.

REFERENCES

Air Line Pilots Association, 1978. Human Factors Report on the Tenerife Incident. Aircraft Accident Report. Available from: <http://www.project-tenerife.com/engels/PDF/alpa.pdf> (accessed 01.12.15.).

Barnett, M., Garfield, D., Pekcan, C., 2006. Non-technical skills: the vital ingredient in world marine technology? Proceedings of the International Conference on World Marine Technology. Institute of Marine Engineering, Science and Technology, London.

Beaubien, J.M., Goodwin, G.F., Costar, D.M., Baker, D., 2004. Behavioural observation scales. In: Stanton, N.A., Hedge, A., Brookhuis, K., Salas, E., Hendrick, H. (Eds.), Handbook of Human Factors and Ergonomic Methods. CRC Press, Boca Raton, FL.

Burgoon, J.K., 1994. Nonverbal signals. In: Knapp, M.L., Miller, G.R. (Eds.), Handbook of Interpersonal Communication, second ed. Sage, Thousand Oaks, CA, pp. 229–285.

Chief Counsel's Report (ePub eBook), 2011. Macondo: The Gulf Oil Disaster. Chapter 5. National Commission on the BP Deep Water Horizon Oil. Available from: <http://www.eoearth.org/files/164401_164500/164423/full.pdf> (accessed 28.11.15.).

CSB, 2007. Refinery Explosion and Fire (15 Killed, 180 Injured), BP Texas City, Texas, March 23, 2005. Report No. 2005-04-I-TX. Available from: <http://www.csb.gov/assets/1/19/csbfinalreportbp.pdf> (accessed 19.11.15.).

Driskell, J.E., Salas, E., 1992. Collective behaviour and team performance. Human Factors 34, 277–288.

Grabowski, M., Roberts, K.H., 1999. Risk mitigation in virtual organizations. Organ. Sci. 10 (6), 704–721.

HSE, 1996. Effective Shift Handover—A Literature Review. Report Reference No. OTO 96 003. Available from: <http://www.hse.gov.uk/research/otopdf/1996/oto96003.pdf> (accessed 04.12.15.).

HSE, 2006. Improving Communication at Shift Handover. Available from: <http://www.hse.gov.uk/humanfactors/topics/shifthandover.pdf> (accessed 04.12.15.).

HSE, 2010. NSRMU Report Ref. No. (HF/GNSR/10) entitled "Preventing Person to Person Communication Failures in the Operation of Nuclear Power Plants". Available from: <http://www.hse.gov.uk/humanfactors/topics/persontoperson.pdf> (accessed 01.12.15.).

HSE, 2015. Safety Critical Communication. Available from: <http://www.hse.gov.uk/humanfactors/topics/communications.htm> (accessed 04.12.15.).

Lardner, R., Maitland, J., 2009. To Err is Human. A Case Study of Error Prevention in Process Isolations. Symposium Series No. 155, IChemE, Hazards XXI. Available from: <https://www.icheme.org/~/media/Documents/Subject%20Groups/Safety_Loss_Prevention/Hazards%20Archive/XXI/XXI-Paper-077.pdf> (accessed 19.11.15.).

Mitchell, J. 2008. The Necessity of Trust and 'Creative Mistrust' for Developing a Safe Culture, Symposium Series No. 154, IChemE, Hazards XX: Process Safety and Environmental Protection, Harnessing Knowledge, Challenging Complacency, 15–17 April, 2008.

Orasanu, J.M., 1993. Decision-making in the cockpit. In: Wiener, E.L., Kanki, B.G., Helmreich, R.L. (Eds.), Cockpit Resource Management Academic Press, San Diego, CA, pp. 137–168.

Reason, J., 1997. Managing the Risks of Organisational Accidents. Ashgate, Farnham, ISBN 978-1-84014-105-4.

Rock, D., 2008. SCARF: a brain-based model for collaborating with and influencing others. Available from: (accessed 20.12.16).

Tannebaum, S.I., Smith-Jentsch, K.A., Behson, S.J., 1998. Training team leaders to facilitate team learning and performance. In: Cannon-Bowers, J.A., Salas, E. (Eds.), Making Decisions Under Stress: Implications for Individual and Team Training. American Psychological Association, pp. 247–270.

Watcher, J.K., Yorio, P.L., 2013. Human tools performance. Engaging workers as the best defense against errors and error precursors. Professional Safety Feb, 54–64.

Wise, J., 2011. What Really Happened Aboard Air France 447? Available from: <http://www.popularmechanics.com/flight/a3115/what-really-happened-aboard-air-france-447-6611877/>.

Managing fatigue

22

K. McCulloch

LIST OF ABBREVIATIONS

ATSB	Australian Transport Safety Bureau
FRMS	Fatigue Risk Management System
GPS	Global Positioning System
HAZOP	Hazard and Operability (study)
HOS	Hours of Service
KPI	Key Performance Indicator
SMS	Safety Management System

TECHNICAL TERMS

Fatigue	*A state of impaired mental or physical performance capability, resulting from inadequate sleep quantity or quality, extended wakefulness, time of day, or workload*
Fatigue Risk Management System (FRMS)	*A system of controls to minimize the impact of fatigue on workforce alertness and performance, and ssociated risk when performing safety sensitive tasks*

This chapter explores the nature of fatigue-related risk within workplaces, and provides practical management strategies for organizations to consider. Fatigue is defined as *a state of impaired mental or physical performance capability, resulting from inadequate sleep quantity or quality, extended wakefulness, time of day, or workload*. The sensation of *"sleepiness"* is considered one indicator of fatigue-related impairment (Gander et al., 2015).

Fatigue is well-recognized as a significant threat to workplace health and safety, and has been cited as a causal factor in numerous industrial disasters including BP Texas City, Three Mile Island, Chernobyl, and Bhopal (Akerstedt et al., 2002). A case study example of an Australian shipping incident is presented in Box 22.1.

While fatigue-related risk is well accepted, few organizations outside the transport sector have implemented targeted management strategies. This is generally

---

BOX 22.1 CASE STUDY: GROUNDING OF SHEN NENG 1

At 1705 on April 3, 2010, a Chinese bulk carrier Shen Neng 1 grounded on the Australian Great Barrier Reef, seriously damaging the ship's hull and fuel tanks, with potential to cause an environmental disaster. The investigation concluded the grounding occurred because the first officer failed to alter the ship's course at the designated position. During the shift, his monitoring of the ship's position was ineffective, and he made a succession of errors leading up to the incident. The errors included:

- failure to preemptively establish the time for course alteration, instead relying on his intuition based on a glance at the chart in use;
- failure to walk around the bridge to familiarize himself with the navigational task at hand;
- failure to fix the ship's position after shift hand over;
- reliance on receiving a Global Positioning System (GPS) waypoint alarm to alert him to alter course. He forgot that the second officer had told him the waypoints were not entered in the GPS route plan.

These failures are believed to be fatigue-related. In particular, the following issues were revealed:

- the first officer told investigators that he was tired when he started the watch;
- the culture did not encourage "sharing" personal feelings of fatigue;
- in the days leading up to the event, he had accumulated a large sleep debt of approximately 16.5 hours;
- there was a procedure for minimum rest periods, but no practical guidance on how to manage personal fatigue levels of watch-keepers;
- there was a culture to under-record work hours. He had recorded hours on watch, but not total hours worked. This provided an inaccurate reflection of time available for sleep and rest;
- with the lack of formal guidance, individual crew members were responsible for managing their own fatigue levels.

Australian Transport Safety Bureau (ATSB) Transport Safety Report No. 274 MO-2010-003 (2010)

---

because without rigorous trend analysis or incident investigation, fatigue can be difficult to diagnose as an operational concern or as part of incident causation. Instead, it can be likened to a pervasive fog that may be difficult to see, but can exacerbate other hazards, and reduce effectiveness of controls.

One of the primary goals of managing fatigue is simply to enable and encourage individuals to get enough quality sleep. Sleep is severely undervalued in modern society. Feeling "tired" has become normal, to the extent that individuals struggle to identify their own fatigued state. Globalization has made work, home, and play activities accessible 24 hours a day, 7 days a week. Increasingly, people prioritize these activities over sleep, opting instead for heavy caffeine, energy drink hits, or other stimulants, to get them through the day. It is a growing cultural norm that thwarts the best efforts of regulators and organizations in managing fatigue. Therefore, responsibilities for fatigue management strategies rely equally on influencing both organizational and individual behavior.

Performance after having less than 6 hours sleep, or being awake for longer than 18 hours is comparable to having a blood alcohol concentration of 0.05%, the legal driving limit in most countries (Dawson and Reid, 1997). Habitual sleep restriction to 6 hours or less is also related to adverse health outcomes such as cardiovascular disease (Gonzalo et al., 2004), Alzheimer's disease (Osorio et al., 2011), and various cancers (Gallicchio and Kalesan, 2009). It is in everyone's best interest to raise awareness of this topic, and implement effective management strategies.

MANAGING FATIGUE

Fatigue has traditionally been managed through prescriptive limitations on work and rest times (Dawson and McCulloch, 2005). These limitations have typically been industrially negotiated to provide work–life balance and remuneration benefits, with little concern for managing fatigue-related risk. Indeed, it is now widely acknowledged that such limitations on their own are ineffective at managing fatigue (Gander et al., 2011a,b).

While work hour limitation rules can provide minimum windows of sleep opportunity between shifts, they rarely account for "time of day" differences in fatigue, the type of work being conducted, or individual differences. Work hour limitation regimes that have tried to account for these elements have proved extremely complex and operationally restrictive. Instead, organizations and regulators are increasingly adopting multi-layered risk-based approaches for fatigue management.

Fatigue Risk Management Systems (FRMS) are a scientifically based and flexible alternative to rigid work time limitations. They rely on multiple layers of control to minimize, identify, and mitigate the effects of fatigue-related impairment, and subsequent impact on alertness, performance, and safety (Dawson and McCulloch, 2005).

Unlike other "impairment"-related threats such as drugs and alcohol, there are no absolute management strategies for fatigue. That is, one can abstain from drugs and alcohol, and tests can verify impairment levels, whereas everyone experiences some level of fatigue, and there is no test that categorically determines whether someone is, or is not, impaired to dangerous levels. Because there are no absolute controls to eliminate fatigue, no management strategy will be perfect. In a similar manner to Reason's Swiss Cheese model (Reason, 1990), using multiple layers of control to manage fatigue makes it likely each ensuing layer will capture any risk that falls through the "holes."

A FRMS, implemented as part of a broader safety management system (SMS), applies multiple layers of control to identify and manage fatigue-related risk. This is depicted in Fig. 22.1, whereby multiple layers of events precede a fatigue-related incident, for which there are identifiable controls. Each of the layers are discussed throughout this chapter. Conceptually, if an organization provides individuals with sufficient sleep opportunity, they should get enough sleep. If individuals get enough sleep, they should not exhibit fatigue-related symptoms. If they are symptom free, they should not make fatigue-related errors, and similarly, not contribute to fatigue-related incidents.

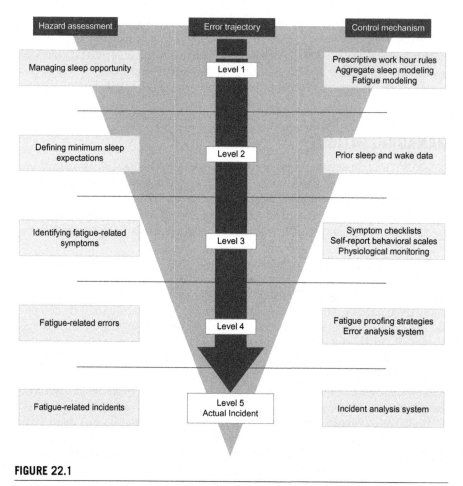

FIGURE 22.1

Fatigue-risk trajectory (Dawson and McCulloch, 2005).

From Drew Dawson and Kirsty McCulloch. Managing fatigue: It's about sleep.
Sleep Medicine Reviews (2005) 9, 365–380

FATIGUE "RISK" MANAGEMENT

As fatigue "risk" management has grown in popularity, understanding of the "risk" aspect has often been poorly understood. Fatigue is not a hazard in the traditional sense. A hazard is any source of potential harm for something or someone under certain conditions at work. With the exception of long-term illness (e.g., sleep disorders or cancers associated with shiftwork), fatigue itself is rarely the direct cause of harm in the workplace. Rather, it acts as a threat to make other hazards more likely to be realized, and can quickly permeate many barriers if not checked.

If we take the example of falling asleep whilst driving, the hazard is "driving," or "moving vehicle," and fatigue is a threat that may increase the likelihood of an incident. Similarly, forgetting a step in an isolation procedure when tired would be risk assessed with the hazard of "working with hydrocarbons" or "pressure," and fatigue

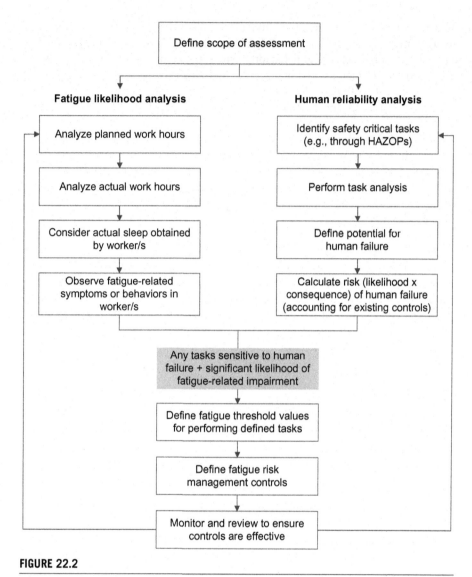

FIGURE 22.2

Process flow for fatigue risk assessment.

is a threat that may make procedural error more likely. Many workgroups attempt to do "pure" risk assessments on fatigue as a workplace hazard itself. This normally results in a vague assessment of fatigue, defining broad sweeping, nontargeted controls that can become a burden, and are often ineffective.

As shown in Fig. 22.2, to structure a risk-based approach to managing fatigue, each workgroup needs to consider:

1. the likelihood of fatigue-related impairment in individuals; and
2. the risks associated with tasks being performed across the work period.

The first assessment, the fatigue likelihood analysis, encompasses the first three layers of the fatigue-risk trajectory (Fig. 22.1), namely, work hour limitations, minimum sleep requirements, and behavioral symptoms. Some organizations find these rules sufficient to identify and manage fatigue. However, as soon as an individual registers any level of fatigue, they would be stood down from duty. This works well for roles such as truck drivers, where there is little variation in risk-based tasks across the shift.

The second assessment is risk-based, using human reliability analysis. This is typically used by organizations that want more flexibility to allow moderately fatigued individuals to work in controlled environments. Many safety-conscious organizations may have already performed human reliability analysis on core safety critical tasks, which can be used as the basis for this assessment. Risk assessment and human reliability analysis has already been discussed in Chapter 6, Human Factors in Risk Management, and is not explored further within this chapter.

The first assessment provides an indication of how tired a given person is. It does not provide any guidance on what tasks that person can or cannot do. Unlike alcohol intoxication, where a person either is or is not under the influence, most people arrive at work each day with a level of fatigue. The use of a risk-based approach helps to determine what tasks can reasonably be performed by each person. For example, a moderately fatigued person may be able to perform routine checks on pressure gauges. However, it is unlikely they would be fit to perform unsupervised complex isolations.

FATIGUE LIKELIHOOD ANALYSIS

The fatigue analysis aims to minimize the likelihood that individuals will be impaired by fatigue, and identify anyone who becomes impaired. Controls should be in place to:

1. ensure that rostered work hours provide sufficient sleep opportunity;
2. record actual work hours to ensure they do not deviate significantly from planned hours;
3. define minimum sleep expectations for individuals; and
4. identify behavioral symptoms that may indicate fatigue-related impairment.

MANAGING SLEEP OPPORTUNITY—PLANNED WORK HOURS

For decades, fatigue research has focused on finding "the perfect roster"; a roster that maximizes safety, productivity, employee health, and satisfaction. Researchers have debated the merits of 8, 10, and 12 hour shifts, best times of day for shift hand overs, how many shifts should be worked in a row, and break length to allow sufficient recovery. During this time, some definitive findings have been made related to extreme work hours. For example, in a study of fatal truck crashes, it was found that working beyond 12 hours more than doubles the relative risk of an incident (Gander et al., 2011a,b). Similarly, driving between 11 pm and 6 am produces up to six times the relative risk of driving between 9 am and 10 pm. However, when considering shift length of 12 hours or less, or 6 am versus 7 am starts, it is clear that there is no such thing as a perfect roster, and optimal work hour planning depends on organizational, environmental, cultural, and individual considerations (Ferguson and Dawson, 2012).

Under FRMS, organizations have the flexibility to structure rosters however they choose, provided there is scientific defensibility to justify why it is safe. This can be achieved through compliance with industry codes of practice or standards that define work hour restrictions or bio-mathematical fatigue modeling. Several bio-mathematical models have been commercially developed, and evaluate the likelihood of fatigue-related impairment based on the sleep opportunity afforded by the roster (Cabon et al., 2012). The models typically use shift length, break length, time of day, number of days worked in a row, and physiological limitations for recovery as key inputs to their calculations. For a more direct justification of a roster, organizations can survey and aggregate sleep patterns of the work teams to ensure that they are getting enough sleep between work shifts. This can be done using sleep diaries, or more objectively using wrist activity monitors. Active engagement of employees to collect this type of data not only helps to validate the roster, it can be a valuable awareness raising tool for employees to become more aware of their sleep needs and habits.

An additional consideration to planned work hours is commute time. Commute times vary significantly depending on the geographic location of the workplace, and where people live. In "fly in fly out" industries, commutes to site can produce periods of 6–12 hours or more time awake before starting a shift (International Association of Oil and Gas Producers, 2015). In this scenario, following a 12-hour shift, the individual would have been awake for up to 24 hours. The performance impairment for this shift would be comparable to having in excess of 0.1% blood alcohol concentration (i.e., not fit to drive). This performance impairment can impact subsequent shifts, and therefore needs to be a significant consideration when designing a FRMS.

MANAGING SLEEP OPPORTUNITY—ACTUAL WORK HOURS

Rostered work hours seldom represent actual work hours. For many shifts, particularly where there is a back to back day-night shift, most workers will arrive earlier and/or stay later to allow a hand over period. In addition, call outs, overtime, overcycle, for example, staying on shift for an additional week to cover absenteeism, and emergency drills within rest periods (when workers sleep at the work-site) all impact on sleep opportunity.

Shift predictability is also important to allow individuals to prepare for their first shift. Individuals may be changed from days to nights (or vice versa) when they first arrive on site to cover absenteeism. Part of the workforce education should include strategies to prepare for work, ensuring they are not awake for too long at the end of their first shift. Unexpected changes to shifts can limit individual ability to prepare and cope with the first shift, and may increase fatigue levels for the following shifts. In many industries, overtime rates encourage people to work excessive work hours, and employees have learned to rely on the additional payment (Gander et al., 2011a,b). In other cases, people simply work extra hours because they feel it is expected of them, or they want to impress their boss.

At times, there may be operational requirements for employees to work additional hours to continue safe operation of plant and equipment. Where this is the case, and an individual is required to work, for example, 16 hours to maintain a safe operational state, it is essential to manage fatigue on subsequent shifts. It may not be reasonable to simply expect that the standard "10 hours off between shifts" will provide enough recovery. Rather, the individual should be instructed to return for their next shift, after they have had sufficient sleep (more detail in the following section), and feel well rested. Supervisors should minimize high-risk tasks performed in the subsequent shift.

Regardless of the reason, any work in addition to the roster will impinge on sleep opportunity, and increase fatigue likelihood. It is important that overtime and callouts do not become a routine workaround to make up for inadequate staffing levels or to cover a serious absence problem. They should be kept to a minimum and ideally tracked as a Key Performance Indicator (KPI) of percentage compliance with rostered hours (IPIECA/OGP, 2012).

DEFINING INDIVIDUAL SLEEP EXPECTATIONS

Whilst the management of work hours can provide opportunity for sleep, managing fatigue-related risk assumes that employees use their time off to actually obtain sleep. From a health and well-being perspective, individuals should obtain 7–9 hours' sleep per 24 hours. From a safety and performance perspective, research has shown that performance becomes impaired when individuals average less than 6 hours sleep per 24 hours, or are awake for longer than 18 hours (Dawson and McCulloch, 2005).

As a minimum, it is important to educate employees on the need for sleep, and expectations of how much sleep they should aim to obtain. Work groups that have unpredictable rosters often rely more heavily on this layer of control, and require employees to report their sleep quantity on a daily basis as a fitness for work declaration. For this layer of control, it is important to establish clear procedures for agreed action when individuals fail to obtain enough sleep prior to starting their shift. In other words, it may require development of a mature culture, and encouragement for individuals to feel (and be) "safe" to declare their unfitness, with agreed strategies available to deal with this (e.g., putting them on light and nonhazardous duties).

Reasons for not obtaining enough sleep may include: poor sleeping accommodation (temperature, noise, bed size, whether sleeping at the workplace or at home); stress or anxiety; poor individual sleep habits, such as, caffeine before bedtime, watching television instead of sleeping, use of portable devices immediately before sleeping; or other interruptions.

Regular reports of insufficient sleep across a whole group of employees may be an indication that the roster does not provide sufficient opportunity for sleep, or that actual work hours may be exceeding the plan. Such reports need to be tracked as a KPI.

CONSIDERATION OF SLEEP DISORDERS

FRMS should consider managing risk associated with sleep disorders, including education and awareness, referral strategies for diagnosis and treatment, and providing clear expectations for return to work (Gander et al., 2011a,b). The main sleep disorders to consider include sleep apnea, insomnia, and narcolepsy. Many organizations prohibit workers with chronic sleep disorders from continuing safety critical work with low error tolerance for fatigue-related impairment, such as truck driving. Where this approach is taken, it is important to avoid penalizing the individual with reduced salary or demeaning tasks. Individuals tend to avoid, or hide diagnoses if they believe they will be unfairly treated. Instead, the focus should be on rehabilitation back to a fit state.

IDENTIFYING FATIGUE-RELATED SYMPTOMS

Fatigue-related symptoms can be split into mental, physical, and emotional categories, and different levels of severity. An overview of typical fatigue-related symptoms is provided in Fig. 22.3.

There may be additional symptoms based on the type of work being performed, or the work environment that can be added to this list. Individuals exhibiting minor symptoms would typically be allowed to continue work, with self-management

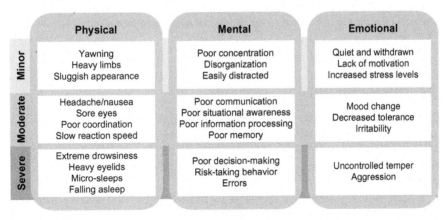

FIGURE 22.3

Fatigue-related symptoms defined by three categories, across three levels of severity.

strategies such as, self-selected breaks or the use of caffeine; and would be monitored by themselves and colleagues to ensure symptoms do not escalate. Individuals exhibiting moderate symptoms would be assessed on their ability to continue with safety critical tasks, and discuss and document appropriate management strategies with a supervisor. Any individual exhibiting severe symptoms should be stood down from work, including driving home, and rested until they are fit to continue.

Providing a symptom guide for employees is an effective method to raise awareness of the risk, and to include in behavior observation programs. It also provides a means to differentiate whether an individual should or should not continue working.

Many organizations simply rely on employees to self-declare when they are unfit due to fatigue, in which case work stops for that individual. Research indicates that tired individuals lack insight when predicting the point at which tiredness has affected their performance to the point of compromising safety (Williamson et al., 2014). They understand they are tired, but do not recognize the point at which they should stop work or modify duties. Using a severity index provides more clarity for employees in detecting the early signs of fatigue, and encourages others to intervene when they see signs in colleagues. Other strategies to identify fatigue-related symptoms include self-report behavioral scales, such as the Karolinska Sleepiness Scale (Akerstedt and Gillberg, 1990), or the Samn Perelli scale (Samn and Perelli, 1982), or physiological monitoring using performance tests.

Observation of symptoms is an effective method to capture any weaknesses in layers 1–3 of Fig. 22.1. For example, if rostering parameters are inadequate, if actual hours of work exceed the plan, if shifts expose individuals to extreme work environments or work tasks, then symptoms may be an indication of the need for tighter control going forward. Similarly, if individuals habitually do not get enough sleep, if there are problems with the sleeping environment, or problems with sleep quality, such as a sleep disorder, observation of symptoms may be a prompt to address any concerns or refer the individual to the employee assistance provider.

FATIGUE PROOFING

There may be occasions beyond the control of the organization, or when the risk of a person "not working" outweighs the risk of them continuing, where individuals may be working with a level of fatigue-related impairment. Fatigue proofing provides a strategy to consider and redesign work tasks to make them more resilient to fatigue-related error (Dawson et al., 2012).

Fatigue proofing strategies should be defined wherever the fatigue likelihood assessment indicates any level of impairment based on work hours, sleep history, or behavioral symptoms. Human reliability analysis identifies the need for tasks sensitive to human failure.

Many organizations find benefit from providing a "black and white" definition of acceptable fatigue threshold values for performing defined tasks. For example:

- some tasks may not be permitted for individuals who have exceeded rostered hours;
- fatigue modeling scores may be structured, based on evidence, to categorize work tasks. Many models have been benchmarked against the task of driving (Darwent et al., 2015). Similar benchmarks can be established for specific work tasks by collecting performance data;
- some tasks may not be permitted for individuals who have obtained less than 6 hours sleep in 24 hours, or been awake for longer than 18 hours. Anecdotal evidence may indicate that greater than 6 hours sleep is needed for some tasks that require high concentration or vigilance;
- the severity of behavioral symptoms may help to provide a definition of what tasks can be performed taking account of the category of impairment.

Whenever these threshold values are reached, or an individual simply declares that they feel tired, fatigue-proofing strategies should be implemented and documented.

The basic premise of fatigue-proofing strategies is to inform supervisors and colleagues of potential risk, which will trigger higher levels of support, observation, and supervision; and reduce pressure, for example, by providing additional time to complete and verify the task was completed correctly. Strategies may be generic and applicable to any "fatigue" situation, or task-specific to either prevent or detect the occurrence of human failure. Ideally, strategies will be developed at the local work group level, to draw on existing informal strategies, and ensure that the strategies are practical and workable.

Examples of generic fatigue proofing strategies include:

- self-selected rest breaks ("time-outs"), so that individuals can rest when they are tired, rather than at scheduled times;
- coworker monitoring;
- supervisory monitoring;
- task rotation;
- use of checklists or double-check systems;
- task reallocation;
- napping opportunity before continuing work;
- cessation of work until fit to continue.

MONITORING AND REVIEW

Any effective FRMS is data-driven and evidence based (Dawson and McCulloch, 2005). Therefore, it is important to record and monitor fatigue risk management controls. Records should be kept for:

- rostered work hours, including management of change assessments (see Chapter 19: Managing Organizational Change);

- actual work hours and actual shifts worked, which should be analyzed to determine percentage compliance with the roster;
- reports of insufficient sleep;
- fatigue-related symptom observation;
- instances where individuals have continued to work when fatigued, including fatigue-proofing strategies; and
- any other reports of potential fatigue-related impairment.

Trends of reports in any one of these elements may be an indication that a higher level control is ineffective (see Fig. 22.1). For example, regular insufficient sleep may be indicative of work hour issues. Regular observation of fatigue-related symptoms may be an indication of individual sleep or work hour issues. By monitoring this data, organizations can tweak their rules and threshold values to more accurately manage fatigue-related risk.

Collecting fatigue-related data is also essential for effective incident investigation. Often, investigations fail to highlight the potential contribution of fatigue, simply because organizations do not have easy access to the information. Any incident investigation should consider the contribution of work hours, sleep history, and symptoms, in addition to any medical issue or substance affecting the performance of the individuals involved.

Occupational health and safety departments play a vital role in establishing surveillance and monitoring of fatigue in high risk workers (e.g., shift workers, or anyone who performs tasks that are susceptible to human failure). This may include screening for sleep disorders, conducting regular surveillance on the effectiveness of fatigue risk management controls, and providing ongoing awareness to leaders and employees.

EDUCATION AND AWARENESS

When designing and implementing a FRMS, it is important to have an informed workforce. Education is the main strategy for helping to mitigate the increasing cultural norm of "less sleep is better," and addressing the potential impact of personal activities on fatigue (Gander et al., 2011a,b). Education is also essential for people who manage or work alongside people who work shifts, to establish the seriousness of the issue and appropriate management strategies. Education should include:

- causes of fatigue;
- specific tasks/hazards that may be exacerbated by fatigue (e.g., driving);
- fatigue management controls including hours of work, personal sleep expectations, and identification of fitness of work issues;
- appropriate fatigue-proofing strategies;
- expectations for reporting fitness for work concerns;
- expectations of how fitness for work concerns will be treated by management; and
- the resources available to assist with personal fatigue concerns.

Education can promote self-reflection to improve personal sleep habits, and provide suggestions for organizational improvement. It is also highly effective to provide education or awareness materials to spouses and families to promote better understanding and coping strategies when recovering from shifts, or preparing to start work after a break.

SUMMARY: MANAGING FATIGUE

Fatigue is a significant threat to any organization that employs people. Effective fatigue risk management involves multiple layers of control. The details of each layer should be developed through consultation, and documented in a formal policy or procedure. The nature of fatigue and management expectations should be communicated through workforce training. Successful management of fatigue is a "win-win" for both organizations and employees. Organizations achieve a safer and more productive workforce, and employees feel more alert and energetic, with less recovery time required during time away from work. In order to achieve these benefits, it is essential to establish KPIs at each level of the fatigue-risk trajectory, and continually monitor and review fatigue-related trends to identify potential improvements. As a minimum, this should be done annually as part of the audit process, but less formally on a more regular basis through ongoing workforce, supervisory, and management communication.

The next chapter discusses managing performance under pressure. The symptoms of stress can be similar to fatigue-related symptoms. Similarly, stress can significantly impact on sleep, resulting in elevated fatigue levels. Therefore, it is important to consider effective stress management strategies in tandem to any fatigue risk management initiative.

KEY POINTS

This chapter has discussed the nature of fatigue-related risk within organizations. It has argued that actions can be taken by organizations, supervisors, and individuals to proactively prevent and manage fatigue-related risk.

- Fatigue is a risk to safety and employee health.
- Work hours should be planned to provide sufficient sleep opportunity between work shifts.
- Individuals should obtain at least 6 hours' quality sleep per 24 hours to minimize performance impairment, and ideally achieve between 7 and 9 hours for long-term health and well-being.
- Symptom checklists can be effective to encourage peer observation and intervention for individuals who may be experiencing fatigue-related impairment.

- Fatigue-proofing strategies can be specifically targeted to manage potential task-based human error.
- Monitoring and review processes are essential for validating FRMS controls and identifying potential improvements to existing controls.
- Establish clear arrangements to manage exceptions to work hours, sleep, or symptoms.
- Provide education and awareness training to at-risk workers, their families, supervisors, and managers.

REFERENCES

Akerstedt, T., Gillberg, M., 1990. Subjective and objective sleepiness in the active individual. Int. J. Neurosci. 52, 29–37.

Akerstedt, T., Fredlund, P., Gillberg, M., Jansson, B., 2002. A prospective study of fatal occupational accidents—relationship to sleeping difficulties and occupational factors. J. Sleep. Res. 11, 69–71.

American Petroleum Institute, 2010. API Standard RP 755—Fatigue Prevention Guidelines for the Refining and Petrochemical Industries, s.l.: American Petroleum Institute.

Australian Transport Safety Bureau, 2010. Independent Investigation into the Grounding of the Chinese Registered Bulk Carrier Shen Neng 1 on Douglas Shoal. Australian Transport Safety Bureau, Queensland, Canberra.

Cabon, P., et al., 2012. Research and guidelines for implementing fatigue risk management systems for the French regional airlines. Accid. Anal. Prevent. 45s, 41–44.

Darwent, D., et al., 2015. Managing fatigue: it really is about sleep. Accid. Anal. Prevent. 82, 20–26.

Dawson, D., Reid, K., 1997. Fatigue, alcohol and performance impairment. Nature 388, 235.

Dawson, D., McCulloch, K., 2005. Managing fatigue: it's about sleep. Sleep. Med. Rev. 9, 365–380.

Dawson, D., Chapman, J., Thomas, M.J., 2012. Fatigue-proofing: a new approach to reducing fatigue-related risk using the principles of error management. Sleep. Med. Rev. 16, 167–175.

Ferguson, S.A., Dawson, D., 2012. 12-h or 8-h shifts? It depends. Sleep. Med. Rev. 16, 519–528.

Gallicchio, L., Kalesan, B., 2009. Sleep duration and mortality: a systematic review and meta-analysis. J. Sleep. Res. 18 (2), 148–158.

Gander, P., Graeber, C., Belenky, G., 2011a. Operator fatigue: implications for human-machine interaction The Handbook of Human-Machine Interaction: A Human-Centered Design Approach. Ashgate Publishing Ltd., Surrey, UK. pp. 365–382

Gander, P., Hartley, L., Powell, D., Cabon, P., Hitchcock, E., Mills, A., et al., 2011b. Fatigue risk management: organizational factors at the regulatory and industry/company level. Accid. Anal. Prevent. 43, 573–590.

Gander, P.H., et al., 2015. Effects of sleep/wake history and circadian phase on proposed pilot fatigue safety performance indicators. J. Sleep. Res. 24, 110–119.

Gonzalo, G., Alvarez, M., Ayas, N., 2004. The impact of daily sleep duration on health: a review of the literature. Prog. Cardiovasc. Nurs. 19 (2), 56.

International Association of Oil and Gas Producers, 2015. Fatigue in Fly-in, Fly-out Operations: Guidance Document for the Oil and Gas Industry. IPIECA/IOGP, London.

IPIECA/OGP, 2012. Performance Indicators for Fatigue Risk Management Systems: Guidance Document for the Oil and Gas Industry. London: OGP Report No. 488.

Osorio, R., et al., 2011. Greater risk of Alzheimer's disease in older adults with insomnia. J. Am. Geriatr. Soc. 59 (3), 559–562.

Reason, J., 1990. Human Error. Cambridge University Press, Cambridge.

Samn, S., Perelli, L., 1982. Estimating Aircrew Fatigue: A Technique with Application to Airlift Operations. USAF School of Aerospace Medicine, Texas.

Williamson, A., et al., 2011. The link between fatigue and safety. Accid. Anal. Prevent. 43, 498–515.

Williamson, A., Friswell, R., Olivier, J., Grzebieta, R., 2014. Are drivers aware of sleepiness and increasing crash risk while driving? Accid. Anal. Prevent. 70, 225–234.

Managing performance under pressure

23

C. Amati, K. Gray and J. Foley

LIST OF ABBREVIATIONS

BEST	Behavioral Emotional Somatic Thinking indicators of acute and somatic stress
EU-OSHA	European Agency for Safety and Health at Work
GAPAN	Guild of Air Pilots and Air Navigators
HSE	Health and Safety Executive
OH	Occupational Health
PSF	Performance Shaping Factors
STAR	Stop, Think, Act, Review—cognitive checking technique for acute stress
STOP	Stand back, Take stock, Overview, Procedures—cognitive checking technique for acute stress

---

TECHNICAL TERMS

Acute Stress	*Reaction to very intense but typically short-lived stressful situation(s)*
Chronic Stress	*Result of exposure to prolonged stress resulting in impaired functioning and potential for longer-term harm to health.*
Pressure	*Positive state of engagement and challenge*

This chapter explores the experience of stress at work, its impact, and actions that individuals and organizations can take to minimize the risk to performance and health.

Stress is typically defined as an imbalance between the demands placed on the individual and the resources available to support coping. Some definitions focus exclusively on the impact of high demands; for example, EU-OSHA states that workers experience stress when the demands of their job are greater than their capacity to cope with them (EU-OSHA, 2015). However, it is useful to also consider that individuals can feel stressed even when they have too few demands, such as when they are out of work or in the wrong job.

It is important to differentiate between different types of stress that individuals at work might experience, as Table 23.1 clarifies.

Table 23.1 Different Types of Stress

	Acute Stress	Chronic Stress
Overview	Caused by a particularly stressful but temporary event or situation that triggers a surge in the release of stress hormones such as adrenaline and cortisol.	Experienced when a person feels under sustained pressure, repeatedly finding themselves in situations they feel they cannot cope with and cannot see a way out of.
Impact	Leads to a range of physical changes "in the moment" that can have either positive or negative effects on performance; includes increased heart rate and muscle tension.	Leads to "strain" on the body due to experiencing ongoing "acute stress" states; a range of psychological and physical consequences are felt, with associated changes in behavior. Exposure to chronic stress for lengthy periods of time can lead to serious physical and psychological ill health.
Recovery	The body is designed to cope well with this type of stress and recovery is fairly fast, taking anything from an hour to a few days, and rarely requiring professional help.	Recovering from chronic stress is a slow process and can require professional input from both medicine and psychology.

This chapter explores both acute and chronic stress, and their impact on individuals' performance as well as their potential for longer-term harm to physical and mental health. It provides practical advice for individuals, managers, and organizations on how to manage the associated risks and concludes with a review of the concept of "resilience" and how managers can act proactively to increase the team's well-being.

THE BODY UNDER DURESS

In the course of their day-to-day employment, employees may be exposed to situations that have the potential to elicit an acute stress reaction, including conditions perceived as potentially life threatening. This subsection explores this experience, discussing evidence of how physiological changes under such circumstances can impact on clarity of thinking and judgment. It then explores implications for the design and implementation of effective preventative and management measures.

ACUTE STRESS: THE EXPERIENCE "IN THE MOMENT"

The human brain is a complex organism of parts that work both in isolation and in interaction with each other. Perceiving something as a real and immediate personal threat generates a reaction in the areas of the brain that manage emotions and house the center of the threat (fight/flight) response. Specific activity is seen in the amygdala and the structures of the brain often collectively known as the limbic system that play a

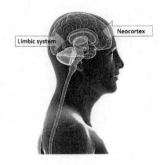

FIGURE 23.1

Hind brain and frontal cortex.

Reproduced with permission of Sebastian Kaulitzki ©123RF.com.

central part in emotional functioning. The threat response is a critical survival mechanism and its main effect is to prepare the body for action. The changes include acceleration of the heart rate, increased blood supply to the major muscle groups, suppression of the immune and digestive systems, and narrowing of attention to the specific threat (e.g. Brann, 2013). These reactions are fast, instinctive, and largely automatic.

The human brain has also developed more advanced neurological functions; this evolution created the neocortex, including the frontal cortex, where conscious thought, working memory, planning, and decision-making is performed (see Fig. 23.1). This process of evolution works from the core outwards and one of the consequences is that there are more connections from the deeper limbic (feeling) brain to the neocortex (thinking brain) than in the other direction. Therefore, the more primitive instinctive and emotional functions of the brain can exert greater influence over our thinking than thinking does over feeling. As a result, instinctive responses, such as the fight/flight response, are difficult to influence through rational thought alone. Therefore, when the person is in a high threat situation, the automatic response triggered by the amygdala can readily "hijack" our conscious thought process (e.g. Brann, 2013).

THE IMPACT OF ACUTE STRESS: IMMEDIATE AND LONGER-TERM

Immediate

In itself, the acute stress reaction is not harmful. The body has evolved this response to enable humans to cope with threats to survival and it is particularly effective in situations that require quick reactions and a physical, active response.

However, in situations that require more complex thinking and access to the working memory required to "think our way out," the way the brain is wired can work against effective behavior. In these instances, individuals can effectively "freeze," fail to take appropriate action, and/or "speed up," taking multiple, rapid, disconnected actions without apparent logic or structure. In these situations, individuals often know what they need to do, but there is a problem with the communication pathways that

Table 23.2 BEST Indicators of Acute Stress

B—Behavioral	E—Emotional	S—Somatic	T—Thinking
• *Fight/flight:* hyperactivity; irritability; argumentativeness; jumpiness • *Freeze:* withdrawal; disengagement; apathetic	• Fear and vulnerability; • Anxiety; • Panic; • Fear of failure	• Increased heart rate; • Dry mouth; • Sweating; • Heightened sensitivity; • Muscle tension	• Impairment of memory, including confirmation bias (failure to take in information that contradicts existing decision); • Reduced concentration, including perceptual tunneling; • Difficulty in decision-making

Adapted from Flin et al. (2008). Safety at the Sharp End: A Guide to Non-Technical Skills. Ashgate Publishers, Farnham.

trigger the appropriate goal directed actions or behaviors (e.g., Brann, 2013). In some people, a fixation of attention also develops, when the individual's concentration is too focused on the narrow task or threat (know as "perceptual tunneling"), rendering them unable to take in the broader situation. The indicators of acute stress reactions can be described according to the BEST acronym (Flin et al., 2008), as illustrated in Table 23.2.

The examples in Boxes 23.1, 23.2 and 23.3 help to illustrate the multiplicity of effects of the acute stress state.

BOX 23.1 AIR FRANCE 447 DISASTER: AN EXAMPLE OF ACUTE STRESS

On June 1, 2009, an Air France Airbus 447 crashed into the ocean. Initial investigations ascribed a possible mechanical cause to the disaster, but later recovery of the plane's black boxes revealed the main issue to be human error. Transcripts of the conversations amongst the copilots and the captain suggested at least one of the crew was suffering from "acute stress," continuing to repeat an incorrect action that, ultimately, caused the plane to stall and crash. The pilots seemed unable to clearly articulate their thinking, access information about the plane's functioning, take in the information from their display, and interact with each other meaningfully. Instead, the transcript appears to reveal signs of panic amongst the crew, possibly exacerbated by the repeated loud alarms, the copilot's inexperience and seeming unwillingness to accept the nature of the situation they were in (Wise, 2011).

BOX 23.2 FORMOSA PLASTICS EXPLOSION: AN EXAMPLE OF ACUTE STRESS

On April 23, 2004, an explosion and fire destroyed the Formosa Plastics plant in Illinois, USA. The accident was caused by an operator overriding a safety valve interlock. However, the investigations found that fatalities could have been reduced and possibly prevented if the staff had evacuated the site, rather than attempt to manage the incident. The stress that the staff were under may have affected their ability to take a step back, pause, consider the situation as a whole, and take more reasoned action (CSB, 2007).

BOX 23.3 THE "HUDSON MIRACLE": MANAGING ACUTE STRESS

On January 15, 2009, Captain Sullenberger and First Officer Skiles successfully landed an Airbus A320 on the Hudson river after a multiple engine failure. In the incident named the "Hudson Miracle," the crew and especially the Captain were commended for their ability to react to the emergency situation in a calm and reasoned manner. In interviews, the Captain referred to the atmosphere in the cockpit being "busy, business-like" and to his confidence in his and his First Officer's ability. These statements indicate a situation of pressure being well-managed, rather than descending into one of acute stress. The Guild of Air Pilots and Air Navigators (GAPAN) awarded the entire flight crew a Master's Medal for outstanding aviation achievements: "The reactions of all members of the crew, the split second decision making and the handling of this emergency and evacuation was 'text book' and an example to us all" (GAPAN press release, 2009).

Post event

In the majority of cases, once the perceived threat has passed, the body will be restored to its normal functioning and the effects of the acute stress response dissipate without adverse longer term consequences. However a sustained stress response can have longer term harmful consequences, as discussed later in this chapter, as can exposure to single, particularly stressful events.

Interviews with people who have survived extremely stressful events comment that the first 3 days are critical to the person regaining sufficient control and equilibrium. Whilst most people will recover, those who fail to do so become more vulnerable to further deterioration of their mental state and could effectively just "give up." Feelings of considerable fatigue, guilt, and sadness will persist and healthy recovery is dependent on the person's support and environment within the first 4 weeks after a traumatic event. If symptoms persist, then Posttraumatic Stress Disorder (PTSD) may occur and this would require the individual to seek expert help (see Box 23.4).

BOX 23.4 POSTTRAUMATIC STRESS DISORDER

PTSD is a psychological reaction following exposure to a traumatic event involving self or others. Common to all sufferers of PTSD is a subjective emotional response that involves an intense fear, helplessness, or horror (DSM-5). Sufferers also report flashbacks and/or nightmares, intrusive distressing memories relating to the traumatic event, and a persistent avoidance of any stimuli that arouse recollections of the traumatic event.

From an organizational point of view, it is most important to recognize the symptoms of PTSD so that action can be taken to ensure the individual receives appropriate treatment. Further, an individual suffering from PTSD is unlikely to be functioning to the level required to ensure accurate judgment and safe behaviors and should therefore be removed from hazardous work environments until their psychological health has been restored.

There are recommended forms of therapeutic intervention to be carried out by a qualified and registered psychologist or other psychological therapist. Individuals being supported correctly can expect to recover.

Note: DSM-5 relates to a specific diagnosis within the Diagnostic and Statistical Manual of Mental Disorders (DSM).

THE CAUSES OF ACUTE STRESS

Acute stress can be brought about by any situation perceived as a threat, such as emergencies, injuries, complex problems, tightly coupled and dependent systems, and high risk situations; in emergency situations, often multiple features can present at the same time. Many of the features above are also recognized as Performance Shaping Factors (PSFs), which have been discussed in Section II, Managing Human Failure. The features of a situation that can cause acute stress include the following:

- *New or unusual events*—where the emerging situation is not well understood because the progression of the event is not known, or appropriate actions are not immediately known;
- *Where expectations are violated and/or plans and expectations do not produce required outcomes*—things fail to develop in the expected direction or systems fail to work as expected, thereby introducing uncertainty (e.g., Box 23.1 Air France 447);
- *Where critical information is missing or wrong*—when making predictions more difficult and outcomes less certain;
- *Where multiple goals are present with conflicting actions*—such as emergencies or complex scenarios in which it is critical to address more than one goal at a time, or where goals may be in conflict with one another (e.g., Box 23.2 Formosa Plastics);
- *If dealing with casualties, situations of personal danger with a high fear of the consequences of failure*—such as emergencies or where the stakes are particularly high;
- *Where there is information overload*—such as the activation of several alarms at the same time (e.g., Box 23.1 Air France 447);
- *Where there is equipment and/or machinery failure*—this might result in an incorrect understanding of the situation or an inability to resolve events quickly;
- *In situations of extreme time pressures*—in many emergency situations there is a limited time before an "event" occurs which adds even more pressure.

MANAGING ACUTE STRESS: MANAGERS AND INDIVIDUALS

There are a number of actions individuals and managers can take to avoid the negative effects of an acute stress reaction. It is important to consider preventing high pressure situations from generating an acute stress reaction as well as managing the effects when individuals are under stress.

As individuals

- Know your limits and do not take on too much—shed less important tasks, where possible, to focus attention on the critical few;
- Be aware of situations that might trigger an acute stress reaction—understand your job and your personal limits and fears; develop your competence and confidence and raise related concerns appropriately within your organization;

- Be aware of how you and others react when under pressure—be prepared to draw the attention of the team leader to any signs of stress in the team; seek help if you feel yourself affected;
- Take action when under stress—effective actions are those that help either:
  a. to regulate physiological/emotional reaction: for example, deep breathing that has a direct physiological impact, reducing the heart beat and calming you down;
  b. to increase cognitive control and behavioral change: for example, self-checking techniques (e.g., STOP: Stand back, Take stock, Overview, Procedures; STAR: Stop, Think, Act, Review) regulate emotions and distracting thoughts to allow you to retain concentration on the task at hand;
- Practice the skills and behaviors you will need if/when you come under acute stress—you are more likely to be able to retrieve and use knowledge that has been so well-learned that it comes automatically; practice before you need to make use of the coping technique;
- Rehearse emergency procedures thoroughly and repeatedly.

As managers

- Understand the work your team are undertaking and possible triggers for acute stress reaction—take preventative action where possible to reduce the likelihood of occurrence of the stress reaction;
- Encourage people to use self-checking or cognitive control techniques, such as STOP (Stand back, Take stock, Overview, Procedures)—take a couple of minutes to run through the process regularly with the team;
- Encourage the practice of deep breathing—support training interventions that teach the basic skills involved in relaxation and their transfer to the workplace;
- Switch those exhibiting severe negative reactions to the situation to nonessential tasks—remove those exhibiting signs and symptoms from high stress tasks and keep them under observation. Engage with them calmly, helping them to understand their reaction as 'normal' or understandable and if possible, keep them within the team;
- Introduce opportunities to over-learn and over-plan: practice skills in a simulated stressful environment—rehearse and practice skills to such a degree that they become automatic and accessible even under stressful conditions. Over-learning and over-planning involve training beyond the level of proficiency normally required. For this to be effective, include training in a simulated stressful environment. This could involve stressors such as noise and time pressure as there is evidence that skills learned in one type of stressful environment can be transferred to another;
- Become familiar with PSFs more generally—understand how high pressure and stress situations can influence behavior and increase the likelihood of human error (see section: Managing Human Failure).

SUMMARY: ACUTE STRESS

Anyone can experience events at work that trigger an acute stress reaction. The way the body reacts has evolved to be helpful but, in some instances, can disrupt thinking, lead to unhelpful behavior and potentially longer-term harm to individuals' health. Individuals and their managers need to be alert to the signs of a stress reaction and consider taking preventative action where possible as well as supporting better management of reactions in these situations.

MANAGING PRESSURE AT WORK: A COMPREHENSIVE APPROACH

The previous section discussed how the body reacts under acute stress and what individuals and managers can do to improve coping in these situations. Although potentially high risk, these situations are not overly common in organizations. A more frequent occurrence is individuals feeling exposed to "chronic stress"; that is, day-to-day situations of high pressure that, when particularly prolonged or significant, start to impact the individual's ability to cope and, in the longer term, negatively impact their physical and mental health. This section explores this experience and what action organizations can take to effectively prevent or manage this type of stress.

CHRONIC STRESS: DEALING WITH PRESSURE OVER TIME

As discussed in the previous section, a pattern of physiological changes occur when individuals are exposed to situations of increased pressure. If dealt with satisfactorily, the body will then return to a "normal" state and recovery occurs. However, in modern workplaces, individuals often need to deal with multiple sources of pressure that cannot necessarily be completely resolved. The result is that many individuals experience a sense of being under pressure for prolonged periods of time, a form of "chronic stress."

To understand this experience, it is useful to understand the impact of stress on performance; this is often referred to as the "Yerkes Dodson Pressure-Performance Curve," shown below and as already discussed in earlier Chapter 20, Staffing the Operation, in this book (Teigen, 1984). Evidence suggests that a certain amount of pressure is positive and is needed to achieve "optimal performance"; such pressure may be achieved through goals and tasks that "stretch" the individual and keep the levels of alertness and readiness "topped up." However, having too much or too little demand has an increasingly negative impact on performance and, over time, health (e.g. Flin et al., 2008). If there is insufficiently challenging or interesting work, the individual may feel bored, listless, and, over time, fall into "rust-out" with a negative impact on their confidence, mood, and energy. If demands are perceived as too high, this can initially lead to increased efforts to sustain performance but, over time, can lead to mistakes, exhaustion, anxiety, and "burn-out" (Fig. 23.2).

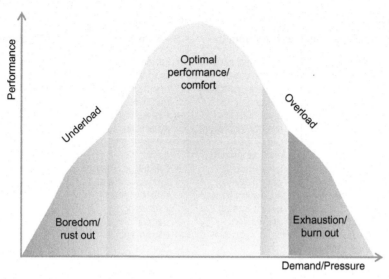

FIGURE 23.2

Performance under pressure: Yerkes-Dobson curve.

THE IMPACT OF CHRONIC STRESS

It is likely that everyone will experience parts of their working life in the "amber" zones of the pressure-performance curve and might also shift from one side of the curve to another during a normal working day. Short periods of time in either amber zone need not be harmful in the longer term, as long as there is an opportunity to recover. However, spending protracted periods of time in the "amber" zones, or being exposed to particularly severe forms of stress, such as bullying, even for shorter periods of time, can start to affect physical and mental health.

Experiencing chronic stress can lead to a number of changes in physical and psychological functioning, although not all of these will be experienced by everyone. Some of these will also not be evident to others, so it is important to focus on the more visible behavioral indicators that might indicate the individual is feeling under stress. Changes to behavior and health can be a direct result of prolonged exposure to stress or a side effect of attempts to manage the symptoms of increased pressure, for example, by "self-medicating" with caffeine, tobacco, or alcohol. The indicators of chronic stress are defined in Table 23.3 using the BEST acronym, as discussed in the previous section (Flin et al., 2008).

Teams that are under pressure can also start to display patterns of behavior that might indicate the presence of stress. These include increased absence (especially short-term); deterioration of relationships and communication; increases in turnover and decreases in performance (Flin et al., 2008). Chronic stress has also been shown to affect safety performance, both directly, in terms of the effect of physical and

Table 23.3 BEST Indicators of Chronic Stress

B—Behavioral	E—Emotional	S—Somatic	T—Thinking
• Increased irritability and aggressive behavior; • Increased withdrawal, isolation, and uncommunicative behavior; • Increased errors and decreased quality of work for similar input; • Increased need for reassurance; reluctance to take responsibility	• Increased sense of worry and anxiety, potentially escalating to panic attacks, anxiety disorders; • Decrease in confidence and lower mood, potentially escalating to symptoms of depression	• Increased incidence of headaches, migraines; • Altered digestion, potentially escalating to gastric conditions; • Difficulties sleeping; • Increased heart rate, potentially escalating to palpitations and cardiovascular disease; • Increased susceptibility to infections	• Lack of concentration • Reduced attention; • Difficulty with memory; • Impaired decision-making; • Failures in planning

Adapted from Flin et al. (2008). Safety at the Sharp End: A Guide to Non-Technical Skills. Ashgate Publishers, Farnham.

psychological symptoms on decision-making and behaviors, and indirectly, for example, due to the knock-on effect of stress-related absence (Amati and Scaife, 2006).

If not managed appropriately, chronic stress symptoms will worsen over time; early intervention is therefore essential, as illustrated in Fig. 23.3. However, individuals who are stressed often remain unaware of how their thinking or behaviour has changed or, even if aware, feel unable to stop and change. This situation can be compared to the continual overriding of a brake mechanism until eventually there is a blindness to the actual speed of travel. The implication therefore is that individuals might need help from colleagues or their manager to become aware of the changes and to take action. Early supportive conversations can help the individual to decide to access further support.

THE CAUSES OF CHRONIC STRESS

The UK Health and Safety Executive (HSE) commissioned research in the mid-2000s to determine the most common and significant sources of work-related stress (Rick et al., 2002; MacKay et al., 2004). The result was the definition of six "risk factors" of job design that, if not managed appropriately, have the potential to negatively affect health and well-being, productivity, and absence. These can be referred to as psychosocial hazards or "badly designed work." The HSE six risk factors are outlined in Table 23.4.

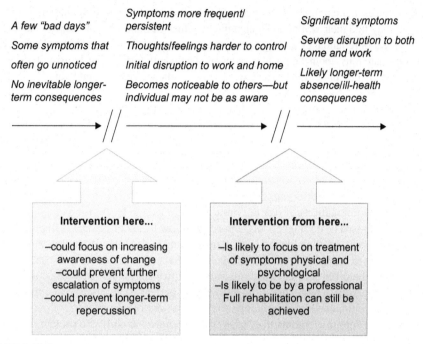

FIGURE 23.3

The benefits of early intervention for chronic stress.

Table 23.4 HSE Six Risk Factors: Work Stressors

Demands	Control	Support	Role	Relationships	Change
Excessively high or low quantitative or qualitative workload	Insufficient autonomy and discretion	Lack of access to support from colleagues, line management, and the organization	Poorly defined or conflicting roles	Exposure to conflict, unacceptable behavior, bullying, and harassment	Poorly managed change

Individuals may also come under pressure from nonwork events, typically referred to as "life events." These can include, for example, difficulties with relationships, needing to support a dependent, or moving house. Although some 'life events' might be regarded as "positive" (e.g., getting married, having a baby) they also create more demands for the individual and therefore have the potential to affect their ability to cope more generally.

There are differences between individuals that influence their experience of pressure and stress; the pressure-performance curve does not look identical for everyone. These individual differences might be due to personality, experience, and/or social circumstances. For example, a confident individual who knows they have their manager's support may find their workload challenging; another with the same demands who feels under threat of redundancy and is experiencing interrupted sleep due to a new baby may feel more anxious about being able to cope. These differences will be explored later within this chapter under the heading of "Stress, Resilience, and the Individual."

These three broad categories of potential causes of stress (work stressors, life events, and individual differences) have so far been discussed separately but there are important patterns in the ways in which these affect individuals (Rick et al., 2002), as outlined below.

- The risk of experiencing stress is typically cumulative; exposure to multiple stressors is more likely to produce harm. For example, experiencing high demands at a time of organizational change will be more likely to cause harm.
- Specific combinations of risk are more likely to lead to harm; for example, being exposed to high levels of demand with low levels of control is likely to cause more harm than having high demands and high levels of control.
- Exposure to each risk factor is not likely to affect all individuals in the same manner; the difference between pressure (positive state) and stress (negative state) is largely dependent on the individual's perceived ability to cope; this might be influenced by personality, experience, and other factors.
- There are certain elements that, when present, have the ability to protect against the risk of stress; in particular, access to high levels of support can help individuals deal with pressure from other stressors more effectively. The role of the manager is also particularly important in this respect and will be explored later within this chapter under the heading of "Stress, Resilience, and the Individual."

MANAGING CHRONIC STRESS: THE ORGANIZATION'S ROLE

A simple three-tier framework can be usefully applied to the management of stress and pressure at work (e.g., Flin et al., 2008). This approach suggests that activity needs to be taken to achieve three separate but interrelated aims: primary interventions have a main focus on stress prevention, secondary (or secondary prevention) interventions on improving individuals' management of stress, and tertiary (or tertiary prevention) interventions focus on treating individuals who are currently suffering from stress, as illustrated in Fig. 23.4.

Organizations can use this framework as a structure to help both identify the focus of their current interventions and identify gaps or areas where further action would be useful. The framework also highlights the risks of taking action at some levels and not others.

Primary prevention—risk assessment

Interventions aimed at prevention need to focus first on identifying potential psychosocial hazards, or sources of possible work-related stress, and then on taking

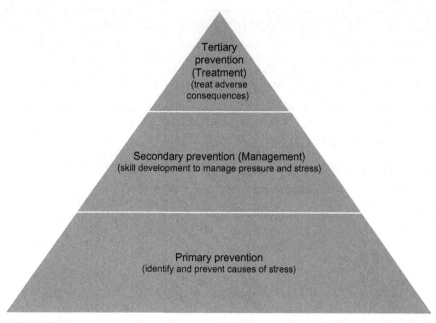

FIGURE 23.4

Three levels of intervention for chronic stress.

Table 23.5 Primary Prevention in Practice

Example Interventions	Critical Success Factors
• Stress risk assessments, stress surveys, focus groups; • Follow-up action plans and reviews	• Staff involvement in assessment and action-planning; • Integration of findings with other organizational data; • Focus primarily on organizational interventions (e.g., work design/redesign)

action to prevent or manage these to minimize the likelihood of harm. These interventions often start with a form of structured assessment such as a survey or stress risk assessment, a series of structured focus groups or a combination of the two; this should identify current levels of stress (harm) and current stressors (hazards). Information to support this phase can also be found in other available data, such as absence records, accident rates, or engagement surveys. The assessment is typically followed up with an action plan with a focus on organizational risk reduction, as recommended by the risk management approach to stress (Cox et al., 2000). There is some evidence that interventions, such as work redesign, workload management, and improved communications are effective in reducing stress (e.g., Rick et al., 2002) (Table 23.5; Box 23.5).

BOX 23.5 THE EU-OSHA "HEALTHY WORKPLACES MANAGE STRESS"

The EU-OSHA is currently running a 2-year initiative aimed at improving the management of psychosocial hazards at work. Combining practical tools, research publications, case-studies, and awards, the related website is a valuable source of information and support for organizations wanting to improve their management of stress. For more information, see https://www.healthy-workplaces.eu/en.

Table 23.6 Examples of Secondary Interventions Aimed at "Management"

Example Interventions	Critical Success Factors
• Activities to promote awareness of stress and its impact; • Educational interventions to develop effective individual coping strategies (e.g., group training, workshops, one-to-one coaching); • Provision of access to services to support coping (e.g., exercise, relaxation, etc.)	• Interventions linked to clear needs analysis: support is tailored to specific needs of the individuals or teams; • Effectiveness evaluated systematically

Secondary prevention—stress management

Secondary prevention, sometimes referred to as "management," refers to provisions of support to enable individuals to better identify and manage stress for themselves. The assumptions made are that some stress will be unavoidable, unpredictable, and/or not easily prevented therefore it is useful for individuals to have the skills they need to face this type of pressure, without risking their health and well-being. There is a wide range of activities that can fall under the generic "stress management" label: from training in coping techniques to massages and meditation (see Table 23.6). Quality evidence as to the effectiveness of individual stress management interventions is not always available (Graveling et al., 2008). However, interventions aimed at increasing physical activity and training/education on cognitive coping have been shown to have a positive impact on well-being (Graveling et al., 2008); there is also some evidence that cumulative interventions are likely to be more effective than interventions in isolation (Flin et al., 2008).

Tertiary prevention—stress treatment

The third level of intervention relates to activities broadly aimed at "treatment," that is the provision of additional support for those suffering from increasingly significant stress-related difficulties (see Table 23.7). Interventions might include the provision of one-to-one clinical support, as well as the management of related absence. Interventions at this level are typically carried out by individuals with professional expertise, such as psychologists, counselors or Occupational Health (OH) nurses, and doctors. Most support is available for the individual to access voluntarily, although some employees might also be referred to a professional by the organization.

Table 23.7 Examples of Tertiary Interventions Aimed at "Treatment"

Example Interventions	Critical Success Factors
• Provisions of access to appropriate professional support (e.g., OH, counseling, clinical services); • Developing of policy and process to effectively manage stress-related absence	• Balance of support for the individual (e.g., support for recovery) and outcomes for the organization (e.g., prompt return to work); • Communication between all people supporting the individual (case-management approach);

Table 23.8 Definitions of Individual and Organizational Resilience

Individual Resilience	Organizational Resilience
"a sense of adaptation, recovery and bounce back despite adversity or change." The ability to adapt and cope with challenges, including bouncing back from setbacks, adversity, and change.	"how well the organization can 'weather the storm' or adapt to the challenges it faces." This includes both the resilience of individuals within the organization and the broader process, competencies, structures, and culture of the organization.

Adapted from Lewis et al. (2011a). Developing Resilience: Research Insight. CIPD Publications, London.

SUMMARY: CHRONIC STRESS

Chronic stress can be understood as the strain experienced by the body and the mind when under extended or excessive pressure. This section has argued that a comprehensive approach to managing chronic stress will involve taking action at three levels: prevention, management, and treatment. In practice, this means organizations need to consider identifying and managing sources of work-related stress first, before or alongside interventions focused on individuals and their coping.

FROM MANAGING STRESS TO PROMOTING RESILIENCE

In the time-frame roughly coinciding with the time of global recession, there has been an increase in interest in the concept of "resilience" and how it might help individuals and organizations better understand and manage pressure and stress at work. This section explores the links between resilience, stress, and pressure and the practical implications for organizations and their managers.

RESILIENCE: RECOVERY FROM SETBACKS

Resilience is generally defined as the ability to cope with the pressures or difficulties encountered in life, including the ability to recover from setbacks. The term can be used to refer to the ability of individuals and of organizations, as defined in Table 23.8.

RESILIENCE: THE "NEW" ANSWER FOR MANAGING PRESSURE AT WORK

Resilience could be understood as the "flip-side" to stress. There are clear similarities between being resilient and being able to effectively manage stress and pressure. Both concepts refer to a broad set of skills, behaviors, and thinking strategies that help individuals to cope more proactively and positively in the face of adversity, whether this is defined as stress, change, unexpected demands, or setbacks. These include individuals controlling their emotional reactions, managing their thinking in situations of increased pressure or challenging and building greater confidence and optimism (Reivich and Shatte, 2003)—see Box 23.6.

However, interventions aimed at improving resilience tend to focus more on individuals and their personal coping skills "in the moment." Efforts to improve resilience are more likely to be effective if they also include attention to those workplace factors that might influence an individual's resilience and longer-term coping. In addition, there are clear links between those factors that improve performance and well-being when present and negatively affect them when absent (for a review, see Amati and Donegan, 2012). The greatest benefits are therefore likely to come from adopting a proactive approach that goes beyond just improving individual resilience, building on understanding of how to design jobs and workplaces to improve psychological health and performance, reduce workplace stress and create "good work" (for an outline of this approach, see Bevan, 2012).

MANAGING STRESS AND BUILDING RESILIENCE: THE MANAGER'S ROLE

The synergy between building resilience and managing stress is also clearly visible in discussions of manager behavior and team well-being. The interaction between the individual and their manager is arguably the most important relationship at work; managers shape and control the employee experience of their workplace, from the jobs that they do, the rewards that they receive and their impression of the company overall.

BOX 23.6 THE SEVEN LEARNABLE SKILLS OF RESILIENCE

Dr. Karen Reivich has written extensively about resilience and the ability to learn how to become more resilient. The following are the seven skills suggested:

1. *Emotion awareness or regulation*—being able to identify what your feelings are and, when necessary, the ability to control your feelings.
2. *Impulse control*—being able to sit back and look at things in a thoughtful way before acting.
3. *Optimism*—understanding your strengths and how you have dealt with challenges in the past.
4. *Causal analysis*—thinking comprehensively about the problems you confront from multiple perspectives.
5. *Empathy*—being able to read and understand the emotions of others.
6. *Self-efficacy*—being confident in your ability to solve problems.
7. *Reaching out*—being prepared to try new things without fear of failure.

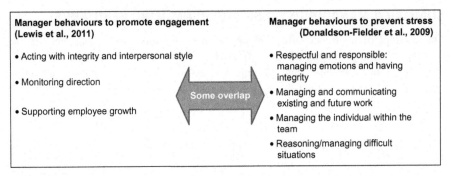

FIGURE 23.5

Manager behavior to prevent stress and promote engagement.

Table 23.9 The Role of Managers

Promoting Resilience

- Provide jobs that are meaningful, motivating, and fairly rewarded
- Manage resources and demands to provide motivating, challenging but reasonable workloads
- Provide opportunities for autonomy within jobs and involvement in decision-making
- Manage relationships and behavior at work to create supportive teams
- Support employee personal and professional development and career progression

Preventing Stress (Primary)	Managing Stress (Secondary)	Stress Treatment (Tertiary)
• Ensure their own behavior does not cause undue pressure or stress • Support individual and teams in identifying stress-related concerns and taking preventative action	• Review the skill mix and provide support to increase coping/resilience skills where needed • Create a culture where pressure and stress can be easily and openly discussed without stigma • Respond appropriately to individuals who are experiencing acute stress	• Act promptly to support individuals displaying signs of stress • Provide access to support when available • Contribute to the management of stress-related absence

Research has shown that, by their actions, managers can both prevent stress from becoming problematic and help individuals manage it more effectively (Donaldson-Feilder et al., 2009)—see Fig. 23.5 (right-hand side of the diagram). These behaviors show a clear overlap with the competencies that can sustain employee engagement and performance (Lewis et al., 2011b)—see Fig. 23.5 (left-hand side of the diagram). This means that, by developing the competencies described, the manager can expect both increases in engagement and a reduction in work-related stress.

By following the steps in Table 23.9, managers can both support engagement (positive emotional reactions to work) and manage stress (negative reactions to work).

SUMMARY: MANAGING STRESS AND PROMOTING RESILIENCE

The concept of resilience offers an opportunity to think more proactively about shaping work to positively affect performance and health. With a focus on providing "good jobs," alongside management of stressors and stress, managers can act to increase engagement in their teams, support resilience and coping.

STRESS, RESILIENCE, AND THE INDIVIDUAL

The discussion in the chapter so far has focused on generic stress reactions and on what action an organization or its managers can take to manage stress and pressure. This section will explore the role of individual differences in the stress reaction.

INDIVIDUAL DIFFERENCES: STRESS AND RESILIENCE

Differences between individuals can affect reactions to situations of high pressure. Individuals differ in their temperament and in the way they react to situations and interpret the challenges they face. Critical factors include the degree of personal self-confidence and emotional intelligence. Self-confident individuals will generally be optimistic and proactive when they face setbacks or obstacles; once doubt their ability to cope starts to emerge, the situation will be perceived as increasingly challenging and the experience of positive pressure may become stress and strain. Individuals who are more able to deal with emotions in themselves and others will be more resilient; individuals who have difficulty controlling their emotions may struggle with what seems like relatively minor challenges to others. Some of these differences between individuals and the way they react to situations are due to personality and up-bringing; there is also evidence that individuals can learn more effective coping strategies, develop greater confidence, emotional intelligence, and become more resilient (Reivich and Shatte, 2003; Graveling et al., 2008).

Individuals also differ in the extent to which they have access to support when they need it and in their readiness to access it. This will be influenced by a number of factors, including the individual's socioeconomic status; their cultural or family background; their personal and work circumstances. Research on joblessness illustrates this well (e.g., Waddell and Burton, 2006): for example, when unemployed, individuals lose income as well as support received through work (colleagues, professional services, etc.); this means they have less "free" resource to access and fewer means with which to access other support.

Finally, different jobs will expose people to different stressors. This works in two ways. Individuals who are able to choose jobs that suit their motivation are more likely to be resilient; individuals who are not as able to choose their job or career path may find their work intrinsically more stressful and less enjoyable. In addition, lower-paid, more controlled, less secure jobs are more likely to cause stress and less likely to provide access to support. Jobs with these characteristics pose a higher risk to individuals, regardless of their own personal resilience levels, as was

Table 23.10 The Role of the Individual

Promoting Resilience
<ul><li>Understand own strengths, motivations, and interests and build these into job choices as much as possible; communicate these to others;</li><li>Understand own personal sources of pressure and stress and take action on issues that are within personal control;</li><li>Identify, practice, and consolidate specific behavioral habits to help build effective coping and resilience (see also Table 23.9).</li></ul>

Prevention	Management	Treatment
<ul><li>Raise work-related issues appropriately and promptly;</li><li>Engage openly with activities related to stress prevention.</li></ul>	<ul><li>Raise development needs promptly;</li><li>Engage with activities provided by organization and elsewhere to improve coping skills.</li></ul>	<ul><li>Access support provided when needed;</li><li>Respect others who might need additional support, without being judgmental.</li></ul>

shown by the large-scale, longitudinal studies of the UK Civil Service, referred to as the "Whitehall Studies" (for more information, see http://unhealthywork.org/classic-studies/the-whitehall-study/).

BUILDING RESILIENCE AND MANAGING STRESS: THE INDIVIDUAL'S ROLE

Given the impact of individual differences, individuals also have a clear role to play in managing their own well-being in relation to stress and resilience. The three-tier approach can also be applied to understand the actions an individual might take to improve their own management of stress and pressure, as depicted in Table 23.10.

SUMMARY: STRESS AND RESILIENCE AND THE INDIVIDUAL

The impact of individual differences on the experience of stress is multifaceted. In practice, individuals can take action to better understand how they react to stress and pressure and to improve their coping strategies. However, for this to be truly effective, individuals also need to be supported by having access to "good jobs" and supportive managers, as discussed in the previous sections.

SUMMARY: MANAGING PERFORMANCE UNDER PRESSURE

Individuals may experience acute or chronic stress at work which may impact on performance or cause harm to their physical or mental health. The causes of stress were discussed alongside comprehensive approaches involving action at three levels: prevention, management, and treatment. The concept of resilience was introduced as a proactive opportunity to support this approach.

KEY POINTS

This chapter has discussed the stress reaction, both as acute "in the moment" and chronic. It has argued that actions can be taken by individuals, managers, and organizations to prevent and manage stress and has suggested that this is integrated with a more proactive approach to create "good jobs."

- The body's stress reaction can be helpful under situations of threat but can also interfere with the ability to "think our way" out of more complex threatening stations;
- Under prolonged pressure, individuals can start experiencing the negative impact of being under stress; over time, these changes can result in longer-term damage to mental and physical ill health;
- Differences between individuals can influence exposure to stressors, reactions to specific situations, and coping styles generally;
- Organizations can manage stress by acting to prevent it where possible, help individuals better manage their reactions, and provide treatment for individuals who are suffering;
- Managers play an active part in managing the well-being of their teams; through their actions they can not only prevent stress but also support engagement with work;
- Individuals can become more aware of how they react under stress and develop more effective coping strategies, building their resilience.

REFERENCES

Amati, C., Donegan, C., 2012. Emotion at work. In: Francis, F., Holbeche, L., Reddington, M. (Eds.), People and Organisational Development: A New Agenda for Organisational Effectiveness. CIPD Publications.

Amati, C., Scaife, R., 2006. Investigation of the Links between Psychological Ill-Health, Stress and Safety. HSE RR488. Crown Publishers.

Bevan, S., 2012. Good Work, High Performance and Productivity. The Work Foundation.

Brann, A., 2013. Make Your Brain Work. Kogan Page Ltd.

Cox, T., Griffiths, A., Barlowe, C., Randall, R., Thomson, L., Rial-Gonzalez, E., 2000. Organisational Interventions for Work Stress: A Risk Management Approach. HSE RR 286/2000. Crown Publishers.

CSB, 2007. Official Site of the CSB's Investigation into The Formosa Plastics Accident. http://www.csb.gov/formosa-plastics-vinyl-chloride-explosion/.

Donaldson-Feilder, E., Lewis, R., Yarker, J., 2009. Preventing Stress: Promoting Positive Manager Behaviour. CIPD Insight Report. CIPD Publications, London, (Available on the CIPD website).

DSM-5, 2013. Diagnostic and Statistical Manual of Mental Disorders, fifth ed. American Psychiatric Association.

EU-OSHA, 2015. European Agency for Safety and Health at Work—Campaign for Managing Work-Related Stress. https://www.healthy-workplaces.eu/en.

Flin, R., O ins://, P., Crichton, M., 2008. Safety at the Sharp End: A Guide to Non-Technical Skills. Ashgate Publishers, Farnham.

GAPAN, 2009. Lists of Awards & Press Statements. http://www.gapan.org/about-the-company/trophies-and-awards/award-winners/the-masters-medal/.

Graveling, R.A., Crawford, J.O., Cowie, H., Amati, C., Vohra, S., 2008. A Review of Workplace Interventions that Promote Mental Wellbeing in the Workplace. Edinburgh Institute of Occupational Medicine.

Lewis, R., Donaldson-Feilder, E., Pangallo, A., 2011a. Developing Resilience: Research Insight. CIPD Publications, London, (Available on the CIPD website).

Lewis, R., Donaldson-Feilder, E., Tharani, T., Pangallo, A., 2011b. Management Competencies for Enhancing Employee Engagement: Research Insight. CIPD Publications, London, (Available on the CIPD website).

MacKay, C., Cousins, R., Kelly, P., Lee, S., McCaig, R., 2004. Management standards' and work-related stress in the UK: policy background and science. Work Stress 18 (2), 91–112.

Reivich, K., Shatte, A., 2003. The Resilience Factor: 7 Keys to Finding Your Inner Strength and Overcoming Life's Hurdles. Harmony.

Rick, J., Thomson, L., Briner, R.B., O'Regan, S., Daniels, K., 2002. Review of Existing Supporting Scientific Knowledge to Underpin Standards of Good Practice for Key Work-Related Stressors—Phase 1. HSE RR. Crown Publishers.

Teigen, K., 1984. Yerkes-Dodson: a law for all seasons. Theory Psychol. 4, 525–547.

Waddell, G., Burton, A.K., 2006. Is Work Good for Your Health and Well-Being? TSO Ltd.

Wise, 2011. What Really Happened Aboard Air France 447? http://www.popularmechanics.com/flight/a3115/what-really-happened-aboard-air-france-447-6611877/.

FURTHER READING

EU-OSHA, 2015. European Agency for Safety and Health at Work—Campaign for Managing Work-Related Stress. https://www.healthy-workplaces.eu/en.

The Centre for Social Epidemiology's website—provides access to information and resources on work-related stress. http://unhealthywork.org/.

The Good Work Commission—forum of leaders dedicated to discussing challenges of work in 21st century in light of need to increase "good jobs" or "good work." http://www.goodworkcommission.co.uk/.

Index

Note: Page numbers followed by "*b*," "*f*," and "*t*" refer to boxes, figures, and tables, respectively.